INTRODUCTION TO
NANOCOMPOSITE
MATERIALS

Properties, Processing, Characterization

Thomas E. Twardowski, Ph.D
Principal
Materials Workshop
Philadelphia, PA

DES*tech* Publications, Inc.

Introduction to Nanocomposite Materials

DEStech Publications, Inc.
1148 Elizabeth Avenue #2
Lancaster, Pennsylvania 17601 U.S.A.

Printed in the United States of America
10   9   8   7   6   5   4   3   2   1

Entry under main title:
    Introduction to Nanocomposite Materials: Properties, Processing, Characterization

A DEStech Publications book
Bibliography: p.
Includes index p. 533

Library of Congress Catalog Card No. 2006935304
ISBN No. 978-1-932078-54-1

HOW TO ORDER THIS BOOK

BY PHONE: 866-401-4337 or 717-290-1660, 9AM–5PM Eastern Time

BY FAX: 717-509-6100

BY MAIL: Order Department
DEStech Publications, Inc.
1148 Elizabeth Avenue #2
Lancaster, PA 17601, U.S.A.

BY CREDIT CARD: American Express, VISA, MasterCard

BY WWW SITE: http://www.destechpub.com

Contents

SECTION III—NANOCOMPOSITES CHARACTERIZATION SCIENCE AND TECHNOLOGY

Preface

The field of nanocomposites is growing by leaps and bounds. A few of the recent commercial applications include sport utility vehicles, furniture, and appliances. Fields interested in reaping the material property advantages of nanocomposites range from agriculture to space science. Many materials, natural and synthetic, capitalize on the behavior of nanoscopic size scales, sometimes by design and sometimes not. The goal of this textbook is to provide a solid foundation for understanding, and beginning to answer, the questions posed by nanocomposites.

I took on this textbook at the recommendation of several colleagues. I worried that the textbook everyone wanted would focus on the clay-based composites that seemed to occupy most of the attention of the late 1990's and early 2000's. (In fact, a book has been issued since then, in two volumes, discussing the state of the art in polymer-clay nanocomposites: *Clay Containing Polymeric Nanocomposites* by L. A. Utracki.) I felt a more general, educational textbook would also be useful. The textbook in your hands focuses instead on teaching the science, processing technology and characterization approaches for studying and making all types of nanocomposites, including low-tech fumed silica and carbon black, high-tech carbon nanotubes, and, yes, clay.

This is a book for teaching nanocomposites, and perhaps for other types of composites: what they are, or can be; how they are, or will be, manufactured; and how they are, or might be, characterized. Nanocomposites look differently depending on who is using them. A surfactant scientist might worry primarily about particle interactions. A chemical engineer could build a career around the challenges of nanoscopic particles in multiphase flows. One mechanical engineer might be most concerned about the structural performance of a support beam while another may be more interested in the surface hardness of a computer casing.

While this text was not meant to be everything to everybody, there were several motivations for the content of this book. For teaching, the text provides fundamentals for structure-property relationships, processing, and characterization in one coherent, nanocomposites-centered presentation. For research and production, the text provides the basics for eventually developing systematic studies and uses of nanocomposites. As Utracki pointed out at the end of his two-volume, detailed review in the area of clay-containing composites, knowledge about processing and characterization remains poorly developed. While there have been a few excellent, substantive treatises, there is a lot of work that is enjoying the euphoria of exploring new combinations of materials without diving too deeply into the whys and wherefores. Nanocomposites have evolved out of composites, and composite technologies are a fusion of materials, chemical, and mechanical engineering, with healthy doses of industrial practice, chemistry, and physics.

This text should help the novice develop a sound foundation for future work and a broader understanding of composite materials. Reading one chapter of this book may leave a reader wondering why there isn't more nano-specific information. Keep reading! The entire book is designed to bring the reader to a systematic understanding of nanocomposites structure, properties, processing and characterization. Many specific subjects of serious consequence to nanocomposites are scattered throughout the text, ranging from the difficulties in achieving complete dispersion of particulate reinforcements, to the dramatic increases in liquid viscosity on adding nanoparticles. Each topic is presented in a logical, supportive context.

Organization of the Text

This is intended as an introductory text, appropriate for both new and experienced scientists, engineers and technicians seeking a deeper understanding of nanocomposites, especially in the broader context of composites. The burgeoning field has benefited from an influx of fresh perspectives. Nanocomposites have drawn researchers from the fields of chemistry, physics, chemical engineering and others. Nevertheless, Utracki, in a two-volume book on clay containing, polymer matrix nanocomposites, states that a need remains for more detailed study of processing and characterization. The intent of this book is to help the student at any stage of a career think about nanocomposites in an integrated manner. The approach in this text is to provide a systematic introduction to composite materials, balancing coverage of materials structure and properties with processing and characterization. As foundational, the text is intended to be supplemented with readings from the now extensive nanocomposites literature.

The first section of the text is a discussion of the physical chemistry of materials leading to nanocomposites. Composite performance is more than the sum of the behaviors of the individual components, but understanding the total behavior begins by exploring the contributions from the components and building toward full understanding. The sequence of presentation in this section is traditional, beginning with a discussion of the structures of materials (Chapter 2), followed by the properties of materials (Chapter 3), and finally by specific features of particles and nanoparticles (Chapter 4). Chapter 4, which focuses on particles, also makes sense as a starting point, rather than Chapter 2.

The second section of the text is devoted to processing. Many parts made historically, in quantity, such as tires, are, in fact, nanocomposites.

Processing content is included because nanocomposites, as novel, must eventually find a way into standard part production. The processing section is not organized in as traditional a fashion as the first section. In Chapter 5, viscosity is discussed as a unifying feature of composites processing. The processing techniques are philosophically divided, rather than grouped into one chapter. This has permitted using different approaches for solution and melt processing without transition within a single chapter. Chapter 6 covers low viscosity processing, which pertains to solution and *in situ* polymerization techniques. Chapter 7 covers high viscosity processing, including melt processing. In general, pure polymers, nanocomposites and composites are produced using the same processes, but in some specific ways, including increased viscosity (Chapter 5), settling (Chapter 6), and mixing (Chapter 7), nanocomposites process somewhat differently.

The third section discusses characterization techniques, loosely organized to follow the format of the first section. Chapter 8 discusses unifying concepts in characterization, including experimental design, visualization of results, and presentation. Chapter 9 discusses structure characterization, while Chapter 10 discusses the characterization of properties. Again, pure polymers, composites and nanocomposites all can be analyzed using the same techniques. However, the small size of particles in nanocomposites has an impact on what techniques can be used to study the particle distribution. For example, wide angle X-ray and standard SEM are generally less effective for analyzing nanoparticles than are electron diffraction and TEM. Also, since one must examine such small regions of a specimen to observe the particles, people are skeptical as to how that small region represents the whole sample. To provide a clearer understanding of the limits and requirements for analysis, Chapter 8 considers experiment design, data analysis, stereology and the eye. In Chapter 9, SEM and x-ray are covered, as well as TEM and AFM, two techniques that provide more reliable pictures of nanoparticles.

Acknowledgements

I have drawn heavily on some particularly useful sources. These are documented at the end of each chapter. As with any text, there are many people to thank. I am grateful to my wife, Meredith, who calls this "A Really Big Book for Solving Very Small Problems." This whimsical observation has had a formative impact on the final product. I am also grateful to Professors Phillip Geil and Richard Gaylord, at the University of Illinois, and Professor Ole Kramer and Dr. Walther Batsberg at Risø Danish National Laboratory, who set me on my current path, and provided immense support over many years. I hope I have finally done something to live up to the investments they made in me. Roger Corneliussen and Chris Pastore provided impetus for embarking on this project. Steven Niezgoda served, in a sense, as the first student to be taught using the book, and provided a valuable student viewpoint during the review process. Gratitude is due to Elizabeth Evans, who took all the AFM photographs. Additional SEM and optical images were acquired by Steven Niezgoda and Elena Petrovicova. Thanks also to Dr. Richard Knight for my initial involvement in nanocomposites and introducing me to thermal spray. All the readers and editors deserve my gratitude and praise for recommendations and improvements.

List of Symbols

Chapter 2 (Pages 11–50)

$\overline{P}_a$ = arithmetic average value of a property for a sample
n = total number of items
N_i = number of objects with a given property
P_i = value of property for objects in interval i
E_A = attraction energy
A = ion dependent constant
r = separation between ions
Z_1, Z_2 = ion valences
e = electronic charge
ε_0 = permittivity of vacuum
E_R = nuclear repulsion
B, n = ion dependent constants
P = polydispersity
$\overline{M}_n$ = number average molecular weight
$\overline{M}_w$ = weight average molecular weight.
L = contour length of the chain
n = number of steps
l_0 = bond length
r = separation between chain ends
θ = bond angle
φ = bond rotation
l_{eff} = effective bond length
α = swelling coefficient
Θ = solvent
r_g = radius of gyration

FCC = face-centered cubic
HCP = hexagonal closest packing
BCC = body-centered cubic
a, b, c = lengths of crystal unit cell sides
α, β, γ = angles of crystal unit cell
T_g = glass transition temperature
T_m = melt transition temperature

Chapter 3 (Pages 51–124)

w = width of specimen
F = applied load
s = span between supports
b = sheet thickness
l_m = length of moment arm
b_n = sheet thickness in plane of crack
COF = coefficient of friction
P_i = value of a component property
a, b = two components
x = component mole fraction

$$\frac{1}{P} = \frac{x_a}{P_a} + \frac{x_b}{P_b} = \text{harmonic mean}$$

$$P = \sqrt{P_a^{x_a} P_b^{x_b}} = \text{geometric mean}$$

k = spring constant
x = position
A = cross-sectional area
k' = modified spring constant
ν = Poisson's ratio
w = width
h = height
l = length
Δx = change in position
Δw = width change
Δh = height change
Δl = length change
ΔV = volume change
A = cross-sectional area
L = sample length
ΔL = the change in length

l = bond length

θ = bond angle

Δl = local change in bond length

k_l = spring, or force, constant for the bond

$\Delta\theta$ = local change in angle

k_θ = valence angle force constant

k_a = transformed angle force constant

ΔG = change in Gibbs free energy

ΔH = enthalpy

T = absolute temperature

ΔS = change in entropy

σ = engineering stress

ε = engineering strain

σ_{true} = true true stress

σ_{ij} = component stresses

τ, σ_s = shear stress

ε_{true} = true strain

ε_{ij} = component strains

γ, ε_s = shear strain

M_i = component modulus

E = Young's modulus, tensile modulus

G = shear modulus

B = bulk modulus

P = hydrostatic pressure

V = sample volume

σ_y = tensile yield stress

τ_y = shear yield stress

σ_t = stress at the crack tip

σ_0 = applied stress a crack length

r = radius of the crack tip

σ_f = failure stress

γ = surface energy

G_c = Griffith fracture energy

F = external work

U = internal energy

b = part thickness

P = generalized load

D = generalized displacement

C = compliance modulus

F_c = critical fracture force

r, θ = point coordinates in polar coordinates

K = stress intensity factor

K_I = mode I stress intensity factor
K_{Ic} = critical value for crack growth
σ_c = stress at the onset of crack growth
Q = geometric parameter
r_y = plastic zone radius
m_p = plastic constraint factor
R = increase in crack length
σ_p = crack growth stress, related to the yield stress
$\varepsilon(t)$ = time-dependent strain
σ_0 = is the constant force, and t is time
τ = relaxation constant
E = instantaneous modulus
η = internal solid viscosity
$\sigma(t)$ = time-dependent stress
σ_0 = initial stress
t = time
τ = relaxation time
ω = period of the oscillating stress or strain
δ = phase shift
T_g = glass transition
F = elastic force
E = internal energy
x = displacement from the equilibrium position
H = Helmholtz energy
S = entropy
k = Boltzmann's constant
Ω = number of possible states
Ω_1 = number of states for state 1
V_1 = volume of state 1
a = diameter of a gas molecule
N = total number of gas molecules
P_N = probability for a given end-to-end separation of an undeformed chain
$\bar{R}$ = end-to-end separation of an undeformed chain
N = number of chain segments
ℓ = segment length
λ_x = deformation in the x direction
K = natural logarithm of the normalization front factor
ν = number of network chains
$\langle \bar{R}_0^2 \rangle$ = average end-to-end separation
$\langle \bar{R}_i^2 \rangle$ = component separations in cardinal direction i

λ_x = directional strain

A = cross-sectional area

l = length

f = force

λ = strain

ν = number of network chains

k = Boltzmann's constant

l = length

ABS = acrylonitrile styrene butadiene

CTE = coefficient of thermal expansion

ϕ_f = volume fraction of reinforcement

$\tan\delta_c$ = composite loss modulus

$\tan\delta_p$ = polymer loss modulus

B = correction parameter

R = mean radius of dispersed particles

ΔR = thickness of the immobilized layer

$M_c = M_p\phi_p = M_f\phi_f$ = rule of mixtures

M_c = a general modulus

M_p = modulus of the matrix polymer

M_f = modulus of reinforcement

ϕ_p = volume fraction of polymer

F = force

N = normal force friction coefficient.

μ = steepest or shallowest angle

Q = mass flux

M = mass

A = cross-sectional

t = time

$$Q = -D\frac{dC}{dx} \quad \text{Fick's first law}$$

C = concentration

D = diffusion constant

x = distance

$$\frac{dC}{dt} = D\frac{d^2C}{dx^2} \quad \text{Ficks' second law}$$

P_i = polymer permeability

P_0 = permeability of amorphous or unfilled polymer

V_i = volume fraction impermeable phase

Chapter 4 (Pages 125–170)

$M.D.$ = Martin's diameter
D = packing density
S = specific surface area
$\bar{d}$ = arithmetic mean
p = number of counting intervals
f_i = number of observations in an interval
d_i = median diameter of an interval
n = total number of measurements
m_2 = variance
s = standard deviation
sk = skewness
m_3 = fourth moment
pk = sharpness or tendency to form a peak
m_4 = fourth moment
$\bar{d}_g$ = geometric mean
$\bar{d}_h$ = harmonic mean
$\bar{d}_s$ = mean surface diameter
$\bar{d}_v$ = mean volume diameter
$\bar{M}_n$ = arithmetic mean of the molecular weight
$\bar{M}_w$ = geometric mean of the molecular weight
$\bar{d}_l$ = linear mean diameter
$\bar{d}_a$ = surface mean diameter
$\bar{d}_w$ = weight mean average diameter volume-averaged diameter
p = aspect ratio
B = breadth
L = length
T = thickness
S = surface area
V = volume
Q = flux of particles
ΔC = change in particle concentration
Δx = distance traveled
D = diffusion constant
k = Boltzmann's constant
T = absolute temperature
d = particle diameter
η = viscosity of suspending liquid
v = velocity of particles
ρ_p = particle density

ρ_L = liquid density
g = gravitational constant
v = liquid velocity
p = bed porosity
ΔP = applied pressure differential
g = gravitation constant
L = bed thickness
η = liquid viscosity
ρ = particle density
S_w = specific surface
k = dimensionless constant
ϕ = angle of flow relative to the perpendicular of the particle orientation
E = energy
C = constant dependent on milling material
X_f = feed stock particle size
X_p = desired size
TEOS = Tetraethoxysilane
pH = scale for measuring H^+ ions (acidic or basic)
$AlCl_3$ = aluminum chloride
Na_4SiO_4 = sodium silicate
OH = hydroxyl group
Ti = titanium
Mg = magnesium
Fe = iron
Cr = chromium
Mn = manganese
Li = lithium
Na = sodium
K = potassium
Ca = calcium
Si = silicon
SiO_2 = silica
Al_2O_3 = alumina
Mg^{2+} = ionized element (magnesium)
V_0 = bulk volume
a, b = material-dependent constants
P = pressure

Chapter 5 (Pages 173–211)

Re = Reynolds number

ρ = density

v = liquid velocity

d = characteristic length

η, μ = viscosity

De = Deborah number

We = Weissenberg number

λ = relaxation or retardation time

t_{flow} = characteristic flow time

τ = shear stress

Δv = steady-state velocity gradient

Δy = distance away from the plate

γ = shear strain

$\dot{\gamma}$ = shear strain rate

$A dy$ = differential volume

$F_s dx$ = energy dissipated through shear

E = energy per unit volume lost to viscosity

$\partial v/\partial y$ = incremental velocity profile as a function of lateral distance from a plate

g = gravitational constant

P = pressure

v = liquid velocity

V = plate velocity

x = distance from plate

h = liquid layer thickness

Q = flow rate

w = plate width

τ_{ij} = shear stress component

T = shear stress tensor

V = volume

T = temperature

R = ideal gas constant

n = moles

M = molecular weight

$Ni(\dot{\gamma})$ = Normal stress difference

Ψ = normal stress coefficients

$\dot{\gamma}$ = strain rate

$\underline{v}$ = velocity in uniaxial elongational flow

$\dot{\varepsilon}(t)$ = strain rate in uniaxial elongational flow

$\underline{\tau}$ = stress in uniaxial elongational flow

$\dot{\varepsilon}_0$ = initial strain rate

l_0 = initial separation between two particles along the z-axis

l = instantaneous separation

ω = angular velocity under flow

$F_{viscous}$ = viscous force

f = friction factor

v = velocity of the molecule or portion thereof

ζ = segmental interaction energy

n = number of segments in rotating chain

E_m = energy loss per molecule

$(r_i)^2$ = two-dimensional radius of gyration

N_A = Avagadro's number

M = molecular weight

l_0 = step length

M_0 = monomer mass

$$\eta = \frac{l_0^2 \zeta \rho N_A}{36 M_0^2} M = KM \text{ Debye viscosity}$$

K = constant

t = time

y = position

ϕ = particle volume fraction

A, B, C and D = curve-fitting parameters

$[\eta]$ = intrinsic viscosity

η_0 = viscosity of pure solvent

η_r = relative viscosity

η_{sp} = specific viscosity

η_{red} = reduced viscosity

η_{inh} = inherent viscosity

C_2 = concentration of polymer

$\sinh x$ = hyperbolic sine function

ϕ_{max} = maximum volume fraction allowing free rotation of particles

A = combined parameter to account for solvation, ellipticity, and experimental error

B = accounts for particle overlap and other concentration-dependent effects

Ω = rotational velocity of spinning plate

v_θ = rotational velocity as a function of position

z = position

h = gap width

$\mathbf{T}$ = total torque

S = stress applied at distance x

$\dot{\gamma}_R$ = rim shear rate

v_ϕ = velocity as a function of radius in a cone and plate viscometer

Θ = upper plate angle

θ = angle with respect to vertical

Ψ_1 = first normal stress coefficient

$\mathbf{F}$ = thrust applied to maintain gap distance

R = outer cylinder radius

κR = inner cylinder radius

L = spinning bob length

τ_R = shear stress at the wall

$\dot{\gamma}_R$ = shear strain rate at the wall

τ_{rz} = local shear stress

ΔP = applied pressure drop

r = position relative to center line

L = length of capilary

R = radius of the capilary

v_z = local velocity

Q = is the flow rate

$$\Delta P = 2\tau_R \frac{L}{R} \quad \text{Bagley correction}$$

$$\eta(\dot{\gamma}_R) = \frac{4\tau_R}{L\dot{\gamma}_a}\left(3 + \frac{d\ln\dot{\gamma}_a}{d\ln\tau_R}\right) \quad \text{Weissenberg-Rabinowitsch correction}$$

$\dot{\gamma}_a$ = apparent viscosity

v_{slip} = slip velocity

$\bar{P}$ = applied pressure

$\bar{h}$ = average height of liquid column

$$\eta = \frac{\pi R^4 \bar{h}\rho g t}{8Vl} \quad \text{Poiseuille's equation}$$

V = liquid volume

l = capillary length

t = time to drain

A = viscometer constant

D = capillary diameter

L = capillary length

MFI = melt flow index

n, m = power law parameters

$$\frac{\eta(\dot{\gamma}) - \eta_\infty}{\eta_0 - \eta_\infty} = [1 + (\dot{\gamma}\lambda)^a]^{\frac{n-1}{a}} \quad \text{Carreau-Yasuda model}$$

η_0 = zero-shear viscosity

η^∞ = infinite-shear viscosity

n, λ and a = fitting parameters

$[C]$ = concentration of holes
h = rate constant
A = front factor
ΔE^* = activation energy
k = Boltzmann's constant
Rate = rate of hole movement
d = distance the hole moves
$\partial v/\partial y$ = shear strain rate
$\dot{\gamma}$ = shear strain rate
λ = distance from first hole required to lose the memory of displacement of the neighboring hole
$\sinh(x)$ = hyperbolic sine
$\exp$ = exponential e^x
β = a collection of constants $\lambda/2dh$
V_h = average volume of a hole
τ_0 = relaxation time, $1/h$
D_t = tube diffusion constant
ζ = segmental friction factor
n = chain length or degree of polymerization
k = Boltzmann's constant
L_t = chain contour length
l_0 = is the monomer length

$$D = \frac{x^2}{2t}$$ Einstein diffusion equation

D = diffusion constant
x = distance traveled
t = time
τ_n = relaxation time for complete relaxation
G = shear modulus
v = is the number of network chains
τ = time required to diffuse along the contour length
t = instantaneous time
t' = integration time
η_0 = zero shear viscosity
τ_0 = longest relaxation time
τ_y = threshold shear
∞ = infinite viscosity
$\dot{\gamma}(t)$ = applied shear strain rate
$\dot{\gamma}_0$ = amplitude
ω = frequency
δ = phase angle

G' = shear storage modulus
G'' = shear loss modulus

Chapter 6 (Pages 213–271)

$\gamma_{SV} = \gamma_{SL} + \gamma_{LV} \cos\theta + \pi_e$ Young equation
γ_{LV} = liquid/vapor interfacial surface tension
γ_{SL} = solid/liquid interfacial surface tension
γ_{SV} = solid/vapor interfacial surface tension
π_e = spreading pressure
$W_A = \gamma_1 + \gamma_2 - \gamma_{12}$ Dupré equation
W_A = thermodynamic work of adhesion
γ_1 = surface tensions of the liquid
γ_2 = surface tension of the substrate
γ_{12} = interfacial tension
γ_i^d = dispersive surface tension component
γ_i^p = polar surface tension component
S = spreading coefficient
γ_1 = total surface tension
$F_{gravity}$ = force exerted by gravity
m = particle mass
g = gravitational constant
F_{bouy} = force exerted by buoyancy
$\underline{V}$ = particle volume
ρ = density of solution
ρ_p = density of particle
$F_{viscosity}$ = resistive, viscous dissipation force
η = viscosity of the solution
R = particle radius
v_s = is the solution velocity
dx/dt = particle velocity
ψ = sphericity
$[P]$ = concentration of particles
v = hindered particle velocity
v_0 = ideal particle velocity
k = Boltzmann's constant
$\bar{v}$ = average particle velocity
T = absolute temperature

$$N(v) = 4\pi N \left(\frac{m}{2\pi kT}\right)^{3/2} v^2 \exp\left(\frac{-mv^2}{2kT}\right)$$ Maxwell speed distribution

$N(v)$ = total number of particles per unit volume

v = velocity

D = diffusion constant

η = viscosity

R = radius of particle

N_A = Avagadro's number

n_{vol} = number of particles in a volume

n_0 = initial number of particles

$V_{particle}$ = average particle velocity

$\rho_{particle}$ = particle density

ρ_{liquid} = liquid density

h = liquid column height

ΔH = change in enthalpy

ΔS = change in entropy

ΔG = change in free energy

ΔS_{mix} = entropy of mixing

R = gas constant

x_i = component mole fraction

ϕ_i = component volume fraction

ΔH_{mix} = heat of mixing

χ = Flory-Huggins interaction parameter

N = number of molecules

$$\Delta H_{mix} = V_{mix} \left[\left(\frac{\Delta E_1^{vap}}{V_1} \right)^{1/2} - \left(\frac{\Delta E_2^{vap}}{V_2} \right)^{1/2} \right] \phi_1 \phi_2 \quad \text{Hildebrand equation}$$

ΔE^{vap} = component internal energy of vaporization

V_n = molar volume

V_{mix} = volume of mixed materials

CED = cohesive energy density

δ = solubility parameter, Hildebrand

$\delta_2^2 = \delta_D^2 + \delta_P^2 + \delta_H^2$ Hansen solubility parameters

f_D = solubility fraction contribution of dispersive forces

f_P = solubility fraction contribution of polar forces

f_H = solubility fraction contribution of hydrogen bonding forces

Hansen solubility parameters

ΔH_{vap} = heat of vaporization

Δh_i = component heat of vaporization

E = energy

F = component interaction forces

V = component volume

χ = Flory chi parameter

b = fitted parameter

Θ = theta condition, temperature, or concentration

f_D = solubility fraction contribution of dispersive forces

f_P = solubility fraction contribution of polar forces

f_H = solubility fraction contribution of hydrogen bonding forces

μ_i = chemical potential of component i

Δn_i = moles of component i

Component 1 solvent

Component 2 polymer

ΔG_m = free energy of mixing

N_1 = number of solvent molecules

n = degree of polymerization

μ_i^0 = chemical potential of pure component i

χ_c = critical Flory chi parameter

$\phi_{1,c}$ = critical volume fraction of solvent

E_s = energy of free surface

γ_s = surface energy

d = particle diameter

k = percent of particle surface occluded by a neighboring particle

f = effectiveness of initiator

$[I]$ = concentration of initiator

$[M]$ = concentration of monomer

k_i = initiation rate constant

$[M^{\bullet}]$ = concentration of the radical monomer species

k_p = rate constant of propagation

k_t = rate constant for termination

$\overline{v}$ = kinetic chain length

$\overline{v}_{tr}$ = average kinetic chain length in the presence of a chain transfer agent

k_{tr} = chain transfer rate constant

$[TR]$ = concentration of chain transfer agent

$[M_i]$ = Concentration of monomer i

$[M_i^{\bullet}]$ = Concentration of radical monomer i

k_{ij} = reaction rate constant between monomers i, j

r_1 = cross-reactivity ratio between monomer 1 and monomer 2

r_2 = cross-reactivity ratio between monomer 2 and monomer 1

X_i = mole fraction of monomer i in polymer

x_i = mass fraction of monomer i in feed

p = extent of reaction

A = instantaneous concentration of one reactant

A_0 = original concentration of reactant
$\bar{n}_n$ = number average chain length
$\bar{n}_w$ = weight average chain length
PD = polydispersity
r = stoichiometric imbalance
A_f = concentration of functional groups in multifunctional monomers
A_{total} = concentration of functional groups in all monomers
ρ = end group imbalance
α = probability that a chain has a multifunctional unit at either end
α_c = critical α for gelation
p_c = critical extent of reaction at gelation
f = number of reactive groups in polyfunctional monomer
$\bar{f}$ = average functionality of all monomers
$[A]$ = concentration of species A
$[B]$ = concentration of species B
k_c = catalyzed rate constant
t = time
$[A]_0$ = initial concentration of species A
k_u = un-catalyzed rate constant
LD50 = lethal dose that kills 50% of the animals to which it was given
TLV = threshold limit value
STEL = short-term exposure limit
MSDS = materials safety data sheet

Chapter 7 (Pages 273–329)

ΔT = range of super-cooling

$\eta(t) = \eta_0 e^{\left[\frac{\Delta E}{R}\left(\frac{T_0-T}{T_0 T}\right)\right]}$ Boltzmann statistics

η_0 = viscosity
T_0 = reference temperature
ΔE = is the activation energy for flow

$\log\left[\frac{\eta(T)}{\eta(T_0)}\right] = \frac{-17.44(T-T_0)}{51.6+(T-T_0)}$ WLF equation

WLF = Williams-Landau-Ferry
T_g = glass transition temperature

T_0 = typically chosen as T_g + 100°C
T_m = melt temperature
ρ = density
C_p = heat capacity at constant pressure
T = temperature
t = time
v_i = velocity in the ordinal x, y or z directions
x, y or z = length in the ordinal directions
k = thermal conductivity constant
η = viscosity
$\Delta\dot{H}$ = point heat generation or loss sources
Q = heat flux
A = cross-sectional area
x = position through the thickness
x_V = volume fraction of the reinforcement
k_T = total thermal conductivity
k_m = conductivity of the matrix
k_r = conductivity of the reinforcement
$\alpha = k/\rho C_p$ a cluster of thermal constants, the thermal diffusivity
x = depth into the solid
T_0 = initial, homogeneous temperature in the solid
T_s = surface temperature
erf = error function
$\dot{Q}$ = heat flow rate
x_l = thickness of the liquid layer
x_s = thickness of an equivalent amount of unmelted solid
K = fitted parameter
R_m = rate of melting
V_0 = velocity of the moving sheet or the moving heat source
I = intensity of the source
$T(x)$ = temperature at any linear position x
T_0 = initial temperature
c = cross-sectional area
h = convective heat transfer coefficient
k = conductive heat transfer coefficient
V_0 = unit volume
x = neck radius
R = particle radius
γ_s = surface tension
t = time
r_0 = initial radius of the holes left in the coalesced film
r = radius

SMC = sheet molding compound
F = central force required to close the mold during compression molding
m, n = power-law constants
$\dot{h}$ = mold separation or charge height
h = mold closing rate
R = final radius of the part after molding
h_0 = initial charge thickness
B-stage = partly cured thermoset material
Q = mass flow rate
Ω = rotational velocity of the cylinder
R = radius of the cylinder
h_r = film thickness at release from the roll
h_0 = twice the gap height at the closest approach between the cylinders
x = distance from the origin, defined as the center of the gap
Ω = rotational velocity of the cylinder
R = radius of the cylinder
x_r = distance to release point upstream of gap
x_g = distance to contact point downstream of gap
d = normalized distance from the gap
d_r = normalized distance to release point upstream of the gap
d_g = normalized distance, downstream of the gap, point of contact with forming surface
$\dot{\gamma}_m$ = aximum shear strain rate
τ_m = maximum shear
P = closing pressure
w = charge width at the pinch point
g = gravitational constant
h = height of a liquid column
P = pressure exerted by a confined, solid column at any depth h
ρ_b = bulk density
D = hopper diameter
h = any depth in the column
H = particle column height
f_w = wall coefficient of friction
K = ratio of horizontal compressive stress to vertical compressive stress
P_{max} = maximum pressure exerted by a confined, solid column
L_0 = initial length of powder charge
D = channel diameter
P = discharge pressure

P_0 = initial pressure

dx/dt = positive displacement of the ram

ϕ = angle between the velocity vector and the force vector

V_0 = velocity of the plate

θ = angle between the flow velocity vector and the velocity vector of the plate

v = liquid velocity

H = channel height

W = channel width

f_{w_1} = friction coefficient with moving surface

f_{w_2} = coefficient of friction with stationary wall

t = average thickness of cubic representation of minor component under shear

L = cube side length

γ = shear

x_V = minor component volume fraction

D = deformation of the droplet

η_1 = continuous phase viscosity

η_2 = secondary phase viscosity

$\dot{\gamma}$ = shear rate

ζ_s = interfacial energy

F = maximum force acting on a pair of particles in contact

r_1 = radius of particle 1

r_2 = radius of particle 2

h = clearance between screw and barrel

δ = clearance between screw flights and barrel

W = channel width

ϕ = average screw pitch or helix angle

e = width of the flights

L_s = screw length

D = barrel diameter

Q_c = maximum flow rate in solids conveying zone

ρ = density of bulk solid

D_s = screw diameter

v_s = screw velocity in revolutions per unit time

V_b = velocity of the barrel

P = pressure on the liquid

z = distance in the ordinal direction perpendicular to the barrel surface

F_d = drag shape factor

F_p = pressure shape factor

t_{plate} = time to cool a thin plate

$t_{cylinder}$ = time to cool a cylindrical part

h = plate thickness

d = cylinder diameter

T_M = melt temperature at injection

T_E = temperature at ejection

T_W = mold wall temperature

TTT = time-temperature transformation curve

$X_{cr} = 1 - \exp(-Kt^m)$ Avrami equation

X_{cr} = percent crystallinity

K, m = constants that depend on the material and conditions of the crystallization process

PE = polyethylene

HVOF = High-velocity oxy-fuel

Chapter 8 (Pages 331–387)

SOPs = standard operating procedures

ASTM = American Society for Testing Materials

NIST = National Institute for Standards and Testing

ANSI = American National Standards Institute

μ = average, most likely value

σ = standard deviation

x = a measured value

y = deviation from the average

$\bar{x}$ = estimated average value

i = individual measurement

n = number of independent values of x used to estimate the standard deviation

$t = \dfrac{(x - \bar{x})\sqrt{n}}{s}$ t-test

$Q = \dfrac{\text{suspect value} - \text{nearest value}}{\text{largest value} - \text{smallest value}}$ Q test

$\bar{x}$ = estimate of correct value

s = estimate of standard deviation

x = experiment value

t, Q = tests for acceptance-rejection

n^{th} = a randomly chosen, regular, integer interval

$2n!$ = $2n$ factorial experiment design

P = sample size

RGB = red, green and blue, the base colors for color image projection

CMY = cyan, magenta and yellow, the base colors for three color printing

CMYK = cyan, magenta, yellow and black, the base colors for four color printing

CIE = International Commission on Illumination (Commission Internationale de l'Eclairage)

CD = compact disc

CTH = controlled temperature and humidity

AFM = atomic force microscopy

OIM = orientational imaging microscopy

OLS = ordinary least squares regression

y = predicted response variable given an input variable

A = Area

d_{max} = maximum diameter

d_{min} = minimum diameter

A_b = Area of a bounding rectangle

D = diameter (Feret's, Martin's, etc.)

L = real object length

P = perimeter

Chapter 9 (Pages 389–439)

ν = frequency

h = Plank's constant

c = speed of light

λ = wavelength

E = energy

eV = electron volts

NaCl = sodium chloride

A = absorbance

T = transmission

T_0 = intensity of incident light

$A = \varepsilon bc$ Beer Lambert law

ε = efficiency of an absorber

b = thickness

C = concentration

^1H = hydrogen isotope with 1 particle in the nucleus

^{13}C = carbon isotope with 13 nuclear particles

δ = shift factor

ppm = parts per million

TMS = tetramethyl silane

θ_1 = angle of incident light resulting in total reflectance

n = index of refraction

α = angle of light collected by the lens

d = smallest distance that can be separated by the lens

L = depth of field

M = magnification

$N.A.$ = numerical aperture

$\lambda = \dfrac{h}{\sqrt{2meE}}$ DeBroglie equation

h = Plank's constant

m = mass of an electron

eE = accelerating voltage

LaB_6 = lanthanum hexaboride

$n\lambda = 2d \sin\theta$ Bragg's law

n = integer whole number of wavelengths

λ = wavelength

d = distance between planes of atoms

θ = angle of the incident light

H, K, L = electron energy levels derived for x-ray spectroscopy

h, k, l = the values of the unit vectors in a unit cell

ATR = attenuated total reflectance

H-NMR = proton nuclear magnetic resonance spectroscopy

C-NMR (CMR) = carbon nuclear magnetic resonance spectroscopy

MRI = magnetic resonance imaging

AAS = atomic absorption spectroscopy

AES = atomic emission spectroscopy

FT = Fourier transform

IR, *ir* = infrared

BSE = back scattered electrons

SE = secondary electrons

CRT = cathode ray tube

SEM = scanning electron microscopy

TEM = transmission electron microscopy

EDS or EDAX = electron dispersive x-ray spectroscopy

XPS = x-ray photoelectron spectroscopy

TEM = transmission electron microscopy

WAX = wide angle x-ray spectroscopy

SAX = small angle x-ray spectroscopy

GPC = gel permeation chromotagraphy
AFM = Atomic Force Microscopy
STM = scanning tunneling microscopy

Chapter 10 (Pages 441–532)

G_{doi} = distinctness of image gloss
R_∞ = reflectance of a unit thickness backed by an infinite thickness of the material
R_1 = reflectance of material backed by perfect reflector
R_0/R_∞ = opacity
T = transparency
I = intensity of transmitted light
I_0 = intensity of incident light
τ = turbidity
z = instantaneous height
L = scan length
l = integration variable for scan length
R_a = arithmetic mean roughness
R_q = root mean square roughness
R_t = distance between highest and lowest points on a rough surface
R_p = distance between the highest point and the mean line on a rough surface
R_v = distance between the lowest point and the mean line on a rough surface
R_{sk} = skewness of a rough surface
R_{ku} = kurtosis of a rough surface
m_n = is the mass fraction of component n
T_n = melting temperature of component n
DSC = differential scanning calorimetry
DTA = differential thermal analysis
DMA = dynamic mechanical analysis
TSC = thermally stimulated creep
TSDC = thermally stimulated discharge current
TGA = thermogravimetric analysis
C_s = unknown sample heat capacity
C_r = heat capacity of a known reference material
ΔT = change in temperature over the region of interest
ΔT_r = change in temperature of the reference over the same range
ΔT_0 = is the change in temperature when there is no sample or reference in the test cells

α = reaction position

k_R = reaction constant

$Q(t)$ = heat generated in an isothermal reaction as a function of time

H_R = total heat of reaction

A = front factor

E_a = activation energy for the reaction

σ_0 = maximum stress

ω = frequency $(2\pi\upsilon)$

t = time

r = radius of hole or fillet

w = width of sample minus the hole

d = characteristic gauge width

D = characteristic grip width

HB = Brinell hardness

HV = Vicker's hardness

HK = Knoop hardness

D = tip diameter

d = indentation diameter

F = applied force

l = length of an indentation side

a = length of the long axis of an indentation

$J'(\omega)$ = is the storage compliance

$J''(\omega)$ = is the loss compliance

E = Youngs' modulus

η = scalar viscosity

δ = phase angle

$J*$ = complex compressive modulus

Λ = log decrement

A = amplitude

λ = wavelength

G' = shear storage modulus

G'' = shear loss modulus

$CaCO_3$ = calcium carbonate

ΔH_{vap} = heat of gasification

q_{rad} = heat released to the environment

q_{in} = input heat flux

CO_2 = carbon dioxide

CO = carbon monoxide

H_2O = water

K = thermal response

ρ = density

C_p = heat capacity

t = thickness

$\Delta T_{ignition}$ = temperature elevation for ignition

γ = free energy

γ_β, β = materials constants

γ_0 = surface tension at 0 K

T_c = critical temperature

γ_∞ = surface tension at infinite molecular weight

k = constant

c = subscript refers to the crystalline state

a = subscript refers to the amorphous state

d = dispersive force interactions

p = polar interactions

$$\gamma_{12} = \gamma_1 + \gamma_2 - \frac{4\gamma_{1,d}\,\gamma_{2,d}}{\gamma_{1,d} + \gamma_{2,d}} - \frac{4\gamma_{1,p}\gamma_{2,p}}{\gamma_{1,p} + \gamma_{2,p}} \quad \text{Young equation}$$

Subscripts 1, 2 refer to two materials in contact.

P = permeability

D = diffusion rate

S = solubility

$P \cdot d = -4\gamma \cos\theta$ Washburn equation

P = applied pressure

d = minimum penetrable diameter

γ = 480 dyne/cm (surface tension of mercury)

θ = contact angle of mercury with the pore wall

$$V \approx \frac{V_m \dfrac{P}{P_0}\left(\dfrac{E_1 - E_L}{RT}\right)}{1 - \dfrac{P}{P_0}} \cdot \frac{1}{1 + \dfrac{P}{P_0}\exp\left(\dfrac{E_1 - E_L}{RT}\right)} \quad \text{BET equation}$$

V = volume of gas adsorbed

V_m = volume of the monolayer

P/P_0 = fugacity (partial pressure divided by the partial pressure above a saturated solution)

E_1 = heat of adsorption of the gas in the monolayer

E_L = heat of condensation for the gas

CHAPTER 1

Nanocomposites Past and Future

The Ben Franklin Bridge was built to cross the Delaware River between Philadelphia, Pennsylvania and Camden, New Jersey. Since being constructed, it has served as a major interstate artery. It now requires a new coat of paint. By some estimates, however, the bridge will cost nearly 10 times as much to repaint as it cost to build! Clearly, refinishing the bridge is a task best done with as durable a coating as possible. Paints and epoxy coatings are not very resistant to surface abrasion and wear from the elements. By incorporating 10% of nanoceramic reinforcement into the epoxy or other plastic coating, to form a nanocomposite, the wear resistance can be doubled and the scratch resistance increased by 50%. Who will design a coating that lasts 50 or 100 years?

In marine applications, wood provides an excellent combination of low density, strength, and low cost. Majestic tall ships still navigate the waterways of the world, and draw large crowds of visitors at each port that they visit. Floating docks serve as the diving platform for swimming on lakes and ponds. Wood makes an excellent deck, where metal and plastic become near-lethally slick when wet. Wooden walkways protect delicate wetlands, allowing passersby to enjoy the scenery without crushing delicate flora and fauna or slopping through mud. Wood, however, deteriorates in wet environments. Yet, there are few alternatives. Metal ship hulls are strong, but without frequent painting will corrode even more quickly than wood. Ceramics are corrosion resistant, but are too brittle to serve in ship hulls. Plastics are wonderfully corrosion resistant and light; polyester-based fiberglass is a mainstay of the modern marine industry. Unfortunately, polymers absorb small quantities of moisture that can decrease the strength of the material and cause layers of

1

the hull to separate after some years in service. The addition of 2% of nanoceramic mica or clay, flake ceramics, will decrease water permeability by up to 50%. Who will design the composite hulls that will be the future of marine manufacturing?

WHAT IS A NANOCOMPOSITE?

A composite is a combination of two or more different materials that are mixed in an effort to blend the best properties of both. A nanocomposite is a composite material, in which one of the components has at least one dimension that is nanoscopic in size, that is around 10^{-9} m. A scaling might be helpful: a coin is on the order of 1–2 mm thick, or 10^{-3} m; a carbon fiber, commonly used as a reinforcement in sporting goods, is approximately 7 μm in diameter, or 10^{-6} m; a carbon-carbon chemical bond, the basic unit of life, is about 1.5 Å, or 10^{-10} m.

The field of nanocomposites is burgeoning. A brief look at new, common commercial uses reveals automotive panels for sports utility vehicles, polypropylene nanocomposites for furniture, appliances, and bulletin board substrates. Advanced technologies implemented include magnetic media, bone cement, filter membranes, aerogels, and solar cells. Nanocomposites represent one area of nanoscale research that has led to marketable products.

Just about every major federal grant agency has dedicated a portion of their available funding to nanoscale research. In 2004, the US Department of Agriculture began a pilot grant program to involve the farm industry in this new field of materials development. The financial investment in nanotechnology, and nanocomposites, is daunting.

The term "nanocomposite" appears to have been coined in 1986. Since then, growth in the field has been geometric (Figure 1.1). According to Google Scholar™, in 1986 there were 2 papers. In 2004, there were 1690 articles. Nanocomposites articles have appeared in just about every materials journal, and now a number of journals are specifically dedicated to nanocomposites. Nanocomposites science and engineering have touched all fields of materials, metals, plastics, ceramics, biomaterials, electronic materials, and more.

Plenty of skeptics believe the "nano" phenomenon to be more hype than productivity. Nanocomposite materials, however, do show improved properties. Materials used for nanocomposites are low technology: soot, ash, and clay. Decades of research have delved into S-glass and graphite fibers or boron nitride whiskers as reinforcements. Viewing the polymer as a mechanical weak link, research has focused on techniques

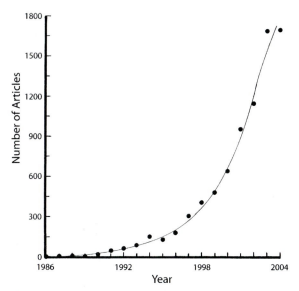

FIGURE 1.1 Number of nanocomposites publications by year, as shown by Google Scholar™.

for improving the volume fraction of reinforcement or improving interaction between reinforcements. Nanoparticles drive research in a different direction.

Whereas traditional composites use over 40% by weight of reinforcement, nanocomposites may show improvements at less than 5%. More importantly, traditional theories do not account for meaningful change in properties when so little material is replaced. Nanoparticles cannot reinforce differently than microparticles, but the continuum theories do not include size dependence. Thus, the mechanics must be understood as arising from load transfer as much as from load bearing.

On the other hand, research describing structures containing nanoparticles seems to rely on methods that are being pushed to the limit of resolution. Few systematic studies have focused on improvements in properties at size scales from the micron scale down to the nanoscale.

Producing nanocomposites also poses very real processing challenges. The list of questions about the fabrication, characterization, and use of nanocomposites is long—despite massive financial and intellectual investment.

The prevailing fascination with all things nanoscopic is a fad. After all, many materials, natural and synthetic, capitalize on the behavior of nanoscopic size scales, sometimes by design and sometimes not. It is a conceit of the information age, with selective amnesia prior to about 1986, that nano is new. Nevertheless, the magnitude of the effects these small particles impart to the bulk properties of a composite are great enough that the science will remain important. A detailed study of nanocomposites, their structure, processing and characterization, will be of value in all walks of the engineering life long after the fad diminishes.

One goal of this textbook is to provide a solid foundation for understanding, and beginning to answer, the questions posed by nanocomposites.

NANOCOMPOSITES PAST

In the middle ages, you are apprenticed to a stained glass artisan. Having earned the trust of the master, you are in charge of handling the precious stores of rare earths used to provide the most vivid reds and yellows. The salts have names containing sounds like aurous or argentous. Maybe, with a reputation for accuracy and care, you are especially favored with conducting routine practical chemistry according to the recipes that start with the base metals, combining them with substances like manure, or quicklime, or bitumen to create the ingredients the master apprentices will need to fill the next order for a far-off cathedral. Perhaps you become the master yourself, deftly adding the proper measure of precious salt, blowing and folding the glass to ensure that the vivid colors that form are evenly distributed. Do you think of yourself as one of the first to synthesize nanocomposites for profit?

Artisans discovered that even minuscule quantities of silver, gold, and copper salts could be used to produce vivid colors in stained glass. The salts had to be insoluble in the glass, and they had to be very finely divided. Adding the insoluble particles unfortunately caused problems with processing, probably requiring quite a bit of talent to insure that the particles were uniformly spread through the glass. There is no evidence that the artisans knew the exact sizes of the particles, or why such small particles gave rise to vivid colors, which were often different from the color of the material before it was added to the glass.

You are a crusading knight, facing an opponent armed with a blade that is unique and unnerving. The sword appears black, and as your foe waves this demonic thing back and forth, the blade shimmers with close-set traceries of black interlacing the dull silver of mere steel, the material making up your own sword. With wild ululations, you and your foe charge, sword meets shield, sword meets sword. Disaster strikes, as your sword fractures when it meets his, blade spinning uselessly away from you, from the fight. His sword does not break. Your last view, before you leave to learn of the life to follow, is of an amazingly strong, sharp, patterned blade.

The smiths of history left no record of how they made Damascene blades, some still in existence today. Modern smiths have rediscovered processes by which steel composites with very thin, carbon-rich layers can be manufactured. Steel making has always included a drive to create iron/carbon composites with finer carbon phases. Did ancient smiths think of themselves as makers of nanocomposites? From the famous Damascus steel swords to modern steel gun barrels, generations have been protected and destroyed by nanocomposites.

In the late 1800s, your company has been making rubber tires for bicycles and other vehicles. The material you produce using the Goodyear process, with sulfur and white lead, produces a beige tire that shows dirt readily. You decide that by adding carbon black to the rubber, the resulting tires will be black and show less dirt. You do so, and find that the tires wear up to five times more slowly than naturally colored rubber. As your company profits from the added value, do you congratulate your skill as a nanocomposites engineer? For a century, the world has moved on wheels made more economical by nanocomposites.

Biological cells are around 20 μm in diameter. The nucleus is around 1/3 of that diameter. Other elements of cell nuclei, such as lysosomes, Golgi apparatus, mitochondria and ribosomes are smaller still. DNA in the nucleus encodes specific proteins; these proteins assemble, by specific chemical interactions, to form the structures necessary to carry out complex functions. Did the cell have to learn nanocomposite technology in order to build these complex features? All life is constructed on nanocomposite building blocks.

MYTHS

Many myths already surround nanocomposite research. Nanocomposite study is viewed as a very new field. Popularizations of the field claim that nanoengineering manipulates structures at the atomic level, and explain that something "different" happens when particles become so small.

New Field

The term *nanocomposite* first appeared in the literature in 1986. The first online articles dealt with ceramics and magnetic composites. The first polymer nanocomposites were being discussed around 1992. These are recent events. The myth that nanocomposites are a new field has been fostered further by online research techniques. Almost everything published from 1986 is on line, while older investigations often are not. If 1986 were truly the birthday of the field, then paper searches through older literature would be less necessary or unnecessary.

There are at least two problems with the new field myth. First, nanocomposite literature really stretches back as far as materials literature does. Research on the effects of carbon black on rubber, or the precipitation-hardening of steel, may provide very useful insight for modern research. Good research is not blind to the past but stands on the shoulders of the giants who have gone before. Second, many techniques are available for understanding, processing, and characterizing modern synthetic nanocomposites. They do not need to be reinvented with each new generation of researchers.

Atomic Engineering

Popularizations of nanoengineering present breathtaking, near infinite vistas composed of flat planes of mica or quartz. On this barren field, atoms are pushed into place to serve as part of a nanomachine. This mental picture is fine, if each nanomachine can accomplish meaningful work alone. For perspective on productivity, consider electrospun fibers. Electrospinning can produce fibers that are 10s of nanometers in diameter. Over the course of the last year, perhaps a kilogram of such material has been produced worldwide. This could reinforce 10 kg of epoxy prepreg at 10% by weight, falling far short of any commercial needs. The reason is simple: small fibers weigh very little. Small devices can do very little work individually. Small wires can carry less current. In order for nanomaterials to pull their weight (or more), technologies for producing much more of them are necessary.

This leads to another aspect of the atom-level production myth. There are nanopreparation techniques that have been employed for a very long time, and can be poorly credited in the nanocomposites literature or popular press. Atoms are angstroms in size, molecules nanometers in size. In other words, chemists work by definition on nanomaterials, in production scales. Biological systems also work with nanoscopic proteins to produce functional devices of varying sizes. The synthesis and characterization tools developed in various fields are likely to be very useful in exploring and developing nanocomposites. A more eclectic, multidisciplinary training can provide a good launching point for nanocomposites engineering.

Properties Change for Nanoscopic Particles

People can draw faulty conclusions from nanotechnological rhetoric. If the first sentence of a report states "material x behaves differently when the particles are less than 100 nm in diameter," some might infer that all previously held understanding of material x is invalid below 100 nm. This is a generalization by the reader, but the author of the report contributed to the misunderstanding.

The properties of a material do not really change as the particle size becomes smaller. A well-trained scientist or engineer knows that the behavior of a material is a function of both the bulk of the material and the surface interactions. The shift occurs in the features that dominate the behavior. For example, particle settling is a balance between gravitational and buoyancy forces. For large particles gravity dominates, while for small particles buoyancy wins the tug-of-war.

NANOCOMPOSITES PRESENT

You are painting your house. You buy a latex paint because you can clean the brush with water. The can guarantees you a coating that will last for 10 years. Do you think of yourself as working with nanocomposites? You are. Paint is a composite of pigment, stiffening and latex binding particles, all smaller than 1 μm in diameter. The water-insoluble paint is stabilized by another nanoscopic feature: the surfactant forms a Langmuir-Blodgett film,. The paint is applied to a surface, the water evaporates, the film fuses over time, and, hopefully, a strong, aesthetically pleasing surface results.

NOMENCLATURE

The focus of this textbook is ceramic or metal nanoparticles in a polymer matrix. The nanoparticles in most common practice, as of the writing of this text, are silicate clay. This can be treated to form nanoscopic flakes. The text will deal with *nanoparticulates* that are particles (1D), fibers (2D), or flakes (3D). Through the text, composites are referred to in the form *matrix material/reinforcing material*, for example, polymer/ceramic or metal/metal. The term *filler* will be used to describe a component mechanically incorporated into the matrix that does not serve as an active component in the composite's performance. *Reinforcement* will denote a type of particle that does interact with the matrix to improve the composite's properties. *Structure* is divided into *primary* structure, which designates chemistry, *secondary* structure, which refers to molecular conformations, *tertiary* or *morphology* to describe crystallinity or other long-range order or disorder, and *dispersion* to describe the organization of the reinforcing phase. No distinction will be made between reinforcement by a soft phase in a hard matrix or a hard phase in a soft matrix.

ORGANIZATION OF THE TEXT

This book draws on the fields of chemistry, physics, chemical engineering and others. The intent is to help the student at any stage of a career think about nanocomposites in an integrated manner. As foundational, the text is intended to be supplemented with readings from the now extensive nanocomposites literature.

Chapters within the book are found in three major sections. The first section is a discussion of the physical chemistry of materials leading to nanocomposites. The sequence of presentation is traditional, starting from a discussion of the structures of materials, followed by the properties of materials, and finally by specific features of nanoparticles. Chapter 4, Nanoparticles, also makes sense as a starting point, rather than Chapter 2, Structures of Materials.

The second section is devoted to processing. The processing section is not organized in as traditional a fashion as the first section. In the introductory chapter, viscosity is discussed as a unifying process feature. Chapters on each solution and melt processing are next presented. Some of the particulars of polymerization are located in the context of creating a nanocomposite using solution processing, rather than as part of a discussion on polymer structure. The processing techniques are philosophically divided, rather than grouped into one chapter. This has permitted using different approaches for solution and melt processing without transition within a single chapter.

The third section discusses approaches to characterization, which loosely follow the organizational format of the first section. An introduction discusses experimental design, visualization of results, and presentation. The next chapter discusses structure characterization, while the third discusses characterization of properties.

Composite Materials

A material is the substance of the world around us. The properties are the purpose, the function of the material. The class of the material determines the basic behavior. There are three basic types of material: polymer, metal, and ceramic. Composite materials combine two or more of these material types. Nanocomposites also combine two different materials, with at least one component having one or more dimensions that are in the nanometer range. However, the composition alone does not determine the properties. A material has a structure that further contributes to the properties.

This section contains three chapters discussing facets of the structure and properties. Chapter 2 provides an introduction to materials and structure. Chapter 3 provides an introduction to the properties of materials, including features of discontinuously reinforced materials such as composites. Chapter 4 provides an introduction to features specific to the reinforcements, including particles, fibers and plates.

CHAPTER 2

Introduction to Solids

A composite is a combination of two or more chemically distinct solid phases. The combination is made in the hope that the best properties of multiple materials will be retained in the composite material. This is not a new or unique idea. Concrete is a combination of an aggregate, rock, cement, and sand. The cement alone is too dense and fragile to serve as a good paving material, but the additional materials provide reinforcement, fill space to make the concrete cheaper, and can impart environmental resistance to temperature or weather changes. Commercial synthetic composites include graphite sporting goods, such as tennis rackets, where graphite fiber provides the strength for the shaft, while an epoxy polymer provides the bonding strength between fibers and the flexibility necessary for proper performance. Composites are also found in nature: wood is a combination of cellulose fibers and a lignin binder. Both components in wood are polymers, but each serves a distinct mechanical purpose.

Nanocomposites can be studied from a molecular or a phenomenological point of view. A nanocomposite can be made from any combination of materials, as long as one of those is nanoscopic in at least one dimension. The nanocomposite is composed of a continuous matrix and a discontinuous reinforcing phase. There are three different types of basic building block materials: polymers, ceramics, and metals. Each of these three classes has specific properties that may be of value in a nanocomposite. Metals are conductive, strong, and ductile. Ceramics are very strong, brittle, chemically resistant, and insulating. Polymers are usually impact resistant, ductile, and insulating. Composite mixtures of these three types of material can result in properties that are a combination of the behaviors of the individual components.

Solid materials, in turn, can be atomic or molecular, crystalline or amorphous. The classification of a solid can therefore be based on atomic composition, molecular size, structure, or on properties.

ATOMIC AND MOLECULAR SOLIDS

Beginning with the primary structure of materials may seem to be putting the cart before the horse in understanding the bulk behavior of nanocomposites. However, many of the properties that materials exhibit will seem less mysterious when considered as resulting from the basic primary (chemical), secondary (molecular shape), and tertiary (crystalline) structures constituting them.

Materials may be atomic, including metals and ceramics, or molecular, including polymers. These two types of solids exhibit very different properties. The single-word explanation for the properties of a material is "bonding": the properties are a function of the bonds that form on the atomic level between atoms and molecules. The simplest way to build an understanding of atoms and bonding is to study the periodic table of the elements.

The Periodic Table

Discovery of the periodic table of the elements is credited to the Russian chemist Dimitri Mendeleev in 1869. Modern scientists understand the periodic table to be an ordering of atoms based on the number of electrons and protons in the atomic structure, subdivided by the number of electrons available to participate in chemical reactions. The periodic table of the elements originally was constructed empirically and was founded on a theory of molecular weights and on observations of physical properties (see Box 2.1). An interesting aspect of the periodic table is that theoretical, unobserved atomic and subatomic particles control the material properties, which can be measured, felt, and manipulated.

One of the useful features of the periodic table is the ability to predict the formation of ions. Recognize that the position and number of the atom on the periodic table describes the number of protons and electrons in that atom. *H*, hydrogen, has one of each, electron and proton. *He*, helium, has two, *Li*, lithium has three, and so forth. In theory, the electrons for each of these atoms can be used to participate in chemical reactions. However, consider the elements in the far right column of the table, the so-called noble gasses, and one finds that if there are eight electrons (well, two for *He*) something has happened to make these atoms very re-

Box 2.1 The Periodic Table of the Elements

The course of the invention of the "periodic table of the elements" is typical of many scientific tools students now take for granted. Scientists and engineers have come to realize that the organization of the elements in rows and columns devolves naturally from the electronic structure of the individual elements. However, the main inspiration for the table arose from consideration of the atomic weight, size, and properties.

The ancient Greeks proposed the first atomic theory. They proposed four fundamental elements: earth, air, fire and water, and invisibly small atoms that composed these elemental substances. They further proposed that the difference between these atoms was the shape. Before being impressed by the foresight of this theory, we should recognize that this was one of a number of competing philosophical theories, which disappeared as philosophers developed new speculative interests.

John Dalton revived the atomic theory in the early 1800s to explain well-known material behavior. Dalton decided that the difference between atoms were weight-based. Various scientists, including Amedeo Avagadro, advanced the theory of atomic weights until, by 1828, there were tables of atoms and atomic weights available. The idea of size was also revived; from the measured density and the proposed atomic weight, atomic sizes could be proposed.

In 1869, Dmitri Mendeleev, a Russian chemist, arranged the atoms in a pattern based on increasing molecular weight that would be understandable based on viewing the modern periodic table. Mendeleev left holes where he believed atomic substances to be missing. Based on this table, Mendeleev predicted the properties of the missing materials. The discovery of three of the missing elements, with the predicted properties, cemented the periodic table of the elements as an invaluable tool for chemistry and engineering.

1 H Hydrogen	2 He Helium																2 He Helium
3 Li Lithium	4 Be Berylium											5 B Boron	6 C Carbon	7 N Nitrogen	8 O Oxygen	9 F Fluorine	10 Ne Neon
11 Na Sodium	12 Mg Magnesium											13 Al Aluminum	14 Si Silicon	15 P Phosphorous	16 S Sulfur	17 Cl Chlorine	18 Ar Argon
19 K Potassium	20 Ca Calcium	21 Sc Scandium	22 Ti Titanium	23 V Vanadium	24 Cr Chromium	25 Mn Manganese	26 Fe Iron	27 Co Cobalt	28 Ni Nickel	29 Cu Copper	30 Zn Zinc	31 Ga Gallium	32 Ge Germanium	33 As Arsenic	34 Se Selenium	35 Br Bromine	36 Kr Krypton
37 Rb Rubidium	38 Sr Strontium	39 Y Yttrium	40 Zr Zirconium	41 Nb Niobium	42 Mo Molybdenum	43 Tc Technetium	44 Ru Ruthenium	45 Rh Rhodium	46 Pd Palladium	47 Ag Silver	48 Cd Cadmium	49 In Indium	50 Sn Tin	51 Sb Antimony	52 Te Tellurium	53 I Iodine	54 Xe Xenon
55 Cs Cesium	56 Ba Barium	57 La Lanthanum	72 Hf Hafnium	73 Ta Tantalum	74 W Tungsten	75 Re Rhenium	76 Os Osmium	77 Ir Iridium	78 Pt Platinum	79 Au Gold	80 Hg Mercury	81 Tl Thallium	82 Pb Lead	83 Bi Bismuth	84 Po Polonium	85 At Astatine	86 Rn Radon
87 Fr Francium	88 Ra Radium	89 Ac Actinium	104 Rf Rutherfordium	105 Db Dubnium	106 Sg Seaborgium	107 Bh Bhorium	108 Hs Hassium	109 Mt Meitnerium	110 Uun	111 Uuu	112 Uub		114 Uuq				

58 Ce Cerium	59 Pr Praseodymium	60 Nd Neodymium	61 Pm Promethium	62 Sm Samarium	63 Eu Europium	64 Gd Gadolinium	65 Tb Terbium	66 Dy Dysprosium	67 Ho Holmium	68 Er Erbium	69 Tm Thulium	70 Yb Ytterbium	71 Lu Lutetium
90 Th Thorium	91 Pa Protactinium	92 U Uranium	93 Np Neptunium	94 Pu Plutonium	95 Am Americium	96 Cm Curium	97 Bk Berkelium	98 Cf Californium	99 Es Einsteinium	100 Fm Fermium	101 Md Mendelevium	102 No Nobelium	103 Lr Lawrencium

FIGURE 2.1 Periodic table of the elements.

sistant to chemical reactions. Further, if the atom is located far to the left or one or two columns just short of the far right, the atom will be much more likely to participate in chemical reactions. Chemical reactions lead to the formation of bonds, and bonds lead to atomic or molecular solids.

What has happened is that, for atoms in the far right column, an entire "orbit" has filled. For atoms in the top several rows of the table, there are two, and only two, orbits, an "s" which holds up to two electrons and a "p" which holds up to six electrons. *He* has the first "s" orbit filled with two electrons, while all the other atoms beneath *He* have both an "s" and "p" orbital filled with eight total electrons. These collections of electrons are not really orbiting the nucleus, and there are far more details about what distinguishes these orbitals and why they behave in this manner, but for the purposes of bonding, this description works. Since the orbital levels are full, the atom is stable. For sound chemical reasons, all atoms have an energetic drive to have a complete energy level.

Therefore, the atoms in the far left column, *Li* and so on, can lose one electron and have a complete shell. In fact, this happens spontaneously, and vigorously, in water. The next column over, headed by Mg, can lose two electrons, and so on. The result is an atom with too few electrons to balance the protons, and a net positive charge. This makes an ion, a cation that will be attracted to a negatively charged species. In addition, since there is a positive charge to draw strongly on the negatively charged electrons, and since the most loosely bound electrons are missing, cations are significantly smaller than the corresponding source atom.

On the other hand, the atoms toward the right, O and F, and so on, can easily gain electrons to complete a shell. The result in this case is an atom with too many electrons for the protons to balance, resulting in a net negative charge. These ions, anions, will be attracted to positively charged species. Again, since there is a net negative charge, which drives the electrons away from each other, and extra electrons, anions are significantly larger than the corresponding source atoms.

The cation, after losing electrons, pulls the remaining electrons in even more tightly, resulting in a smaller net size. The anion, after gaining an electron, does not have the counter-charge to draw the electron to the expected diameter, resulting in a larger net size. The creation of ions is the building block of a major class of materials, ceramics.

A zigzag line divides the atoms that will tend to lose electrons, on the left side of the table, from the atoms, on the right side of the table, which will tend to gain electrons. In chemical terms, this line is the demarcation between atoms that are strongly electronegative, non-metals that gain electrons, and weakly electronegative, metals that lose electrons. Atoms that are close to this line may sometimes gain and sometimes lose elec-

The orbital is not a boundary on which the electron travels. The orbital is the line describing either: (1) the most probable set of locations for two electrons, one with spin "up" and one with spin "down"; or (2) the line containing the location of two electrons, spins up and down, 50% of the time. Another meaning of "shell" is the group of all orbitals "s," "p," "d" etc., along a single row of the periodic table. For the "s" orbital, the electron has the greatest probability of being found right in the center of the atom. However, the particle is very small, and so the vibrational space filled is bigger: a sphere. The electron particle moves very quickly, and within that boundary will exert a force against any encroaching particle, such as another electron. The electron will exert some presence outside of the 50% boundary as well. Especially as the temperature increases, the electron will tend to range out farther from the nucleus. This behavior is easiest to understand for the s shell, but is also true for the "p" orbitals. There are three "p" orbitals, each oriented orthogonally to the other two. The "p" orbital has one node in the vibrational energy, giving the shape shown in Figure 2.2. Additional orbitals, types "d" and "f," have two and three nodes, respectively. These extra nodes add extra energy, and so the 3d shell fills after the 4s and 4p shells do.

When there are more than two electrons, the "s" and "p" orbitals will combine, or hybridize, allowing a four-lobed tetrahedral electron density to form. The reorganization is necessary to explain double and triple bonding. Without hybridization, the "p" orbitals remain at 90° relative to each other, and preclude the overlap necessary to form the "pi" bonds found in multiple bonds between two atoms. The electronic reorganization keeps the charged electrons as far from each other as possible, providing the driving force for hybridization.

The side effect of this perfectly natural reorganization is the tetrahedral bonding structure of carbon. The tetrahedron structure is also compatible with the Lewis dot structures used in basic chemistry to predict chemical structure. However, while the dot structures are drawn at 90°, the real structures form bonds at relative angles of 109.5°, creating a three-dimensional structure.

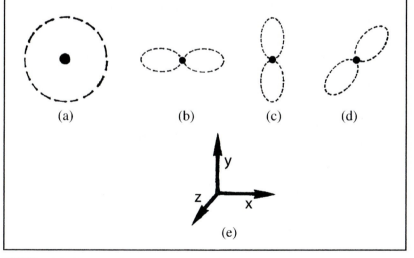

(a)   (b)   (c)   (d)

(e)

FIGURE 2.2 Electron orbitals, also called shells. (a) An "s" orbital. (b)–(d) Three mutually orthogonal "p" orbitals. (e) The x, y, z coordinate system.

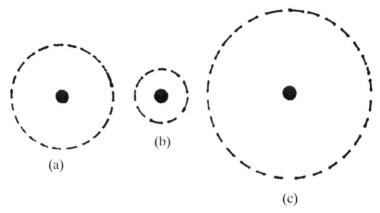

FIGURE 2.3 An atom (a) placed in size scale with its ions (b, c). The creation of ions is the building block of a major class of materials, ceramics. The cation (b), after losing electrons, pulls the remaining electrons in even more tightly, resulting in a smaller size. The anion (c), gains an electron but does not have the countercharge to draw the new electron to the expected diameter, resulting in a larger net size.

trons, atoms farther from the line will usually behave as expected. There is a second line, straight and vertical, that is usually drawn on the periodic table between the atoms in the last column and atoms in the penultimate column, to indicate the noble gases in the far right column. This column is composed of atoms that do not have to gain or lose electrons to have a filled energy level. The strength or weakness of the electronegativity grows with distance from the zigzag dividing line, and the strength of the electronegativity, in turn, determines how likely the atom is to participate in a chemical reaction. Finally, the column that the atom is in determines how many electrons the atom needs to gain or lose in order to achieve a filled shell. The electrons in the d and f shells do not participate in the bonding dance.

There is a column, headed by carbon, which has four electrons and needs four electrons to fill a shell. This creates an odd situation that is responsible for all life: a carbon has to gain or lose too many electrons for ions to form readily. Instead, two or more non-metallic atoms will share electrons instead of gaining or losing them. This forms a covalent bond. When sharing electrons, both atoms can function as if they have a complete electron shell. Covalent bonds are very strong, since both atoms are driven to preserve the bond, plus the stability benefits of having a complete shell. However, since the bond is between two atoms, the strength is only along the bond and not effective in all directions as are the attractions between charged particles bonding ceramics together. Unlike metallic and ionic bonds, which are isotropic, equal in all directions, covalent bonds are anisotropic, between the two bonded atoms only. The attraction between atoms outside of the covalent bond is limited to secondary bonds, which are far less strong than the covalent bonds.

All of the nonmetals are capable of forming covalent bonds, but the most important are carbon [C], hydrogen [H], nitrogen [N] and oxygen [O]. The chemistry of making and breaking covalent bonds, especially with carbon, is organic chemistry. Sulfur [S], chlorine [Cl], fluorine [F] and silicon [Si] are also of some importance in organic chemistry.

The atoms to the left of the zigzag line include aluminum (Al) and tin (Sn), and magnesium (Mg) and calcium (Ca). These materials are metallic and exhibit specific properties because of this. Atoms toward the left column will relinquish electrons readily. These electrons can be shared amongst all neighboring atoms forming a metallic bond. If an electron is removed from one atom, an electron from a neighboring atom can transfer to the vacated electron energy level. This will form a cascade of moving electrons, which can be measured as a flow of current. Thus, metals are conductive.

Two large categories of atoms do not occur in the periodic table until first the fourth and then in the sixth row. These are the transition metals beginning in the fourth row, and the rare earths and actinides, added in the sixth row. The transition metals are the result of electrons occupying "d" orbitals. The rare earths and actinides are the result of electrons occupying the "f" orbitals. Like lithium and arsenic, the transition metals, rare earths and actinides occur to the left of the stair-step line, demarking them as losing electrons and therefore metallic. Unlike items in the left two and right six columns, however, the number of electrons the transition metals may lose is often variable. Transition metals will most commonly lose one, two, or three electrons. The number of electrons that will be lost depends on the neighboring atoms with which the metal bonds.

Transition metal chemistry can be complicated. When forming metallic bonds, the atoms will pack as a sphere with its radius depending on the atom. When forming ions, the positive charge will be variable.

THE ROLE OF STATISTICS IN MATERIALS

Understanding the properties of materials may be less confusing if one can first understand that the properties observed for a solid are an average of the behavior of all of the constituents composing the solid. That is: each individual atom, or group of atoms, or molecule, may exhibit individual behavior differing from the whole, but individual elements contribute only a part of the behavior. The final behavior is the behavior of the average contributor.

Calculating the average value of something can be an interesting task, however, as shown in Do It at Home 2.2. Not all properties depend on the same aspect of an atom or molecule. The standard manner of averaging, in which all of the properties are summed together and normalized, or divided by, the total number of objects having that property is the arithmetic average, or the first moment.

Do It at Home 2.1

Experiment 1

For materials, strength is not limited by the weakest link but is an average of all the links. The elastic modulus of a material can increase through improvements to any contribution in bond strength from any part of the total system.

The strength of rubber bands: Collect some rubber bands of various types. There should be around five rubber bands of each type. Take a bundle of the five weakest rubber bands and pull on them. Consider the force required to do this. Now take a bundle of the five strongest rubber bands. Consider the force required to stretch this bundle. Finally, take a mixture of any five rubber bands. The force required to stretch the bundle should be intermediate between the weakest and the strongest bundle.

Experiment 2

The contents of a container measured by mass: Collect balls of different mass, for example ball bearings, ping-pong balls, beans, or golf balls. Place all of the balls in a shoebox. The mass of the box is equal to the mass of all the items placed inside. The average increase in mass with the addition of each particle is the mass of the whole, divided by the number of balls put in the box. So far, so good: that is, until a ball is removed and one tries to predict the new mass of the box. The new mass depends very much on the mass of the ball removed. Removing one light particle hardly affects the mass at all, removing a heavy particle affects the mass significantly. A single heavy particle is "worth" more than a number of light particles. If a handful of balls are removed, the new mass can be calculated simply using the number of balls removed, because any random handful should include a general representation of the total content of the box.

$$\overline{P}_a = \frac{\sum\limits_{i=1}^{n} N_i P_i}{\sum\limits_{i=1}^{n} N_i} \qquad (2.1)$$

$\overline{P}_a$ is the arithmetic average value of a property for a sample, n is the total number of items in a sample, N_i is the number of objects with a given property, and P_i is the value of that property for those objects. There are other moments of the average. The second moment is described by:

$$\overline{P}_2 = \frac{\sum\limits_{i=1}^{n} N_i P_i^2}{\sum\limits_{i=1}^{n} N_i P_i} \qquad (2.2)$$

A number of moments of various properties can be calculated. The different moments of the average of a property bias the behavior toward the largest values of the property.

PRIMARY, SECONDARY AND TERTIARY STRUCTURE

Primary Structure

The primary structure refers to the atomic structure, and how the atoms interact with each other. There are three types of bonds, characteristic of the three classes of materials. Ionic bonds form in ceramic materials, metallic bonds form in metallic solids, and covalent and secondary bonds are both contributors to molecular solids, of which polymers are the engineering material representatives.

Ceramics

Ceramics are ionic solids. That is, cations and anions mix to form a charge-balanced solid. The ions are approximately spherical in shape, and the charge exerts an attractive force in all directions, or isotropically. Naturally, therefore, the cations and anions will move into positions relative to each other so that each positive + charge will be balanced by a net negative – charge, and vice versa. This spacing will be regular. The alternative to regular spacing would be irregular spacing. If the spacing between ions becomes irregular, the charges may be thrown out of balance.

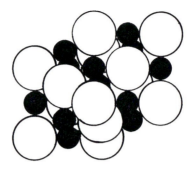

FIGURE 2.4 The unit cell for sodium chloride, NaCl. The large atoms are chlorine anions; the smaller atoms are sodium cations. The structure shown is the "rock salt" unit cell.

Heating causes the atoms to move around relative to the optimal bond length. Atoms dislocated from the regular structure are not the preferred, lowest energy condition.

The type of bond described is an ionic bond. Cations and anions, having opposite charge, will be drawn to each other. The electrostatic charge is strong, and can be described by the mathematical relation:

$$E_A = -\frac{A}{r} \tag{2.3}$$

E_A is the energy of attraction, A depends on the types of ions involved, and r is the separation between the ions. r is used because the attraction is isotropic, independent of direction, and can be approached from any direction, resulting in a sphere of influence. Electrostatics provides a form for the attractive energy.

$$A = \frac{(Z_1 e)(Z_2 e)}{4\pi\varepsilon_0} \tag{2.4}$$

Z_1 and Z_2 are the valences of the ions, e is the electronic charge and ε_0 is the permittivity of a vacuum. The implications of this expression are that the electrostatic charge has a known value, and that the strength of the attraction between ions changes is a function of separation distance. On the other hand, as the ions approach each other, they will begin to experience a hard-core repulsive force.

$$E_R = \frac{B}{r^n} \tag{2.5}$$

E_R is the repulsion experienced once the electrons stop shielding the nuclei of the two ions from each other, and B and n are a function of the ions. This core repulsion will eventually stop the ions from continuing to close the gap between them. As a result, the balance of attractive and repulsive forces creates an energy well with a minimum where the repulsive force starts to overwhelm the attractive force. The ions will form a bond with an average length defined by the separation at the bottom of the well. An interesting observation from this analysis is that an ion does not really have a defined size unless it is approaching another ion, and the size can change with respect to the species it is approaching. For ceramics the energy well describing bonding is deep and symmetrical.

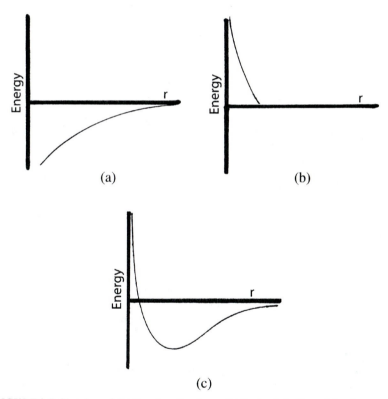

FIGURE 2.5 Sketches of: (a) the attractive force, (b) the hard shell repulsive force, and the combination of these two forces resulting in (c) the energy well describing bond formation. Note the equilibrium energy and bond length corresponding to the bottom of the energy well. Separating two bonded atoms requires supplying enough energy to remove the atoms to infinite separation.

Do It at Home 2.2

Simulating the packing in an ionic solid: Collect identical round magnets. Place half of the magnets face up and half face down. The south and north poles of the magnet stand for the positive and negative ionic charges. Try to arrange the magnets into an ordered structure. The magnets should form a square, with two up magnets in diagonally opposed corners, and the down magnets completing the other two corners. This arrangement of magnets is quite stable.

Ionic bonds are quite strong, balanced and isotropic. This gives rise to the characteristic properties of ceramics. Since the atoms in the solid all have ionized to form a filled shell, and the ions formed by filling those shells are balanced by neighboring and oppositely charged ions, ceramics are insulating and chemically resistant. Ionic bonds are also difficult to break, making ceramics strong and rigid.

Another important feature of ceramics is that the cation and anion usually are very different in size. This has an impact on the ordering in the solid. Since the cations can be considered effectively packed into holes in a regular array of anions, it is difficult to move one ion past another. Further, in order to cause the ions to slip one past another requires enough energy to completely separate two ions. Therefore, ceramics are brittle, breaking instead of deforming when an excessive force is applied to the solid.

Metals

In an atomic solid, the ease of losing electrons translates into the ability of electrons to freely migrate away from the source atom. The sharing of electrons in atomic orbital theory results from the interactions of the orbital energy levels of two bonding atoms, which creates low energy states that are only partially filled. The result is that metals are conductive. In order to create the necessary electron balance, metals will share electrons between all neighboring atoms. This delocalized electron density is known as a metallic bond. Metallic bonding is isotropic. Since the electrons will be in continuous motion, the bond is easily reformed if it is broken or dislocated. The bond is reasonably strong. The electrons can wander over some distance, making a metallic bond a long-range force. Therefore, because of the metallic bonds, metals are strong, withstanding deforming forces well. Metals are also ductile, due to the easily reformed and displaced bonds. The number of electrons available for sharing is determined by the number of electrons that the original atom is capable of easily losing.

There are specific rules for two or more different metallic atoms to mix well. The atoms must be of similar physical size. Even though the atoms do not lose their electrons in metallic bonds, they must be able to share the same number of electrons. The atoms must have approximately the same affinity for electrons or the electrons will not be shared. Finally, the atoms must form the same type of crystals, with the same conditions for structural packing interactions.

Organic Molecules

In nonmetallic atoms such as carbon and oxygen, covalent, shared bonds will form. The electron travels in a joint orbital, creating a strong, directional bond. The electron orbitals of the atom tend to mix in character, resulting in reactive sites that are as far away from each other as possible. This creates a tetrahedral bonding structure with four reactive poles (Figure 2.6). In carbon and silicon, each of these reactive poles can form one bond. In oxygen two of the poles can form bonds, while two electrons each already occupy two of the poles. Nitrogen can form bonds with three of the poles while two electrons already naturally occupy the fourth pole. So, carbon can form four bonds, oxygen can form two bonds and nitrogen can form three bonds. A little geometry shows clearly that two carbons can overlap one, two or three bonds to form single, double or

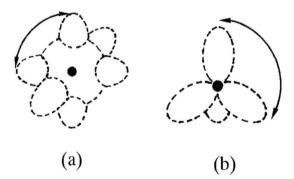

(a)   (b)

FIGURE 2.6 Hybridized orbitals. (a) Overlapping "s" and "p" orbitals. The electron density shows considerable overlap. (b) Hybridization of the "s" and "p" orbitals creates a tetrahedral bonding structure. One, two, or three of the lobes of electron density can overlap with a neighboring atom, creating single, double or triple bonds, respectively.

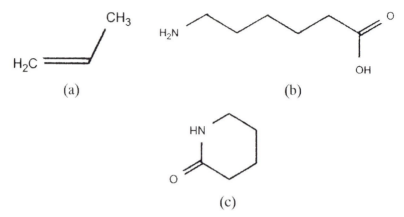

(a)

(b)

(c)

FIGURE 2.7 Chemical structures of some monomers used in nanocomposites, including those with double bonds for addition polymerization, such as (a) propylene, and acid-base functional monomers for condensation polymers, such as (b) 6-amino hexanoic acid, and (c) caprolactam. Caprolactam is the cyclized isomer of 6-amino hexanoic acid, created by elimination of H_2O.

triple bonds between atoms, respectively, as can nitrogen. Oxygen can form single or double bonds with itself. In nature, oxygen and nitrogen are usually found as doublets: oxygen double bonded to oxygen, nitrogen triple bonded to nitrogen.

Of course, the interesting organic molecules are formed from bonds between mixtures of different types of atoms. Figure 2.7 shows sketches of some typical organic molecules that comprise interesting monomers for polymerization. These can be divided up into hydrocarbon and heterochain molecules. Hydrocarbon molecules are composed exclusively of carbon-carbon backbones. Heterochain molecules have another atom in the backbone, usually either nitrogen or oxygen.

Polymers

Polymers are the most diverse set of solids. Polymers are composed of long, repeating sequences of organic molecules. Here, long means in excess of 2000 carbons (or equivalent) in a single backbone. In order to make the low quality polyethylene found in six-pack binders and milk jugs, about 10,000 carbon atoms are polymerized to form a single chain. The milk jug in turn contains thousands of these long chains. The repeat sequence of a polymer can be composed of a variety of different types of organic or silica-based molecules. These long chains interact with each other in some ways similar to other molecules, with some complications

The traditional chemical shorthand uses the chemical element symbols, strung together in specific combinations. However, all of the descriptions in Figure 2.8 are recognizable as polystyrene, and each chemical representation is useful for different reasons. The shorthand chemical formula provides quick information about the elemental structure. The line representation pares away hydrogen, as all bonds that are not shown are implicitly with hydrogen, improving the visibility. The line representation also provides better information about bonding. The stereographic projection furnishes information about the resistance to rotation about a single bond. The planar zigzag representation depicts a lowest-energy, planar projection, providing an echo of the three-dimensional structure. The overhead projection retains some of the features of the planar zigzag, but allows the reader to recognize structural isomers easily.

Doubly or triply connected lines indicate doubly or triply bonded atoms, respectively. Double and triple bonds tend to be mobile, through a process called delocalization, and may be partially stabilized by side groups such as the chlorine in Figure 2.9(c). The delocalized electron density along the chains is part of the conditions necessary for conduction in polymers.

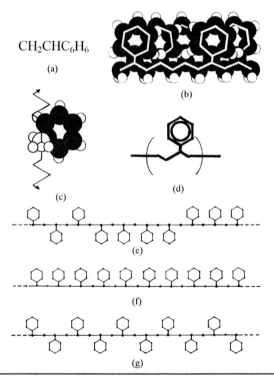

FIGURE 2.8 Several different representations of polystyrene. (a) The shorthand chemical formula, (b) a space-filling, chain extended representation, (c) stereographic projection, (d) line structure, and overhead projections, showing the (e) atactic, (f) isotactic and (g) syndiotactic structural isomers.

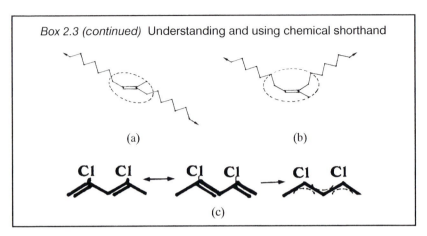

Box 2.3 (continued) Understanding and using chemical shorthand

(a)   (b)

(c)

FIGURE 2.9 Structural isomers of isoprene, natural rubber, and the linear structure of double bond-containing conducting polymers. (a) trans isoprene, (b) cis isoprene, or gutta percha, and (c) two poly(chloro acetylene) repeats with resonance indicated.

due to the molecular weight. These interactions create the properties of polymers, including toughness, rubbery elasticity, adhesiveness, transparency, low density, insulation, flexibility, and so forth.

The bonds between atoms within the molecules, the intra-molecular bonds, are covalent. There are also interactions between molecules outside of the covalent bonds. These interchain interactions are much weaker than the covalent bonds. The interchain interactions are grouped together under the term "secondary bonds", consisting of London dispersive forces, hydrogen bonding and dipole bonding. This mixture of strong and weak bonds is behind the diverse properties of polymers.

Hydrogen bonding typically occurs in heterochain molecules where there is a hydrogen bonded to an oxygen or a nitrogen on one molecule and a nitrogen or oxygen on another molecule. In this situation, the hydrogen can "change allegiance," becoming primarily bound to the original atom but also partially bound to the oxygen or nitrogen from the neighboring chain. In this manner, a hydrogen bridge between the two molecules is formed. This is the strongest of the secondary bonds. Therefore, hydrogen bonding creates the closest, or shortest, secondary bond. This bond is easily broken and reformed under the influence of temperature or force.

Permanent electron dipoles are present predominantly in heterochain molecules. When a covalent bond forms between two atoms with dissimilar electron affinities or electronegativities, the electrons forming the bond will tend to contribute more electron density to the more

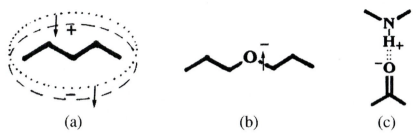

(a)   (b)   (c)

FIGURE 2.10 Types of secondary bonds. (a) In a dispersive bond, a temporary +/− dipole charge is developed when more electron density occurs momentarily on one side of the atom. (b) A permanent dipole on some molecules can result in strong interactions between molecules. (c) A hydrogen bridge occurs where one hydrogen atom is affiliated with one or more neighboring but non-bonded atoms.

electronegative atom. Thus, for example in a carbon-oxygen bond, the oxygen has "more" of the shared electrons, giving the oxygen a slight negative charge and the carbon a slight positive charge. This electronic dipole is a weak but permanent feature of the molecule. The positive head of the dipole on one molecule can be attracted to the negative head of a neighboring molecule, giving rise to a moderately strong attractive force between the two molecules. This is the second strongest form of secondary bond, but is even more easily broken and re-formed than hydrogen bridges.

Finally, dispersive forces can result in very weak secondary bonds between chains. The electron clouds surrounding the atoms forming the molecule may all become biased, for various reasons including pure chance, toward one side of a molecule. The electrons are moving very fast, and interact with each other slightly, so this can happen often enough that the molecules will frequently have a charge imbalance, creating a short-lived dipole. The attraction between these dipoles is very weak, but sufficient to explain the liquid and solid phases of hydrocarbons. Clearly, the bigger the molecule, the more electrons are available to become unbalanced, so there should be an observable size effect. Ethane, with two carbons and six hydrogens, is a gas at room temperature. Octane, with eight carbons and 18 hydrogens, and kerosene, twelve carbons and 26 hydrogens, are both liquid at room temperature. Vaseline, at approximately 20 carbons and 42 hydrogens, is a grease, while paraffin, at approximately 50 carbons and 102 hydrogens, is a wax. While individually, due to the temporary nature of the dipoles and the weak attractive force, dispersive bonds are the weakest of the secondary bonds, they do have significant effects in great number.

Do It at Home 2.3

Overlap about an inch of both sides of a long strip of Velcro™. Try to pull the two strips apart by opposed, tensile, force rather than shear, or peeling, force. Consider how little force is required to do this. Now overlap successively longer segments of the Velcro™ strip. Is there a length at which separating the strips becomes functionally impossible? The bond is still temporary, and can be broken easily by peeling the layers apart. Why is the force required for separation different in peel and shear?

In order for a molecule, or monomer, to participate in polymerization the molecule must have two or more reactive groups. The monomer must be able to connect to at least two other molecules in order to form a continuous chain. Hydrocarbon type chains form by a process called chain polymerization. This type of reaction uses monomers with carbons that are double-bonded. During the reaction, the double bond is half-broken,

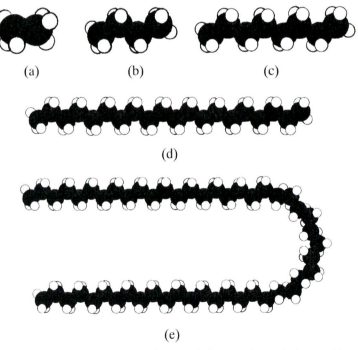

FIGURE 2.11 A homologous series of carbon chains: (a) ethane, (b) butane, (c) octane, (d) twenty carbon chain, e.g. Vaseline™, and (e) fifty carbon chain, e.g. paraffin wax.

leaving a single bond between the original carbons but allowing each of the two carbons in the original pair free to form another bond. Thus, a continuous chain can form. The reaction is very fast and generally clumsy, resulting in disorganized polymer chains. However, there are ways to make the polymers more organized. Rubber trees, for example, use enzymes to produce polyisoprene, natural rubber. Other common chain growth polymers include polyethylene, polystyrene, polypropylene, poly(vinyl alcohol) and poly(vinyl chloride).

A second major type of polymerization is step growth. Step growth occurs by the route most people commonly think of as chemistry. Two different types of monomers, one with an acid functional group and one with a basic group, react to create a new, larger compound. These can react in a continuing series to build up long backbones. Some common commercial step growth polymers are nylon, polyester, polyurethane and epoxy. Nature forms many types of step growth polymers, as well. In animal systems, substances such as collagen and proteins are formed from amino acid building blocks using templates such as RNA or enzymes. Plants form cellulose, the main mechanical building block for plant fibers, from different sugars (saccharides), linking them together to form long chains.

Polymerization is discussed more in Chapter 6.

Molecular Weight

A frequent problem to which statistical averaging applies is the molecular weight of long polymer chains. The periodic table provides the atomic weight of atomic solids. The atomic weight is the mass, in grams, of a mole, or 6.02×10^{23} atoms, of atoms. The molecular weight of small molecules is the sum of the compositional atoms. So, water, H_2O, has 2 hydrogens, at approximately 1 g/mole apiece, and oxygen at approximately 16 g/mole. The molecular weight of water is 18 g/mole. These atomic weights were determined after experiments with changes in mass due to chemical changes in bulk weight after certain types of reaction, such as oxidation.

The molecular weight of a polymer is less simple to define. Polymerization reactions are a statistical process. The chains formed have a structure and length that depends on the technique used for polymerization. Some biological systems and some specialized synthesis techniques can create a polymer with a regular structure or a "monodisperse" distribution of chain lengths. A brief list of examples is given in Table 2.1.

How does one measure the molecular weight of a polymer? The chains are long, but the chain lengths are uncertain. One can measure the mass of a sample, and then count the chain ends. This provides a measure of

TABLE 2.1 A list of molecular weight distributions and structures in some polymers.

Polymer	Synthesis	$\overline{M}_w / \overline{M}_n$	Primary Structure	Secondary Structure
Polystyrene	Free-radical	Ideally 2	Random, branched	Random coil
	Catalyzed free-radical	Ideally 2	Isotactic	Crystallizable
	Emulsion	$1.1 > P > 1.0$	Random	Random coil
Nylon	Condensation	> 2	Alternating	Crystallizable
RNA	Enzyme	1	Alternating	α helix
Collagen	Cellular assembly	> 2	Alternating	Triple helix

the arithmetic average, provided the chain is not branched, since branches create more than two ends per chain. Gel permeation chromatography is the current standard for obtaining molecular weight. However, many techniques, including osmotic pressure, viscosity, and light scattering are used to measure molecular weights.

For most polymers, there will be a distribution of chain lengths. The number average molecular weight is the arithmetic mean, and the weight average is the second moment of the weight distribution. The polydispersity, P, is the ratio of $\overline{M}_w / \overline{M}_n$, and provides an estimate of the shape of the distribution. A monodisperse polymer has a P of 1. A purely normal distribution of chains has a P of 2. The proper moment of the molecular weight to use in calculations depends on the property. Viscosity and other processing parameters depend more heavily on the longest chains, and so $\overline{M}_w$ is the appropriate average. Thermodynamic properties, such as the freezing point depression, boiling point elevation, and osmotic pressure, which governs the dialysis of blood, depend on the total number of molecules present, and so depend on the $\overline{M}_n$.

Secondary Structure

Polymer chains can have a secondary structure caused by the primary, chemical structure. The secondary structure describes the ordering of the chain in space. For example, disordered chains can occupy a large volume defined by the random coil. Helical chains, such as nylon or DNA, take up less volume but are longer. Very rigid chains may be chain extended, making them as long as possible. The arrangement of the individual chains in space can affect the way the chains interact with each other. Therefore, the secondary structure can affect all of the properties ranging from viscosity of flow to mechanical strength of the solid phase.

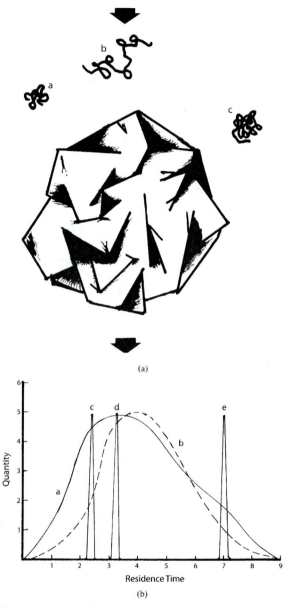

FIGURE 2.12 (a) Schematic of interactions between polymer chains and substrate in a packed-bed GPC column, with a—an example chain, b—a chain expanded due to affinity with the surrounding solvent, and c—a longer chain, which also takes up more volume. Chains occupying larger volume can interact less with a substrate particle than smaller chains, resulting in less residence time in the column. (b) A schematic of a GPC curve. Two molecular weight distributions a. and b. are shown, with superimposed c = high, d= medium, and e = low molecular weight calibration standards.

31

Box 2.4 Gel Permeation Chromatography (GPC)

In GPC, a sample is dissolved in a solvent. The solution is passed through a static, porous bed. Long chains cannot penetrate the porosity, and pass through rapidly. The process is described more fully in Chapter 9. The plot of mass of material vs. molecular weight that emerges is the complete distribution of chain lengths, as shown in Figure 2.12. The data from this plot can be used in the same sort of average calculation described previously.

$$\overline{M}_n = \frac{\sum N_i M_i}{\sum N_i}$$

$\overline{M}_n$ is the number average molecular weight and $\overline{M}_w$ is the weight average molecular weight.

$$\overline{M}_w = \frac{\sum N_i M_i^2}{\sum N_i M_i}$$

The disadvantages to GPC lie in the solution properties of the polymer. If a different solvent is used to dissolve the polymer, the diameter of the polymer coil will be larger or smaller. This will change the apparent molecular weight. In order for the measurements to be accurate, the unknown polymer and the control polymer also must have exactly the same solution diameter character.

Random Coil

The most common shape assumed by the chain is a random coil. A random coil will look something like a poorly stored extension cord, with loops, crossings and spaces. In molecular solids the chains of other molecules or solvent molecules may occupy the spaces between the elements of a single chain. Alternatively, there may be small amounts of trapped space, too small to fit any other component. This trapped space has been called "free volume". The natural molecular motion of the chain, which increases with heat, also will introduce unoccupied volume to a solid.

Some simple math and statistics will allow the estimation of the size of the random coil. The length of the chain can be described by a mathematical construct called the random walk. The walk is comprised of steps of a size corresponding to the bond length, with steps at angles defined by the bond angle and bond rotation, as shown in Figure 2.13. The contour length of the chain, L, is defined by the number of steps, n, and the bond length l_0.

$$L = nl_0 \tag{2.6}$$

The contour length is a characteristic length for a chain of a given composition. However, a more representative characteristic length for the chain is the separation between the endpoints of the chain. In the absence of restrictions imposed by bond angles and rotation, the end-to-end separation is characterized by the random walk in three directions.

$$\bar{r}^2 = nl_0^2 \qquad (2.7)$$

r is the separation, and r is used to indicate that it can be in any three-dimensional direction. The geometric restrictions on rotation due to the bond angle, θ, and bond rotation, φ, cause the end-to-end separation to expand. This imposes constraints that modify the basic mathematic representation.

$$\bar{r}^2 = nl_0^2 \frac{1 - \cos\theta}{1 + \cos\theta} \frac{1 + \overline{\cos\varphi}}{1 - \overline{\cos\varphi}} \qquad (2.8)$$

The average of the cosines of all accessible angles of rotation is used to calculate the expansion for bond rotation. If the temperature is high enough, the chain may rotate freely, and the effect of bond rotation is usually neglected.

The end-to-end separation provides a reasonable estimate of the actual length covered by a chain in a random coil. However, the random walk presents several potential problems. First, random walk statistics include the possibility of an infinite number of steps in a straight line. This would correspond to an infinite polymer chain, which is unrealistic. Second, in the case of a polymer, the walk must be self-avoiding to avoid the chain crossing through itself. This expands the chain slightly. In order to allow for unknown factors, the end-to-end separation also may be described using an effective bond length, l_{eff}, rather than the more specific, geometric form.

$$\bar{r}^2 = nl_{eff}^2 \qquad (2.9)$$

This expression recovers the simplicity of the original expression for end-to-end separation.

A solvent may have an additional impact on the chain size. The solvent may be good, in which case the polymer will absorb extra solvent and swell. The solvent may be poor, causing the polymer to exclude the solvent, shrinking somewhat in response. The swelling can be accounted for by a swelling coefficient, α, which can range from greater than 1 in a good solvent to less than 1 in a poor solvent.

$$\bar{r}^2 = \alpha nl_{eff}^2 \qquad (2.10)$$

If the solvent is just a little poor, so that α is just a little below 1, the shrinking exactly counteracts the swelling caused by the self-avoiding

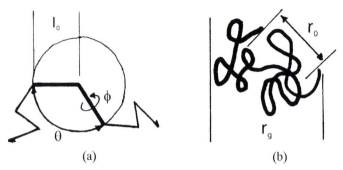

FIGURE 2.13 (a) An illustration of a random walk for a polymer chain. Each step has length l_0, occurs at a bond angle θ, and with a rotation ϕ. (b) An illustration of the random coil structure. The coil is characterized by the end-to-end separation, r_0, and by the radius of gyration, r_g, the radius of a sphere encompassing fifty percent of the volume of the chain.

character of the random walk. This set of conditions, defining a temperature, solvent quality and polymer concentration, is called the Θ conditions. A solvent meeting these conditions is a Θ solvent. The size of a chain in a melt, that is, a polymer chain with the polymer as the solvent, appears to act as if in a Θ solvent.

Another characteristic is the radius of gyration, r_g, the radius of the sphere that is most likely to contain 1/2 of the segments of the chain. The r_g is closely, mathematically related to the end-to-end separation.

$$r_g^2 = \frac{1}{6} \bar{r}^2 \qquad (2.11)$$

The radius of gyration is one of the most useful estimates of the average size of a random coil for calculations.

Chain Extended

Very rigid chains will not coil, but remain chain extended. The primary examples of the chain extended structure are found in conductive polymers and polymer crystals. In a polymer crystal, lengths of the chain are extended in the planar zigzag. This state has a lower entropy than the random coil. Provided enough energy, the chain will relax to the coiled state. However, the increased number of interactions along the straight chain allows a gain in energy leading to stability in the crystalline form for extended chains. Conductive polymers have a rigid backbone, with alternating double bonds. These chains are linear and cannot rotate easily, regardless of interactions with neighbors.

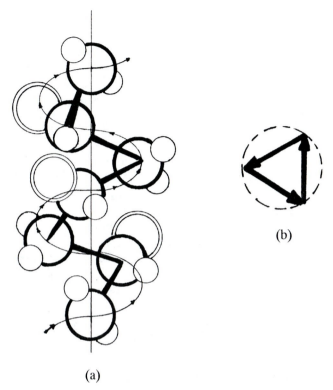

(b)

(a)

FIGURE 2.14 Many polymers assume a helical structure to minimize the interactions of side groups. (a) The polypropylene helix executes one full rotation over three residues. This is the 3:1 helix, which has a planar projection shown in (b).

Helixes

The final common secondary structure is a helix. Some polymer chains have repeating structures with regular side groups. The side groups, such as in polypropylene, will overlap if the chain is fully and completely extended. The overlap causes an increase in energy. The chain can minimize the energy caused by overlap by executing a regular helix with a fixed rotation in each repeat along the chain. The characteristic of a helix is the number of mer repeats and the number of helix rotations that occur before the structure begins to repeat. A helix can be either clockwise or counterclockwise. Polyamino acids, such as proteins and collagen, often assume a helix along some length of the chain. Many synthetic polymers, including nylon, polypropylene and Teflon, also form helixes.

Tertiary Structure

A solid with regularly packed atoms or molecules is "crystalline." When the units, usually molecules, are disordered, the solid is described as "amorphous." Some materials are partially ordered and partially disordered, and these solids are semi-crystalline. Many plastics and polymers are semi-crystalline.

Crystalline Solids

Crystalline solids have long-range order. The crystals in a solid can be quite large or very small. For example, in bronze the crystals are often large enough to be plainly visible without magnification. In high quality gemstones, the crystals are the size of the gem. In poor quality crystals, such as quartz "crystals," the gem may consist of many smaller crystals. These smaller crystals are the right size to interfere with light shining through the stone. The result is a solid that appears cloudy.

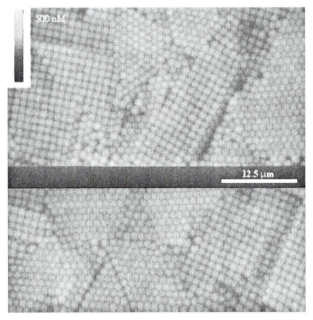

FIGURE 2.15 An image showing long-range packing of particles. The particles pack into regular cubic structures. Each area of packing is a grain, so an individual grain has a single crystal structure. Where the grains meet, the crystals do not necessarily line up, forming a grain boundary.

Do It at Home 2.3

Place ping-pong or golf balls in a shoebox. Shake the shoebox and observe the patterns made by the balls. Is there a discernable pattern? Repeat the experiment with pennies and with kidney beans.

All solids can be classified as ordered or disordered. If the constituents of the solid are similar enough in shape and size, the constituents will adopt a regular, ordered crystal structure. While the charge-based driving force to regular packing in ceramics is obvious, there are similar forces that drive other solids to be regularly packed. In nature, everything will pack into as regular a structure as possible. This includes atoms and molecules, or particles such as nanoparticles or coins. All collections of objects will tend to pack into some regular structure, characteristic of the shape of the atom, molecule or particle, and the types of interactions that exist between these constituents. For some solids, however, the constituents are not regular enough in shape or size to pack well. At some point, packing can be disrupted to a point where there is no discernible order.

In a regular structure, a unit cell can be identified that when moved one unit in any direction will contain a new area of space exactly identical to the space examined just previously. Everything in the new area is completely replicated, including all of the matter and all of the space. There are a limited number of unit cell types, and in all crystalline solids, the atoms or molecules are packed into one of these unit cells.

The easiest ordering to understand is simple cubic (cubic). In this structure, a simple cube can be drawn that contains precisely one set of all unique constituents. If this is an atomic solid, then there is only one atom in the unit cell. If copies of this cube are stacked, the atoms will form a regular pattern. This pattern is characteristic of some metals and ceramics, such as CsCl. The cube can be drawn anywhere so that it contains the set of unique constituents, but the cube is usually drawn so that at least one atom is located at one corner. Therefore, instead of one atom the unit cell really contains eight eighths of an atom, one at each of the eight corners of the cube. This turns out to be easier to visualize, and facilitates calculations of the geometry of the unit cell. This also hints at an important fact: the atom sizes determine the size of the unit cell. For an atomic solid with a simple cubic structure, in which the atoms touch along the edges, the unit cell is $2r$ along all three sides, where r is the radius of an atom. Recall that this radius is an average determined by a bal-

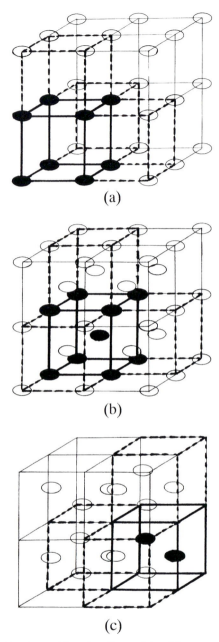

FIGURE 2.16 Stacking in a cubic unit cell containing (a) one atom and (b)–(c) two atoms. Notice that the unit cell can be drawn so that one atom is located in the corner of the unit cell, as in (a) and (b), or not, as in (c). This is true because the unit cell describes any volume of space, not the arrangement of the atoms. Repeating the unit cell will still faithfully reproduce the entire crystal.

ance of attractive and repulsive forces. The geometry of the unit cell and the sizes of the atoms will also determine the density of the material.

While the simple cubic structure is the easiest to understand, other structures are possible and may even be more common. In solids, spherical atoms will generally prefer to pack as closely together as possible, yielding "close-packed" structures that are therefore more common than the simple cubic structure: face-centered cubic (FCC) and hexagonal closest packing (HCP). These structures are the closest together uniform spheres can be packed. They are similar in many ways, but the visualization of FCC is as a cube with duplicate atoms at the center of each face as well as at the corners. HCP is visualized as a hexagonal volume, again with atoms located at the center of each face. Rounding out the cubic unit cells are the more complicated body-centered (BCC) and side-centered structures. These structures can be visualized as cubes with an extra atom located either at the center of the cube, or on one pair of opposing faces, rather than on all three sets of opposing faces. Non-close-packed structures can be forced by the shape of the constituents or by balances of attractive forces. Not all unit cells are cubic: there are rectangular unit cells, including orthorhombic, and tilted unit cells, including monoclinic and triclinic. The unit cells are shown in Figure 2.17.

Crystals are built up of unit cells. Growth of the crystal, however, adds another level of complexity. Each of these crystal structures can grow in different ways, creating the varied forms of crystalline solids. The crystals can grow by adding unit cells in all directions, in which case the crystal growth shape, or habit, will be identical to the shape of the unit cell. Microscopic examination of a crystal of NaCl reveals a perfect cubic crystal. However, crystalline solids can grow preferentially along two faces of the unit cell to yield a rectangular shape, or only one face, yielding a spindle shape. The growth shape is described by the crystal habit, examples of which are shown in Figure 2.18. Finally, the crystal can grow so rapidly that flaws cause an irregular growth pattern. This is called dendritic growth, and is the source of observed patterns in nature, such as snowflakes.

Polymer Crystals

Polymer crystals are a little different from atomic crystals. The chains are too long to constitute particles to pack on a regular lattice. Instead, the mers are the units that pack on the lattice. Therefore, two of the unit cell parameters, usually defined to be a and b, are the result of secondary bonds. a and b are typically of a length scale defined by secondary bonds.

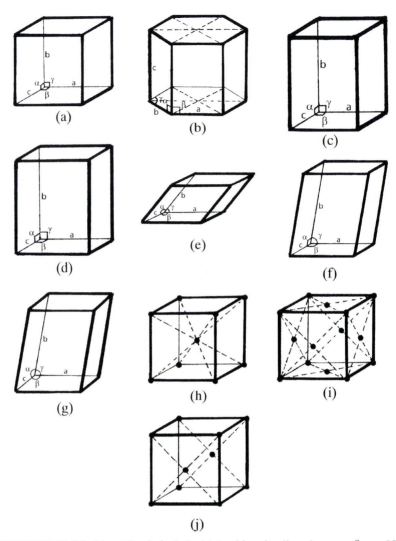

FIGURE 2.17 Primitive unit cells include: (a) A cubic unit cell. $a = b = c$, $\alpha = \beta = \gamma = 90°$. (b) A hexagonal unit cell. $a = b \neq c$, $\alpha = \beta = 90°$, $\gamma = 120°$. (c) A tetragonal unit cell. $a = b \neq c$, $\alpha = \beta = \gamma = 90°$. (d) An orthorhombic unit cell. $a \neq b \neq c$, $\alpha = \beta = \gamma = 90°$. (e) A rhombohedral cubic unit cell. $a = b = c$, $\alpha = \beta = \gamma \neq 90°$. (f) A monoclinic unit cell. $a \neq b \neq c$, $\alpha = \gamma = 90°$, $\beta \neq 90°$. (g) A triclinic cubic unit cell. $a \neq b \neq c$, $\alpha \neq \beta \neq \gamma \neq 90°$. Non-primitive unit cells have the same potential shapes as the primitive unit cells, but contain additional centers of mass, including: (h) The body-centered unit cell, drawn with an atom in the corner position and a second atom in the center of the unit cell. (i) The face-centered unit cell, drawn with an atom in the corner position and an atom in the center of each face. (j) The side-centered unit cell, drawn with atoms located at the corner position and in the center of one face.

The chain direction is traditionally defined to lie along the c-axis of the crystal. The length of c is defined by the length and number of mers in a repeat. Polyethylene packs in a chain-extended form, and so c is the bond length of a single mer. Polypropylene executes a 3:1 helix, and so three repeat units along the chain define the c length. Polymers tend to form orthorhombic, monoclinic, and triclinic unit cells, where a ≠ b ≠ c. The angles of the crystal are influenced by both packing considerations and intermolecular interactions.

The chain ends of the polymer would introduce crystal defects. Therefore, they tend to be excluded to the edge of the crystal. Polymer crystals are generally thinner than the length of a polymer chain. Experiments have revealed that the polymer chain executes regular folding to create a thin plate, called a lamellae. The lamellae will generally grow in dendritic fashion from a central nucleus, resulting in a spherulite. There is space within the spherulite and between spherulites for polymer that has not been crystallized. There is also disorder in the chain folds. Thus, polymers are almost always semi-crystalline.

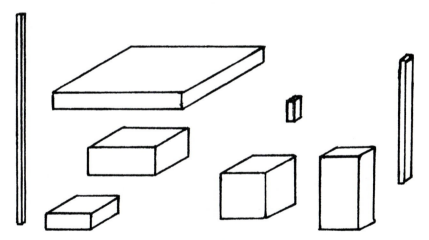

FIGURE 2.18 Some orthorhombic unit cell growth habits. An orthorhombic unit cell has three potential growth faces. Many crystal habits may result from different growth rates at each face of the unit cell. In a lamella, the primary structure in polymer crystals, grows rapidly along two faces but does not grow in the third, resulting in a plate-like structure. A fibrillar habit forms if growth is rapid on one face but is very slow in the other two. An isometric structure grows evenly in all three dimensions. A needle habit grows rapidly along one face and slowly along the other two. An acicular habit grows slowly in all three directions. (After Wunderlich, *Macromolecular Physics Volume I*, Academic Press, 1973.)

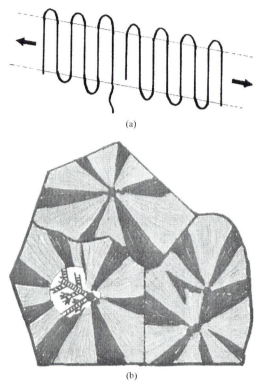

(a)

(b)

FIGURE 2.19 Features of polymer crystals. (a) The polymer chain typically folds back and forth, creating a chain-folded lamella. The chain ends are crystal defects, so they are excluded to the surface of the lamella. (b) New lamella will nucleate from the surface of lamella already formed. The resulting, space-filling structure is called a spherulite.

Polymer crystals grow in two stages, just as do any other type of crystal. The first stage is nucleation. Nucleation can occur either by random fluctuations causing the melt to form a stable nucleus, or by attaching to a foreign additive. Homogeneous nucleation, controlled by random fluctuations in the melt, is spontaneous and happens under the correct temperature conditions. However, the process is relatively slower. Heterogeneous nucleation begins at some nucleation site. Since new nuclei are not required, heterogeneous nucleation is both spontaneous and also relatively instantaneous on reaching satisfactory conditions for crystal formation. The foreign initiation sites may be the leftover nuclei of previous crystals, a nucleation additive, or the particles of a composite. Any particle can serve as a crystal nucleus, even relatively larger particles such as glass fibers. Moreover, if there is a match between one of the growth faces of a polymer crystal growth habit and an edge of the particle, the foreign material may be very efficient at nucleating a crystal. If a polymer, such as polypropylene, has more than one crystal form, a nucleating agent can create bias toward one of the phases over the other. Nanoparticles, since they can be closer to the order of the size of the crystals, and because they may have more edges to serve as nuclei, may result in significantly different crystallization behavior than that generated by larger particles.

From the nucleus, the crystal will grow outward. The rate is controlled by how rapidly fresh material can reach and then settle onto the growth face. Chains in molten liquid can move more rapidly than individual chains diffusing to the crystal. The long polymer chains generally take longer to both reach and attach to the growth face, so the crystallization rate of polymers is typically much slower than for atomic solids. The solid phase in a composite can restrict mobility even more. Therefore, crystallization rates in composites may change from those of pure polymer fluids.

Amorphous Solids

If there is not enough similarity between the constituents of a solid, they will not be able to approach each other closely enough to gain the natural benefits of packing together. As a result, there is no long-range order. Such solids are termed amorphous.

SiO_2, silica glass, is an example of a material that can be either crystalline or amorphous. Quartz is the crystalline form of SiO_2, while glass is amorphous. Quartz can be seen through only if the crystals inside it are so small or so large that they do not interfere with light. Amorphous glass has no structure that can interfere with light, and so is always transparent.

Metals pack as uniform spheres, so metals tend to form cubic structures, especially FCC and HCP. There are specific packing circumstances that can also result in non-closest packed structures such as simple cubic and BCC. However, metals crystallize readily. "Amorphous" metals are usually metals with nanoscopic crystal structures.

The strength of a crystal is determined by both the structure of the crystal and the defects within the crystal. Introducing a defect to the structure deforms a material. If the material is amorphous, such small disruptions have little effect on the structure, just a few rearrangements. In a highly ordered crystal, introducing defects requires breaking bonds. There are several types of defects, shown in Figure 2.20. The vacancy is a hole in the crystal, where a particle is missing. An interstitial occurs where an extra particle is forced into empty space in between regularly packed par-

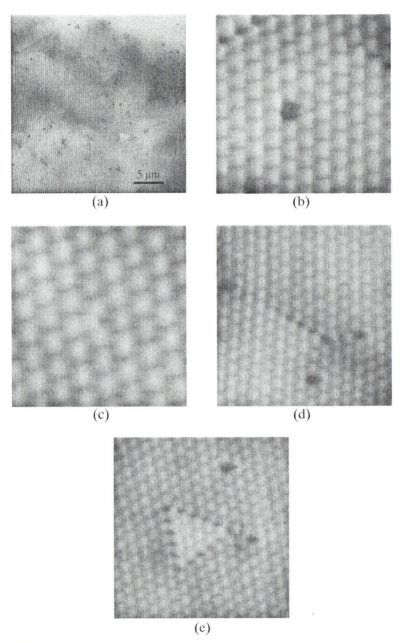

FIGURE 2.20 Crystal packing defects, illustrated with atomic force microscope images of an approximately 300 nm latex bead suspension. (a) the base image, with the beads packed in a long-range, face-centered cubic (FCC) structure with multiple defects, (b) a vacancy, (c) an interstitial, (d) a line defect, (e) the intersection of 3 line defects, creating a pinned defect structure.

ticles. A line defect, or dislocation, occurs where there is a line of mismatched particles within the crystal. The dislocations are collections of point defects. The dislocations move under the influence of deformation. If they run into each other at angles, they can create a pinned structure. This stable feature can also pin other moving defects, thus changing the properties of the solid.

TRANSITIONS

The properties of a solid depend on the structure of the material. Crystalline materials tend to be stiffer and stronger than amorphous materials. Solids with stronger bonds also tend to be stiffer than solids with weaker bonds. Solids with some ability to reorganize the bond structure tend to be more ductile, more deformable, and more fracture-resistant than solids that cannot. Changes in properties match changes in structure. Changes in structure result from increases in constituent mobility caused by increases in temperature.

Perhaps the strongest, and most property-affecting, transition occurs in crystalline materials. The packing of atoms or molecules into a regular array forms crystals. The size of the unit cell that defines the array is determined, as discussed above, by the average size of the constituents or, more appropriately, by the average distance between the constituents. However, this average separation will increase with increasing temperature, until at some temperature the structure will be disrupted entirely. This is the crystal melting transition. Materials have a variety of characteristic transition temperatures, including melting, that depend on the chemical structure.

The melting transition is the temperature at which the crystals in a crystalline solid become disrupted. This may allow the atoms or molecules to move freely past each other. Therefore, many materials are solid at temperatures below the melting transition and liquids at temperatures above it. An alternative way to look at melting is that as more energy is added to a system of bonded constituents, more of the bonds will be disrupted, and, since the energy is high enough, will be unable to re-form. A little deduction reveals that bonds break and crystals are disrupted by the same increase in energy. A defect in a crystal is a local event resulting from the disruption of bonds. Less perfect, smaller crystals melt at a lower temperature, since the crystals already have more disruptions in the form of grain boundaries. Larger, more perfect crystals melt at a higher temperature, as more defects must occur to disrupt the crystals enough to allow melting.

Crystallization occurs when a melt cools to the point that the constituents can approach each other closely enough to form bonds. Since a formed bond has a lower energy than the individual non-bonded constituents, bonds will form whenever possible. Since the packing occurring in a crystal demands that the maximum possible number of constituents be in proximity, a crystal represents the lowest energy state available to a solid. The optimal number of nearest neighbors insures the largest number of bonds possible. If the solid is crystallized very close to the crystallization temperature long enough for the entire solid to crystallize, perfect crystals will form. If the solid is cooled more rapidly, the full optimization does not have time to occur and smaller, less-perfect crystals will result. If the constituents are spherical atoms or simple molecules, crystallization will occur at the same temperature as melting. Water melts and freezes at 100°C at standard temperature and pressure; indium melts and freezes at 153°C. A melting crystal results in a step change in volume, as shown in Figure 2.21.

Some solids containing constituents with complicated shapes are resistant to crystallizing. As a result, crystallization may occur at a temperature below the melting temperature. This is recorded as ΔT, the degree

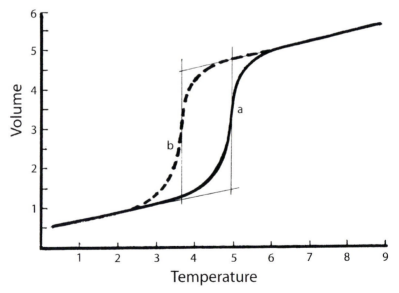

FIGURE 2.21 The volume as a function of temperature showing a crystal melting transition for a semicrystalline polymer. (a) The heating curve. (b) The cooling curve. Note that the heating and cooling curves are asymmetric. Super-cooling is required to drive crystallization.

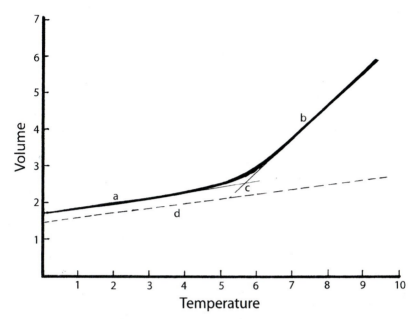

FIGURE 2.22 Volume as a function of temperature showing a glass transition for a polymer. (a) Below the glass transition, the volume increase as a function of temperature is the coefficient of thermal expansion. (b) At the glass transition, the coefficient of thermal expansion increases. (c) The intersection point of the lines describes the ideal glass transition. (d) Theories of glass transition suggest that there is a hard-core volume occupied by the molecules in the solid. There is also some free volume trapped in the solid as well. At the glass transition, the free volume trapped by the movement of chains increases suddenly.

of under-cooling required to crystallize the solid. Long-chain organic molecules, polymers, are examples of solids that have different melting and crystallization transition temperatures. Further, the long chains result in longer times required to allow crystals to form.

A number of solids are not crystalline, or are semicrystalline. For these materials, the transition from solid to liquid is marked by a glass transition (T_g) instead of or in addition to a melting transition. This includes polymeric solids, for which the large chains, often irregular in shape, interfere with the packing of the constituent elements. The properties of the material change at the glass transition. The change in volume with temperature around the glass transition is represented in Figure 2.22. The mechanical modulus drops, the index of refraction decreases more rapidly, and all other properties experience different dependence on temperature above and below this characteristic temperature.

Some think of the glass transition as disruption or melting of the secondary bonds, but this is not valid. Secondary bonds are temporary by nature, anyway. The glass transition has been best characterized by free volume arguments. Below the glass transition, there is a specific amount of space available surrounding the constituents defined by the relatively small space required to contain the moving atoms. This space is on the order of the range of secondary bonds. At certain temperatures, the atoms making up small pieces of the constituent molecules begin to do things other than bounce around. For example, a group of three atoms can start to do things like rock, scissor, wave, or wag, as seen in Figure 2.23. This causes these parts of the molecule to take up more space.

At higher temperatures, larger groups of atoms can move in coordination. At the glass transition temperature, a portion of the chain large enough to be physically indistinguishable from the entire chain begins to move. This has been evaluated experimentally to be approximately 50 carbon units, depending a little, but not very much, on chain chemistry. Apparently, whatever chain motions create the glass transition effect, whether S-shape wiggles along the chain backbone or something else, require at least that long a chain to occur.

At a temperature sufficiently above the glass transition temperature, the solid will flow readily. The difference between the glass transition temperature and the processing temperature depends somewhat on molecular weight and chemistry. Engineers often start evaluating the flow properties at about 50°C above T_g, raising or lowering the temperature

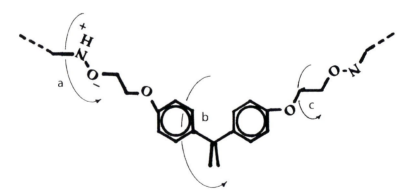

FIGURE 2.23 Molecular relaxations, illustrated with bisphenol-A epoxy cross-linked with an amide hardener. (a) Crankshaft motion causing movement of an electronic dipole. (b) Crankshaft motion of the phenol rings. This is a large molecular moiety that would require significant energy to occur. (c) A crankshaft motion requiring the addition of an intermediate amount of energy to activate.

from there to achieve sufficient flow at reasonable pressures without degrading the solid.

There are other transitions and relaxations of importance in materials. Any change in structural mobility will result in a transition that can be measured in a temperature-dependent property. In naturally occurring biomaterials, like collagen, there is a helix to coil transition that occurs at around 37°C. The helix is the functional form of the protein; the random coil is a heat-denatured form. This transition is irreversible. Some solids have two or more crystal states, and there are specific temperature and pressure conditions at which a crystal-crystal transition may occur. For magnetic materials, the ordering of the magnetic dipoles in the metal is lost if the temperature is heated to above the Curie temperature. In polymers, many of the smaller rotations and vibrations of the atom clusters in the chain occurring below T_g absorb energy, making the solid tougher and more impact-resistant. At temperatures below these small transitions, the solid is more brittle; above them, the energy from the impact can be transformed into molecular vibrations, or heat, instead of causing bond rupture.

The thermal properties of a composite are an interesting class of behavior. At a single temperature, such as room temperature, there are no "thermal properties" except for certain rare materials with a material transition at near room temperature. Still, they are important for determining two features: the processing conditions and appropriate use-temperatures. Further, because they probe heat flow, which is linked to thermodynamic values, thermal analysis can provide data that illuminates the atomic behavior more directly than continuum mechanical measurements.

EFFECT OF SCALE

The packing of small round particles, such as fumed silica, polymer latex, and carbon black would be expected to follow in similar fashion to spherical metal atoms. The dominant factor in particle packing is physical size and shape, but surface energy will also play a role in how readily the particles pack. Irregular particles will pack less efficiently. The shape may force the packed structure to have a more complex unit cell than represented by closest-packed structures. The surface energy will also affect the ability to disperse the particles. Nanoparticles may have a driving force to pack and may require significant energy or a good solvent to disrupt the packing and enable the constituents to be separated and incorporated into a liquid or polymer.

REFERENCES

Asimov, A, *On Chemistry,* Anchor Books Press, Garden City, NY, 1974.

Callister, W, *Materials Science and Engineering: An Introduction,* John Wiley and Sons, Hoboken, 2003.

Hiemenz, P, Polymer Chemistry, the Basic Concepts, Marcel Dekker, Inc., New York, 1984.

Geil, P, *Polymer Single Crystals,* John Wiley and Sons, New York, 1963.

CHAPTER 3

Properties

This chapter introduces and develops the knowledge base for some important physical properties, including bulk and surface mechanical and barrier performance. The focus of learning will occasionally shift in this chapter to what is called the "continuum" level, which considers the behavior of the solid object as a whole. The continuum approach uses the average behavior of the solid to describe the physical properties of the nanocomposite, but forsakes understanding the interaction of structure with performance. The continuum approach often relies on empirical relationships to correlate and predict properties. Using the atomic theory developed in the last chapter as a starting point, the physics of a material can sometimes be extrapolated from statistical arguments, for example using statistical thermodynamics to connect, rather than correlate, structure and properties. Extrapolating atomic behavior from empirical observations is seldom appropriate. For example, continuum mechanics allows assigning the same elastic modulus to two distinctly different physical material states in two different, systematic mechanical experiments. However, the processing history and therefore the microstructure of these two states may be very different. An aim of atomic theory is that many properties can be predicted, once a structure is known. The continuum theory particularly begins to break down in composites, where the second phase can no longer be treated as discrete entities isolated from the matrix. This occurs when the particles are smaller than about 100 nm and for particle volume fractions above 10%. An argument for using the quantum mechanical averaging of the atomic scale to rescue the continuum perspective occasionally will be made.

The properties of a nanocomposite are an aggregate of two phases: ce-

51

ramic and polymer, polymer and metal, ceramic and ceramic. The greatest advantage in nanocomposites arises by combining two materials with very different properties. Polymers have excellent barrier and electrical properties, are light, impact resistant, can be flexible, transparent, and have a host of other desirable features. They also tend to be "soft", that is they have relatively lower moduli than ceramics or metals and are susceptible to scratch and wear. Ceramics and inorganic glasses are good insulators, have relatively higher moduli and are hard. They also tend to be brittle and heavy. Metals are conductive, have relatively high moduli, and contain many examples of ferromagnetic materials. They also are susceptible to micro-damage that can increase permeability, are quite heavy, and susceptible to oxidation. As discussed earlier, each material class has strengths and weaknesses appropriate for solving specific engineering problems. Combining the properties in composites and nanocomposites can provide useful average properties, while ameliorating some of the weaknesses. In nanocomposites, the components are difficult to regard as isolated and individual contributors to the properties. The implications of this will become apparent.

Composite behavior results from some combination of the properties of the individual components. A ceramic reinforcement is added to a soft polymer with good impact resistance and ductility in order to provide improved hardness and scratch resistance. Some understanding of the component contributions to properties may be accomplished empirically, using rules such as the "rule of mixtures." The properties, however, really arise from the chemical composition and the processing conditions used to form the composite. Particularly for nanocomposites, the empirical property models will often provide estimates that are far from the real values.

The rule of mixtures is one method for calculating the property of a combination of materials. It states:

$$P = x_a P_a + x_b P_b \tag{3.1}$$

P is the value of the property, such as the density or a transition temperature. a and b are two components, although the rule of mixtures can apply to any number of components. x is the mole fraction of each component. The rule of mixtures is one way of averaging the properties. An alternative mixing rule is the harmonic mean:

$$\frac{1}{P} = \frac{x_a}{P_a} + \frac{x_b}{P_b} \tag{3.2}$$

Another alternative is the geometric mean:

$$P = \sqrt{P_a^{x_a} P_b^{x_b}}$$ (3.3)

Each of these formulas emphasizes a different aspect of the properties. The real behavior depends on the structure, and how the different structures contribute to the properties.

There are several types of properties of commercial importance. The mechanical properties, surface properties and permeability properties of the polymer are modified by added particles. These properties, and the changes, are introduced and discussed here.

PHYSICS OF THE MODULUS

The bulk mechanical responses of a solid to applied force are among the most important and familiar of properties. Mechanical responses govern the usefulness of a material in varied applications. The forces acting on a solid can be resolved conveniently along a set of three, linearly independent axes. For most calculations, defining these axes as mutually orthogonal is helpful. Most mechanical forces and deformations are applied in combinations of two modes, normal and shear. Normal forces act along one or more axes of a solid. The vector of application results in a linear displacement. Shear force acts perpendicularly to one or more axes of the solid. The vector of application results in a perpendicular component to displacement, and an angle of rotation in the solid. The final consideration is hydrostatic force, which is applied uniformly and negatively along all three axes simultaneously. This situation is shown in Figure 3.1.

The simplest form of mechanical behavior is uniaxial tensile elongation. The ends of a block of material are pulled in opposite directions. The block lengthens, or deforms, in the direction of the applied force. Normal materials also shrink in width and depth, or perpendicularly to the force, as shown in Figure 3.2. Hooke's law states that for small deformations the force is directly proportional to the deformation:

$$F = kx$$ (3.4)

or:

$$\frac{F}{A} = k' \frac{\Delta x}{x}$$ (3.5)

This is akin to observing that a solid material behaves much like a me-

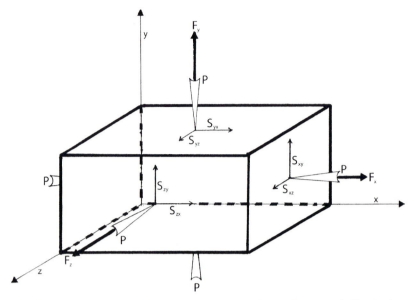

FIGURE 3.1 A prismatic solid showing 3 orthonormal axes. The normal ($F_n \rightarrow$), shear ($S_n \rightarrow$), and hydrostatic force (P >) directions are indicated.

chanically perfect spring. The constant of proportionality, the spring constant, is known as either the Tensile or Young's Modulus. Any energy applied to cause the deformation in this regime can be elastically recovered. The modulus can be affected by temperature, and is different for every material or combination of materials. For polymers, as shall be described later, there are additional dependences on the rate and amount of deformation. The nature of the tensile response forms the simplest basis for mechanical study of a solid. Hooke's law is a phenomenological description at the continuum level. Deviations from Hooke's law and some of the observations of temperature, time, or strain dependence may be explained by deeper analysis.

For unconstrained materials, an additional observation is important. In most experiments, the material will shrink perpendicularly to the force in proportion to the elongation of the specimen. This is called the Poisson effect, and the constant of proportionality is Poisson's ratio, ν:

$$\nu = -\frac{\Delta w / w}{\Delta l / l} = -\frac{\Delta h / h}{\Delta l / l} \tag{3.6}$$

w is the width, h is the height and l is the length in the direction of deformation. There is an equivalent expression in terms of the volume change, ΔV:

$$v = \frac{1}{2}\left(1 - \frac{\Delta V}{\Delta l / l}\right) \tag{3.7}$$

Physically, materials retract to conserve the total volume of the deforming body. There is some additional influence to the deformation from the incompressible nature of the solid. An incompressible material exhibits a Poisson Ratio of 1/2. Materials that can exhibit some compressibility can have ratios of less than 1/2. For an incompressible solid, the sample will shrink in depth and width, each, by 1/2 the increase in length.

Constructing the Energy Well

The modulus, and the Poisson effect, may be understood from consideration of the bonding. From an examination of the physical changes to the sample, several observations may be made. First, as the sample deforms under elongation, atoms in the solid must move farther apart in the length direction and closer together in the depth and width dimensions, in order to maintain their relative positions with each other. Second, the energy of deformation is measurable. This energy can be determined by integrating the force required to achieve a given displacement. An examination of the atomic level changes in the solid may provide some insight into these observations.

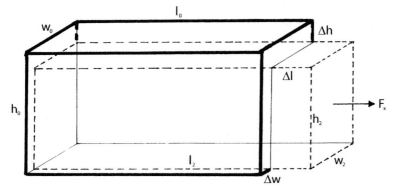

FIGURE 3.2 Simple tensile force. As the sample elongates in the x direction (l_1 to l_2, Δl), conservation of volume requires that the sample shrink along the y (h_1 to h_2, Δh) and z (w_1 to w_2, Δw) directions.

As any two atoms approach each other to form a solid, they experience both attractive and repulsive forces. The energetic forces acting upon any two atoms govern the separation. The two types of force are generally considered pair-wise, as occurring between two isolated atoms. The real force acting on each atom is the sum of all forces acting on that atom. In a metal or ceramic, the attractive forces, electrostatic or ionic, act isotropically around each atom. In polymers, there are two attractive forces: the isotropic electrostatic force, creating secondary bonding forces; and the strong, unidirectional force exerted by a shared electron pair in a covalent bond. Any two atoms will experience the strong repulsion of atoms overlapping as they approach to within atomic distances, angstroms. The balance of the attractive and repulsive forces can be illustrated by drawing a curve showing the individual forces and their sum. The resultant energy curve shows a well, with the minimum energy at some atomic separation determined by the energies specific to the pair of atom types, which was depicted above in Figure 2.5.

The average separation between atoms occurs at the minimum in the energy. For pure metals, the separation distance is the same for all of the component atoms in the solid. This gives rise to materials with simple packing structures and isotropic properties. The attractive force is isotropic, of moderate strength, and can bond with any other atom in the metal. If the atom is displaced far enough to lose the connection of the original bond, the atom can fall into another energy well, or bond, with a new neighbor. The depth of the energy well determines the strength of a metal, while a metal's ductility is determined by how easily the atoms can relocate from one well to the next.

Ceramics are composed of both anions and cations, and have two charged atomic components that must balance each other. The pair-wise interaction is still valid, but the overall force acting on an individual atom is more complicated. The energy for an ionic bond is stronger than for a metallic bond, so the well tends to be deeper. The strength of the ionic bonds leads to strong solids. New ionic bonds are less likely to form due to the proximity of similarly charged and therefore repulsive ions. When failure occurs, a ceramic solid is more likely to fracture than deform.

Polymers have a combination of attractive forces. The covalent bonds have deep wells, forming strong bonds with small separation distances along the covalent bond. There are also weaker, isotropic attractions caused by van der Waals forces. These result in larger average separations between atoms along the van der Waals interactions. The energy well is also shallower, resulting in low energy costs to disrupt the secondary bonds. On the other hand, forming new bonds is energetically favorable. Since each element of the polymer is connected to the chain, displacing one element requires reconfiguring the chain, at a corresponding increase in energy to cause the deformation. This leads to the complicated mechanical properties exhibited by polymers, where some polymers are very ductile, while others are mechanically rigid. The interactions between chains in rubber are very weak, so the deformation in rubber depends only on the energy required to reorganize the chains, as will be discussed in the section on rubbery elasticity.

The average separation between atomic centers, the bond length, predicted by the sum of attractive and repulsive forces becomes more accurate at lower temperatures. At higher temperatures, the atoms have more energy, oscillating and vibrating. The result is that the atoms will occupy a higher level in the energy diagram, as shown in Figure 3.3. This in turn will lead to a change in the average bond length, which usually gets longer, although the position of the atoms relative to each other also becomes more uncertain. When heated, a solid typically will expand, a property known as thermal expansion, which can also be quantified by a coefficient of thermal expansion. The deeper and more symmetric the energy well, as observed in ceramics, the less thermal expansion will occur with an increase in temperature and the smaller will be the thermal expansion coefficient. On the other hand, polymers, with a shallow energy well, will exhibit a larger thermal expansion coefficient. At higher

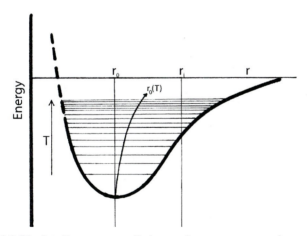

FIGURE 3.3 The bonding energy well. At very low temperatures, the average bond length is r_0. As the temperature increases, the energy in the bond allows larger vibrations, and a larger uncertainty in bond length. The energy well asymmetry results in an increase in the equilibrium atomic separation as a function of temperature. The increasing bond length is the source of the coefficient of thermal expansion. The inflection point in the energy well, r_i, is important in the force diagram.

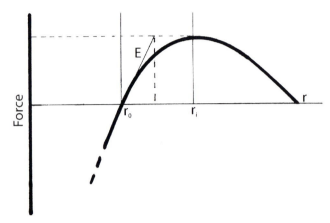

FIGURE 3.4 The force vs. bond length curve for atomic separation is the derivative of the energy well. The displacement force is zero at the equilibrium bond length, r_0, and increases until the maximum in force occurs at r_i. The slope of the force curve about the r_0 is related to the Young's modulus, E.

temperatures, removing individual atoms to infinite separation also becomes easier. The physical consequence of this is softening in the solid.

Predicting the Modulus

As seen in the simple description of tension, when a force is applied to a solid the atoms are forced farther apart. This separation can be described as supplying energy to drive the atoms farther apart. Therefore, the energy curve can be differentiated to determine the amount of force required to change the average separation of the atoms in specific ways, as shown in Figure 3.4.

At the minimum energy, or the average atomic separation, the force required to maintain the length is zero. To move the atoms slightly closer together or farther apart will require a small amount of energy. As the displacement increases, the amount of force required to supply the excess energy to maintain the atomic separation will increase. The integral of the curve locally will be a straight line, requiring a compressive force to push atoms closer together and a positive, extensive force to pull atoms further apart. The curve will pass through zero at the minimum in the energy well. The slope of the curve at this point is also the constant of proportionality between the force and displacement required to separate an atomic pair a given distance, much like a spring constant.

In order to understand the implications of the force curve, it is useful to consider an average of all of the displacements and forces acting within the solid. The average slope can describe the constant of proportionality between force and deformation for a bulk material. In simple force fields, this yields Hooke's law, with tensile properties governed by a constant of proportionality. The force curve also provides an assessment of the average force required to remove an atom to infinite separation, which might be the yield point, if new bonds can be formed, or failure point, if forming new bonds is difficult.

On the laboratory scale, this average is achieved by accounting for all of the bonds over which a force is applied, and summing up all of the deformation experienced by the bonds. The number of bonds is proportional to the cross-sectional area of a part. Therefore, by normalizing the force across the cross-sectional area, a material response is obtained that is independent of sample geometry. This response is called the stress, σ:

$$\sigma = \frac{F}{A} \tag{3.8}$$

The length of the part is the sum of all the bonds laid end to end. If each bond is elongated by 1%, then the whole part will elongate by 1%. As long as each bond is deformed equally, an assumption known as affine deformation, then the length change is also a material response that is independent of the sample geometry. This response is the strain, ε:

$$\varepsilon = \frac{\Delta L}{L} \tag{3.9}$$

L is the length of the sample, and ΔL is the change in length. The affine deformation generally holds for a linearly elastic material. The linear response of a perfectly elastic material is a proportional response between stress and strain:

$$\sigma = E\varepsilon \tag{3.10}$$

E is the Young's modulus, similar to the spring constants described earlier. The modulus is a property of the material. This modulus is most valid at small deformations, where the bond lengths are close to the equilibrium position. Far from the equilibrium separation, some of the bonds may break, and the mechanical response will become non-linear. The deformation, or strain, also may no longer be completely recoverable.

Polymer deformations are not inherently different from those in metals and ceramics. A covalent bond is directional and strong, stronger even than ionic bonds. The simplest case to be described for a polymer is for deformation of a chain-extended conformation, such as found locally

TABLE 3.1 Approximate moduli of various materials.

Material	Modulus (GPa)
Gray Iron (G1800)	66–97
Ductile Iron (60-40-18)	169
Steel (1020)	207
Aluminum (6061)	69
Copper	115
Beryllium-Copper	128
Brass	97
Copper-Nickel	150
Bronze	100
Tungsten	400
Lead	13.5
Alumina	380
Diamond	700–1200
Concrete	25–37
Borosilicate Glass	70
Graphite	11
Silicon Carbide	207–485
Styrene Butadiene Rubber	0.002–0.010
Nylon 6,6	1.59–3.79
Polycarbonate	2.38
LDPE	0.172–0.282
HDPE	1.08
Polyethylene Terephthalate	2.76–4.14
Polypropylene	1.14–1.55
Polystyrene	2.28–3.28
Polymethyl Methacrylate	2.24–3.24
Polytetrafluoroethylene	0.40–0.55
Polyvinyl Chloride	2.41–4.14
Kevlar^{TM} Fiber	131
High Modulus Carbon Fiber	400
E Glass Fiber	72.5

Table derived from Callister.

in a crystal or along the axis of a fiber. Deformation along the chain can occur by both elongating the carbon-carbon bonds, as well as distorting the angle between the bonds (Figure 3.5). The net energy requirement for the deformation is determined by the sum of the contributions. The change in chain length under a force, F, is a function of the bond length l and the bond angle θ:

$$\Delta L = n(\Delta l \cos\theta - l\Delta\theta \sin\theta) \qquad (3.11)$$

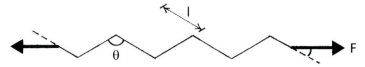

FIGURE 3.5 Distorting a polymer chain in a chain-extended crystal. The bond angle, θ, and length, *l*, will both respond to the application of an axial force *F*.

The local change in bond length, Δ*l*, is produced by the resolution of the force along the bond:

$$\Delta l = \frac{F}{k_l} \cos \theta \qquad (3.12)$$

k_l is the spring, or force, constant for the bond. An expression for the torque applied to the bond is defined by the moment of the force resolved along the bond angle:

$$\Delta \theta = -\frac{1}{4} \frac{Fl}{k_\theta} \sin \theta \qquad (3.13)$$

k_θ is the force constant for the valence angle. This force constant can be transformed into a second linear force constant:

TABLE 3.2 Bonds, angles and corresponding force constants.

		Length (Å)	k (Pa/m)
	C—C	1.53	436
Bond	OC—NH	1.40	780
	CH$_2$—NH	1.47	574
	H$_3$C ‾‾ CH$_3$		350
	H$_3$C ‾‾ NH$_2$		360
Angle	H$_3$C ‾NH‾ CH$_3$		680
	H$_3$C ‾(O)‾ CH$_3$		380

Table from McCrum, Read and Williams

Box 3.1 Deformation can be described by the same energy
equation that governs all state changes: $\Delta G = \Delta H - T\Delta S$

ΔG is the change in free energy induced by a change in state, for example, from relaxed to deformed; ΔH is the enthalpy, or energy, required to cause the change; T is the absolute temperature, and ΔS is change in entropy induced by the change.

Enthalpy is measured by the amount of energy required to cause the change, evaluated here by the number of bonds broken or formed. The enthalpy can be broken into components by considering the internal energy of the system, that is, the bonds themselves, separately from the work. The work can be put in via the deformation:

$$\Delta G = \Delta U - F\Delta L - T\Delta S$$

F is the applied force and ΔL is the change in length.

Entropy can be calculated by counting the number of distinct ways in which to occupy each state, and the change in entropy is the difference between the two states. In an atomic solid, since the atoms are interchangeable there is little entropic change caused by reorganizing the atoms. Only introducing defects such as vacancies causes a change in entropy, resulting in plastic and irretrievable deformation. With long chains, such as in rubber, the number of ways a relaxed chain can cross an intervening distance is large. If the chain must cross a greater distance, as in a deformed state, the number of distinct ways in which to do so decreases. The result is a decrease in entropy, requiring an applied force to supply the energy. This force is recoverable, provided there is no way for the chains to reorganize permanently. If the chain can reorganize over a characteristic time, then the mechanical properties of a solid composed of these chains are time dependent. The time dependence will be dictated by the same forces that resist the initial deformation: locally overcoming van der Waals bonds, while displacement along the chain is further resisted by secondary bonds-by-proxy. Displacement has a nearest neighbor contribution, a next nearest neighbor contribution, and so on. The greater the thermal energy in the system, the more coupling can occur between nearest neighbors, next-nearest neighbors, next-next-nearest neighbors, and so on. That is, at higher temperatures energy is less localized on a specific bond, and easier to displace farther along the chain. Therefore, the energy wells and resulting time dependence may have multiple modes that range from a unit length the size of a monomer up to the full length of the chain.

The resistance to deformation in a glassy polymer, at a temperature well below the glass transition, is an average of the van der Waals interactions between local segments and the covalent bonds along the chain. As the glass transition is approached, more energy may be coupled along the chain to allow some chain reorganization. This results in a non-linear elastic, or recoverable plastic, deformation. Well above the glass transition, there is sufficient thermal energy in the system to allow simple stress release through chain reorganization.

If the material is semi-crystalline, the force can be borne by the stronger average interactions in the crystalline regions, but overall the properties will continue to be time and rate dependent. If the material is cross-linked, the ultimate reorganization of the chains to recover the applied stress cannot occur. If the material is not cross-linked, the stress recovery will allow flow, over however long a period, as dictated by the resistive force of the secondary bond energy.

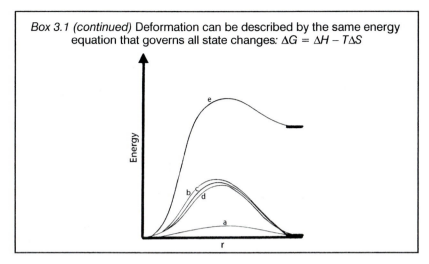

FIGURE 3.6 Schematic representation of the energy required to distort various types of
bond: (a) a secondary van der Waals bond, (b)–(d) three different metallic bonds, and (e) a
covalent bond. Differences in the peak distortion energies of the different metallic bonds
are relatively smaller than the differences between secondary and covalent bonds.

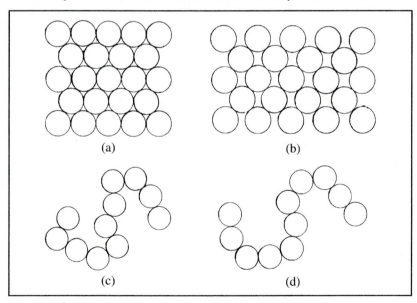

FIGURE 3.7 A constant volume deformation in an atomic solid (a)–(c), compared to reorganization of chains in a rubber (b)–(d). Notice that there is no meaningful change in
the relative atomic positions in the metal, so there is no entropic contribution to the free
energy. Deformation in the rubber occurs through relatively free reorganization of the
chains. Since the reorganization occurs against little resistance, there is no enthalpic contribution to the free energy, but there is a large change in the entropy.

63

$$k_\theta = k_a l^2 \tag{3.14}$$

The modulus for the chain-extended crystal is derived from force and displacement:

$$E = \frac{l \cos\theta}{A} \left[\frac{\cos^2\theta}{k_1} + \frac{\sin^2\theta}{4k_a} \right]^{-1} \tag{3.15}$$

A is the cross-sectional area. Chain-extended polymers can show a high modulus, as observed in Kevlar™ and Spectra™ fibers.

The modulus of a coiled, or at least non-extended, chain is more complicated to predict. If the temperature is above the threshold for easy bond rotation, the stress on a covalent bond can be relieved with little energy cost by rotating a nearby bond. The isotropic secondary bonds contribute additional small amounts to the energy required to displace an individual atom, effectively creating a force field through which interconnected atoms must move. Therefore, the energy well is shallow, and the energy cost of deformation is low. Further, the bond energies of an individual atom are connected to neighboring and distant atoms through the covalent bonds. A small force applied to a neighbor can remove a local force accumulated by the atom. While these changes have a small energy cost, there are energy requirements, usually readily available as thermal energy above the glass transition. Below the glass transition, chain mobility is limited by the difficulty in rotating from one conformational state to another. Below the glass transition, polymers are more rigid, behaving somewhat like ceramics. Internal atomic displacement is barred by an increase in the height of the energy barrier for overcoming both rotation and secondary bonds between chains.

The complex behavior of polymers is described by rubbery elasticity or viscoelasticity considered briefly below.

CONTINUUM MEASUREMENTS

The force is always applied over an area. If the same force is applied over twice as much as area, the sample will deform half as much, whereas if the force is applied over half as much area, the sample will deform twice as much under the same load. Dividing the applied force by the cross-section gives a measure of an average force parameter: the stress. Stress is inherently an averaged measurement of a force field. As long as measurements take place over an area that is representative of all of the microscopic, or nanoscopic, features of the solid, the stress and modulus can be generalized. The force can be normalized using the initial cross-section, providing the engineering stress. This has the advantage of being precisely measurable at the beginning of the experiment. However, because of the Poisson effect the cross-section will not be a constant in an incompressible medium. The force can also be normalized by the instantaneous cross-section, providing the true stress. For incompressible materials at small stresses, that is, assuming a deformation with no volume change, there is a simple correlation between the engineering and true stress of the solid depending upon the mode of deformation:

$$\sigma_{true} = \sigma(1+\varepsilon) \tag{3.16}$$

However, the correlation sometimes breaks down in real testing. Choosing either the true stress or engineering stress as the basis for measurement may be driven by practical considerations.

The stress, again, applied force normalized by unit area of application to transform the response into a material property, can be applied to a solid at any arbitrary angle. To the external observer, it is usually convenient to consider the force as being applied on a three-dimensional axis system. The primary axis can always be defined as the axis along which the primary force is applied. This means that the force can be resolved along three axes, creating nine components. These components can be defined by subscripts with the first subscript describing direction of the normal to the plane, and the second the direction of the applied force: σ_{11}, σ_{12}, σ_{13}, σ_{21}, σ_{22}, σ_{23}, σ_{31}, σ_{32}, and σ_{33}. Therefore, the stress components, σ_{11}, σ_{22} and σ_{33} are coincident with the normal of the plane on which each acts, and are "normal" stresses. The remaining stress components are

TABLE 3.3 Stress, true stress, strain and true strain.

	Definition	Written in terms of the engineering values
Stress	$\sigma = \dfrac{F}{A_0}$	
True Stress	$\sigma_{true} = \dfrac{F}{A_i}$	$\sigma_{true} = \sigma(1 + \varepsilon)$
Strain	$\varepsilon = \dfrac{\Delta l}{l_0}$	
True Strain	$\varepsilon_{true} = \ln\left(\dfrac{l_i}{l_0}\right)$	$\varepsilon_{true} = \ln(1 + \varepsilon)$

shear stresses. Unless the solid is rotating, the shear stresses must balance, forcing $\sigma_{21} = \sigma_{12}$, $\sigma_{13} = \sigma_{31}$, and $\sigma_{32} = \sigma_{23}$. This reduces the number of independent shear stresses to six, and a stress tensor can be written as:

$$\sigma_{ij} = \begin{vmatrix} \sigma_{11} & \sigma_{12} & \sigma_{13} \\ \sigma_{12} & \sigma_{22} & \sigma_{23} \\ \sigma_{13} & \sigma_{23} & \sigma_{33} \end{vmatrix} \tag{3.17}$$

The axes can be chosen such that only the normal stresses are non-zero. In this case, σ_{11}, σ_{22}, and σ_{33} are the principle stresses applied along the principal axes. In uniaxial tension, there is one principal stress, σ_{11}. When there is a free surface, there are two principal axes lying in the plane of the free surface. Note that general tensile stresses are usually written as σ, while general shear stresses are written as either τ or σ_s. The principal stresses are written as: σ_1, σ_2, and σ_3.

The deformation is measured by a change in sample dimensions, positive or negative, in percent change in dimension. The initial length is often a convenient reference point, and is the denominator for the

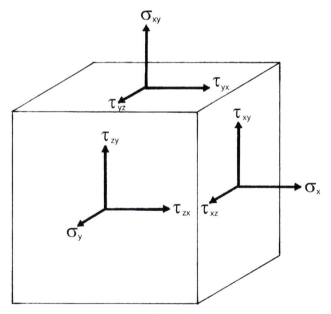

FIGURE 3.8 Force definitions as visualized using a set of three, orthonormal, experimental reference axes, x, y, and z. In the notation system described, the tensile stress σ_{11} is applied along the x-axis, and the shear stresses normal to the x-axis are τ_{12} and τ_{13}.

engineering strain. The strain can also be measured as the instantaneous change in length, providing the true strain. Again, the initial length is simpler to measure, while instantaneous changes are often more conveniently calculated. Therefore, at small stresses and strains and as long as the Poisson ratio is a constant, the true strain is easily obtained by calculation, either by assuming the Poisson ratio to be 1/2 for an incompressible solid, or using a measured value from an affine but compressible solid. The strain calculated according to the equation in Table 3.3 holds for elastic solids.

The total strain, as a differential of the change in length, can be very complicated when resolved for a generalized, three-dimensional deformation. However, for small strains, in the elastic deformation regime, they can be written as a combination of the displacements along each axis. The result is again a combination of nine components, given the same definitions: $\varepsilon_{11}, \varepsilon_{12}, \varepsilon_{13}, \varepsilon_{21}, \varepsilon_{22}, \varepsilon_{23}, \varepsilon_{31}, \varepsilon_{32}$, and ε_{33}. During the process of creating the tensor definition of strain, the off-axis strains are again found to be symmetrical. This yields a three-dimensional tensor:

$$\varepsilon_{ij} = \begin{vmatrix} \varepsilon_{11} & \varepsilon_{12} & \varepsilon_{13} \\ \varepsilon_{12} & \varepsilon_{22} & \varepsilon_{23} \\ \varepsilon_{13} & \varepsilon_{23} & \varepsilon_{33} \end{vmatrix} \tag{3.18}$$

The strain axes can also be chosen so that the principal strains act along the principal axes, with the shear strains equal to zero. Transforming the tensor along the principal axes coincides with the strain transformation for an isotropic solid. Note that general tensile strains are usually written as ε, while general shear strains are written as either γ or ε_s. The principal strains are written as: $\varepsilon_1, \varepsilon_2$, and ε_3.

The simplification of the stress-strain tensors to 1×3 matrices simplifies the algebra of mechanics to matrix form, and the various moduli occupy a 3×3 property matrix:

$$\begin{vmatrix} \sigma_1 \\ \sigma_2 \\ \sigma_3 \end{vmatrix} = \begin{vmatrix} M_1 & M_2 & M_3 \\ M_4 & M_5 & M_6 \\ M_7 & M_8 & M_9 \end{vmatrix} \begin{vmatrix} \varepsilon_1 \\ \varepsilon_2 \\ \varepsilon_3 \end{vmatrix} \tag{3.19}$$

So, $\sigma_1 = M_1\varepsilon_1 + M_2\varepsilon_2 + M_3\varepsilon_3$. The tensile stress lies along 1, and so the elastic modulus can be identified as $E = M_1$, while there are additional shear stress contributions along 1 due to M_2 and M_3. For an isotropic material, M_5 and M_9 also are E, or rather there is a unique value of E independent of the axis of the applied force. The other moduli, including M_2 and M_3 are the shear modulus, G. The bulk modulus generally does not play a role in small deformations.

True stress and true strain are particularly difficult to use for polymers. These values often must be measured rather than calculated. The instantaneous area can be measured using a traveling microscope to record the instantaneous cross-section. Likewise, the displacement can be measured with an extensometer or traveling microscope, at least at moderate deformations. Even these precautions cannot help correct for any non-affine deformation, common in polymers and especially in composites. Resolving the principal axis in anisotropic solids, such as composites, also is more mathematically intensive and may require keeping more components of the stress-strain tensors, which may in turn require the use of a four-dimensional property tensor. Nevertheless, a modulus and yield properties can be measured for both polymers and composites, because the properties are averaged over a large area, if the sample is large enough. Nanoreinforcements, especially when well dispersed, have influence on the local deformation and cross-sectional area over so small a volume that any experimental measurement can assume an average behavior. Thus, continuum level mechanics can be particularly useful in measuring behavior. Another difficulty is that, especially during compression, polymers can experience significant hydrostatic pressure, which is distributed across all axes of the solid.

There are two special cases of stress and strain in mechanical testing that merit some attention. If the test specimen is very thin, then the entire cross section is under the same force. Thus, a very thin film under tensile stress is in plane stress. If the specimen is very thick, a single plane in the sample will experience a uniform deformation. Thick specimens are used to ensure that testing is conducted in plane strain. This is the standard experimental test design.

Other Moduli

The discussion so far has focused on tensile mechanical properties. This has been primarily because people are familiar with sensing what happens when they pull on a material. Uniaxial tensile deformation also provides the clearest definitions and simplest calculations of the necessary parameters. Reaching conclusions regarding shear, bulk, or compressive deformation follows the same general route. Some polymer and polymer composite properties are most easily observed in these deformation modes.

Shear is a common type of deformation. As seen in Figure 3.9, shear in basic form is similar to trying to slide a block across a surface by putting one's hand on the top face and moving it sideways. Unless the block slides or tips, then some of the forces being applied to the block are shear forces. Before the block slides, the surface upon which the bottom face sits will exert a shear force on the bottom face, exactly counter to the manually applied force at the top face. The top of the block will be displaced, while the bottom remains in contact with the surface and undisturbed. Of course, depending on the stiffness of the material composing the block, the displacement may not be visible or easily measured. Just as for tensile force, Hooke's Law defines the shear displacement:

$$\tau = G\gamma \tag{3.20}$$

τ or σ_s, is the shear stress. In sliding a block, the force continues to build, contrary in sign, at both surfaces until the static coefficient of friction is overcome. Then, the applied shear force drops to a limit defined by the dynamic coefficient of friction. Sliding will be considered later in the chapter.

The final separate type of force is hydrostatic. The resistance of the material to hydrostatic compression is defined by the bulk modulus, *B*. The governing law in this case provides the relationship:

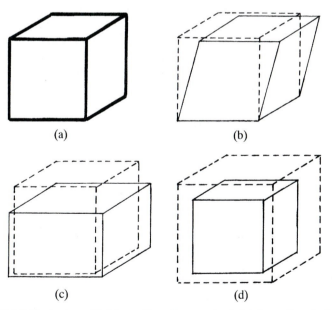

(a)   (b)

(c)   (d)

FIGURE 3.9 Beginning with (a) a cubic volume, potential deformations include: (b) shear, (c) unidirectional compression, and (d) bulk deformation.

$$P = B\left(\frac{\Delta V}{V_0}\right) \qquad (3.21)$$

P is hydrostatic pressure, or the pressure exerted by the surrounding environs, and V is the sample volume. Note that P is measured relative to atmospheric pressure.

The three material moduli and the Poisson ratio are not independent. Tensile deformations are visible from applied shear or bulk forces. Likewise, a tensile force could be transformed by a rotation of axes into a shear force. The relationship between the moduli is evaluated using the Poisson ratio:

$$E = 2G(1+v) = 3B(1-2v) \qquad (3.22)$$

Uniaxial compression is the rough opposite of tensile deformation. How rough the comparison is depends on the material. In compressive deformation, a negative force is applied along one dimension of the bulk specimen. Whereas in tension, the length extends while the width and depth shrink, the opposite is observed in compression. Continuum theory provides a mathematical explanation similar to tension, with the sign reversed. This is reasonable for many materials until plastic deformation occurs. Beyond the plastic deformation, the mechanics of failure take over. Compression, however, can have a significant hydrostatic component as the material resists expanding. If a material has a weak bulk modulus, particularly true of polymers, then the contribution of hydrostatic force to the compressive response can be important. This usually results in the yield stress in polymers being significantly greater in compression than in extension. The hydrostatic force, being isotropic and usually invariant, can be subtracted from the stress using the principal axes, creating the deviatoric stress tensor.

Plasticity

Deviations from the simple elastic deformation described by Hooke's law occur frequently, often at moderate or small deformations. Non-linear behavior can be observed for several important reasons, including permanent, unrecoverable changes to the structure. Energy may be diverted into displacements that cannot be recovered. For example, if an atom of a metal is displaced from the base unit cell, creating a defect, the atom will not return to the original position after the force is released, and the energy cannot be recovered. Similarly, if a plane of atoms is sheared across the plane, the deformation cannot be recovered. Generally, such deformations cause more displacement than would be predicted by an elastic response. These deviations are called plastic deformation. While plastics also behave this way, they do so for different molecular reasons. In the context of mechanics, the words plasticity and plastic refer to the behavior of metals and ceramics.

Some types of non-linear behavior are, however, fully recoverable, and are discussed using non-linear elastic models. Polymeric and plastic non-linear mechanical deformations arise from viscoelasticity or rubbery elasticity. Composite materials, especially those with deformable matrixes, also will often exhibit non-linear elastic behavior. The continuum mechanical plastic deformation will be discussed first.

In certain materials, permanent plastic deformation takes place after some significant initial deformation. In this case, the initial slope provides a useful measure of the modulus. In other materials, plasticity is obvious from the beginning of deformation. In these cases, while there is an initial slope that provides Young's modulus, this is an upper bound on the value of the effective modulus. Often, a secant line is drawn between the initial point, usually (0,0) and the highest strain of interest. In practice, this is frequently the yield point. The slope of this line provides the secant modulus, a lower-bound modulus applicable until yielding oc-

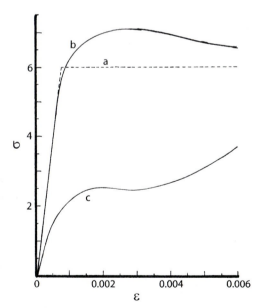

FIGURE 3.10 Examples of plastic deformation. (a) ideal plastic deformation, (b) plastic deformation in a metal, and (c) plastic deformation in a thermoplastic.

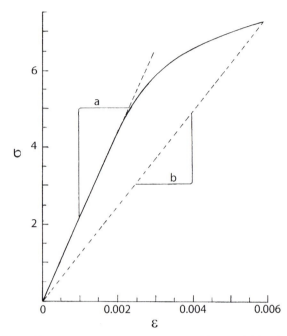

FIGURE 3.11 (a) The elastic modulus is measured taking the initial slope, where the material behaves as a perfectly elastic solid. (b) The secant modulus is evaluated by taking the slope of the line connecting the origin to the yield stress. The secant modulus describes the average behavior of the solid over a larger range of mechanical responses.

curs. Engineering design with the secant modulus provides a safety factor not present when using Young's modulus.

YIELD

Yield has been mentioned several times already. The first mention occurred in describing when the deformation force added enough energy to start displacing atoms. Now, the yield point in a linearly elastic material provides an assessment of the upper stress and strain boundaries to the appropriate use of modulus and safety constraints in design.

There are several phenomenological measurements for the yield point. One common measure is a 0.2% offset. Using this method, a line with the same slope as the initial modulus is drawn with a 0.2% offset from the origin of a stress-strain plot. When the plastic curvature causes the stress-strain behavior to cross this offset line, the material is taken to exhibit plastic deformation. Above this point, the deformation is not fully recoverable. The atoms have been displaced farther than can be accounted for by lengthening the bonds. In polymers, this can be described as an extrinsic yield criterion, because the values of σ_y and ε_y, the yield stress and yield strain, respectively, are derived from an extrinsic criterion.

The Considère construction, shown in Figure 3.13, is another method for extrinsically determining the yield point. Using this method, the true stress is plotted against the nominal strain. The true stress is given in Table 3.3. The deformation for materials below yield is assumed to be incompressible, and therefore at a constant volume. The nominal strain can be transformed using Table 3.3. At yield, the nominal load will cease to rise with increasing strain, as shown by:

$$\frac{d\sigma}{d\varepsilon} = \frac{\sigma}{1+\varepsilon} \tag{3.23}$$

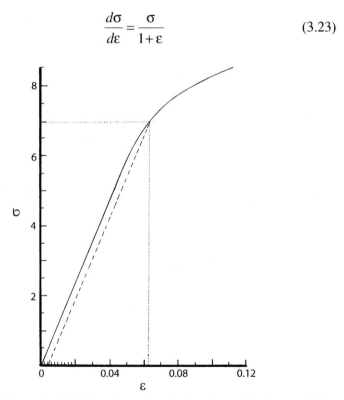

FIGURE 3.12 The extrinsic yield can be evaluated using a line with a slope parallel to the initial elastic behavior, offset 0.2% along strain. Yield is defined to occur where the offset line intersects the deformation curve. 0.2% is a typical value for many metals, but other values are standard for certain metals, such as cast iron or magnesium.

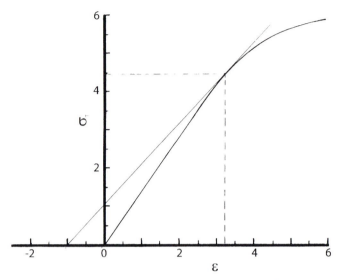

FIGURE 3.13 Considère construct for yield in ductile polymers. The true stress is used instead of the engineering stress. A tangent line to the curve is drawn that intersects at −1 strain. The point of tangency provides another extrinsic definition of yield.

This indicates that a tangent line intersecting the nominal strain at −1 at zero stress is a condition for yielding.

Deformation beyond the 0.2% offset may be fully recoverable in a non-linear elastic material. Thermoplastic polymers often exhibit a local maximum in stress at some strain. This change in behavior consistently occurs at a stress-strain fixed by the material's composition. At this point, the strain may stop increasing, increase more slowly, or decrease. Regardless of the specific manifestation, which is also characteristic of a polymer type and depends additionally on formulation and processing history, the point at which this stress-strain behavior change occurs is considered the yield point. This can be described as an intrinsic yield, because the performance change is driven by largely unrecoverable structural changes in the polymer. There can be unrecoverable strain at loads below this yield point. Further, even if the polymer is strained beyond yield, some, if not all, of the deformation may be recoverable by heating the polymer.

There are several traditional criteria for determining the yield. The Tresca yield criterion states that the yield occurs when a maximum is reached in the shear stress. The principal stresses σ_1, σ_2, and σ_3, can be picked such that $\sigma_1 > \sigma_2 > \sigma_3$:

$$(\sigma_1 - \sigma_3) = 2\tau_y = \sigma_y \qquad (3.24)$$

von Mises proposed that the yield would occur when the shear strain energy reached a critical value, rather than the shear stress:

$$(\sigma_1 - \sigma_2)^2 + (\sigma_2 - \sigma_3)^2 + (\sigma_3 - \sigma_1)^2 = 6\tau_y^2 \qquad (3.25)$$

In two dimensions, for any combination of tension, compression, or shear, the yield envelope can be drawn. Both Tresca and von Mises predict that the yield criterion for tension and compression should be the same. This is not observed for polymers, however, where the yield stress is higher in compression than in tension. Polymers are also observed to have a higher yield stress under increased hydrostatic pressure.

Phenomena

The primary model for plastic deformation is the formation and movement of dislocations. During simple tension, the atoms will displace slightly, relative to each other, to maintain constant volume and sustain all the original bonds. This deformation is completely elastic and recoverable. Eventually, enough energy will be added through the force to overcome one or more bonds. When this level is reached, the energy can be stored more effectively by introducing defects in the crystal structure,

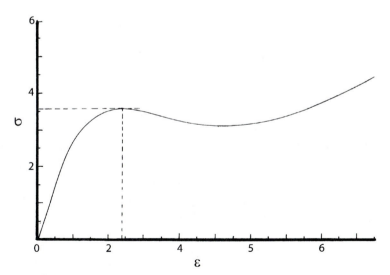

FIGURE 3.14 Thermoplastic polymers often exhibit a local maximum in stress, characteristic of the polymer. This provides an intrinsic definition of the yield.

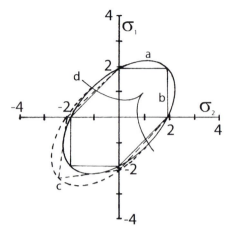

FIGURE 3.15 Various criteria for yield in a material: (a) the Tresca criteria, and (b) the von Mises criterion. (c) Yield is often distorted in materials with a significant bulk response, such as polymers. (d) Brittle polymers may exhibit crazing before failure through yield defined by the Tresca or von Mises criteria.

causing dislocations. However, the energy used to create these defects is dissipated rather than stored, and the deformation is no longer recoverable. Point defects can be created to disrupt single bonds: a vacancy, or hole, in the structure, creates one atom's worth of new surface area, distributing the energy to the newly created surface. An interstitial atom will cause a disruption of the simple shape of the unit cell, distributing excess energy to the displacement of the neighboring atoms. Note that an interstitial atom will also make moving neighboring atoms physically more difficult.

Line and plane defects can be induced as well. Such defects introduce slightly more disorder, and slightly greater energy into the solid, again absorbing some of the applied force into unrecoverable deformation. In metals, after the displacement occurs, bonds can re-form without causing the solid to fail. In polymers, deformation typically occurs resisted by the van der Waals forces, which can also re-form. In ceramics, the displaced atoms move into contact with atoms of opposite charge. Displacement requires too much energy, leading to fracture. The ease with which defects can form, the types and numbers of defects that can form in a particular structure, and the resistance to displacement of these defects control the plasticity of a material.

Slip is the primary description of how the molecular movement leading to plastic deformation occurs. Point defects and dislocations will move under the influence of the applied force, causing material flow. As dislocations flow in one direction, the mass will shift in the opposite direction, resulting in a plastic deformation. In atomic solids, there is no restoring force retained during this displacement, and plastic deformation is unrecoverable. In polymers, the molecular interconnectivity along the chain can store some displacement energy that may later be recovered under the proper circumstances.

Sometimes, moving dislocations will run into each other. Two intersecting dislocations can bounce off and around each other, combine and annihilate, or combine into a different type of dislocation or a larger flaw. Unless the defects annihilate, the result is continued, if modified, deformation by dislocation slip. However, when three dislocations meet they can trap each other, creating a stable defect. This stable feature can trap new dislocations, absorbing them. The result is strain hardening, introducing another mechanism for preventing large-scale deformation.

Slip has several physical manifestations in extreme cases of yielding. One type of visual disruption is kink banding. In a kink band, pairs of dislocations cause a redirection of a crystallographic plane. The result is a kink in the crystals, observable using microscopy. Another manifestation is the creation of new grain boundaries through an accumulation of dislocations, making the crystals smaller and less perfect.

A grain boundary is also an accumulation of dislocations. Twinning boundaries are an example of a low energy grain boundary. In twinning, part of the crystal rotates along a slip plane. The result is a new crystal that shares a crystallographic edge with the parent crystal where the two different crystal system lattices match. This match means there are few holes and less increase in free surface.

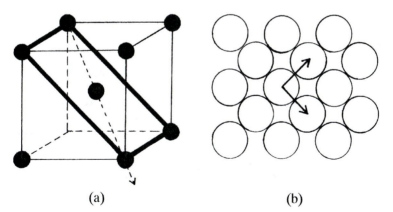

(a)   (b)

FIGURE 3.16 (a) the (110) slip plane in a body-centered cubic unit cell, and (b) the atomic organization of the slip plane.

Box 3.2 A simple model of flow comes from Eyring.

Flow can be modeled as a hole moving through a solid. Modeling the hole is easier and more general than modeling the movement of atoms or molecules because there are fewer unknowns about the hole. In order to move, the hole must have sufficient energy to cross some energy barrier to movement.

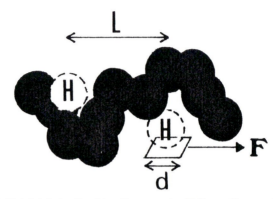

FIGURE 3.17 (a) A hole, H, with a diameter, d, will flow a distance L under an applied force F.

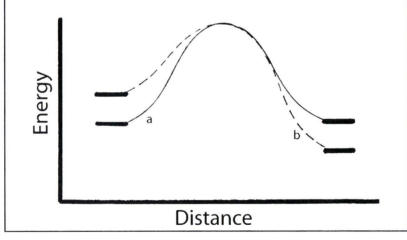

FIGURE 3.18 (a) In the absence of a force, the hole has equal energy in either the initial position or the displaced position. Flow may occur with sufficient thermal energy, but the hole may return to the initial position the same way. (b) Under an applied force, the energy for the hole is lower downstream, leading to flow.

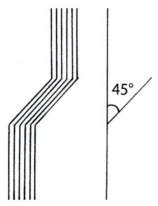

FIGURE 3.19 A kink band is a typical failure mode, when a crystal buckles under compressive force. The kink band is typically at 45°.

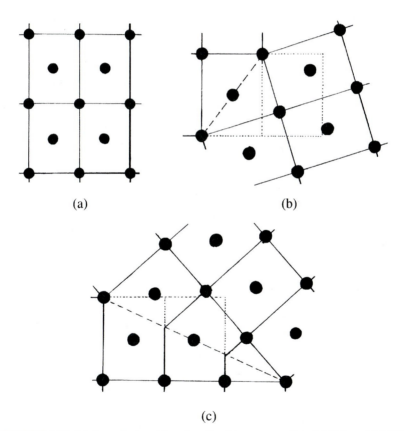

(a)

(b)

(c)

FIGURE 3.20 (a) An orthorhombic unit cell, with corresponding (b) 110, and (c) 310 twin planes.

(a)   (b)

FIGURE 3.21 Schematic of polycrystalline material with (a) small crystals and (b) large crystals.

On the other hand, crystals can meet at a high-energy boundary with little registration of crystal lattices. There is a store of energy in the free surface created at such boundaries. Since there is already free surface at these boundaries, there is less energy barrier to motion along these boundaries. Therefore, grain boundaries are a convenient slip line in most materials. Changes in the crystal structure that increase the number and energy level of grain boundaries influence the mechanical properties. Annealing can eliminate grain boundaries and cause crystals to grow and become more perfect. Deformation can break grains apart.

During crystallization, the lowest-energy crystal structure is usually formed. Some materials are capable of crystal-crystal transformations between stable structures and meta-stable, high-energy configurations. Crystal-crystal transformations can occur at defined temperatures to accommodate thermal displacement of atoms, such as occurs in Teflon™. In strain induced crystal transitions, alternatively, the low-energy favored structure can be distorted by stress to the point where a meta-stable crystal phase forms, releasing more of the stored energy. This transformation may dissipate energy irreversibly, so the structure will not always return to the favored state when the stress is removed. Yttria is a ceramic with a metastable, stress-induced phase. Zirconia may be stabilized by adding Yttria, which absorbs some stress by transforming without disrupting the overall structure.

In polymers, there can be an accumulation of amorphous material at boundaries between crystals. There is amorphous content in the fold structure as the lamellae are formed. There can be additional amorphous material between the lamellae in a spherulite. There is additional amorphous material at the boundaries between spherulites.

Polymer deformation will usually occur through the reorganization of chains in the amorphous regions, up until the average amorphous chain becomes fully extended. Chain reorganization is resisted by the energy wells created by the neighboring van der Waals interactions, the van der Waals interactions between neighbors of the covalent partners, as well as any energy well associated with local crystallinity. Therefore, deformation occurs preferentially between spherulites, and then along the boundaries between lamellae as the spherulites are disrupted. The additional slip mechanisms result in the ability of polymers to undergo large-scale reorientation. This reorganization results in the formation of a stable neck. The amorphous material around the spherulites is drawn and oriented, relocating spherulites. Then the spherulites will break up to form an oriented shish-kebob structure. The lamellae can unfold, as well, but the evidence for this occurring during cold drawing is mixed.

In composites, the reinforcement can nucleate polymer crystal growth, leading to more crystallinity. This will occur if the particle is larger than

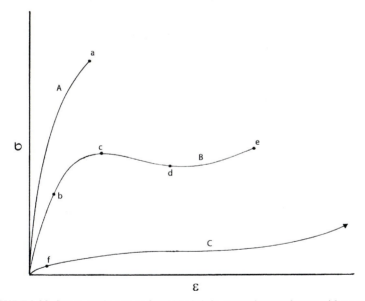

FIGURE 3.22 Stress-strain curves for: (A) a brittle, amorphous polymer with a non-linear elastic response leading to a fracture (a). (B) a semicrystalline polymer with (b) a non-linear elastic response, leading to (c) yield, (d) crystal reorientation and finally (e) fracture. (C) a rubbery elastic polymer with (f) a typical rubbery elastic response.

necessary to serve as a crystal nucleus, or if one dimension is the same length as a crystal lattice parameter. Specifically, nanoparticles may not be large enough or otherwise suited as nuclei. Generally, reinforcements that do not nucleate crystal growth act in crystallization the same as impurities. Such reinforcements will occupy positions along spherulite boundaries, and may prevent crystals from disrupting as necessary during deformation.

Yield will occur as the chains begin to reorganize, where the local segments readily move from one energy well to another. Some resistance to unrestricted deformation in this region is provided by the energy-by-proxy induced through the interconnectivity of the chain. Resistance is particularly strong in the environment of any crystalline structures, consisting of more systematic, concentrated secondary bonds between segments as well as the primary bonds along the chain. When sufficient force is applied, the crystal can be disrupted in part, resulting in the reorganization of the crystals parallel to the direction of the force.

Below the glass transition, restriction of motion is provided by the van der Waals energy-by-proxy through the interconnectivity between segments along the chain, rendering the net energy required for displacement very large. Sufficiently above the glass transition temperature, the average energy in each segment is sufficient to overcome the secondary bond energy. Above the threshold for easy bond rotation, the stress on a

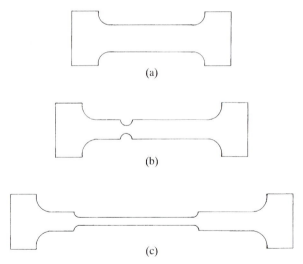

FIGURE 3.23 Polymer necking: (a) A dogbone specimen is tested in tension. (b) At yield, a stable neck forms. (c) Under further deformation, more material is drawn into the neck.

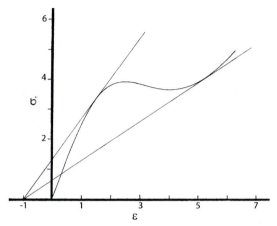

FIGURE 3.24 A material capable of stable necking will exhibit two tangent lines in the Considère construct.

covalent bond can be relieved with little energy cost by rotating a nearby bond, reducing the energy contribution to stress.

Some polymers and metals will exhibit necking. The material will begin at a certain stress and strain to form a narrow point along the gauge length. This will manifest on a stress/strain curve as if the strain is increasing at a constant or decreasing stress. The force is now being applied over a reduced area, causing increased force in the neck zone. The neck can be unstable, in which case the narrowing will continue until the test sample breaks. The neck can be stable, in which case the polymer will narrow until a certain point, then stabilize. The neck region will then grow, with more material being drawn into the neck. After the entire gauge length is transformed, the stress will usually increase again until the sample breaks.

The Considère construction provides a means to determine the conditions for forming a stable neck. A stable neck can form if two tangent lines can be drawn from the $(-1,0)$ point. In metals where work hardening occurs, the dislocations can pin each other, leading to increased strength within the work zone. In polymers, the chains are observed to align, resulting in applying the force to the primary bonds. In both cases, at higher strains, a second mechanical mechanism takes over, and this mechanism results in the formation of the stable neck. Composite reinforcements can impact the second mechanical mechanism. For example, if nanoparticles suppress chain reorientation, then the secondary mechanism, primary bond deformation, is not brought into play and the material may not neck. If, on the other hand, tensile deformation reorganizes the reinforcement, the composite material may form a neck when the matrix ordinarily would not.

Finally, brittle polymers are capable of a final type of yield not available to metals or ceramics. The polymer can undergo microscopic cracking, or crazing. Under hydrostatic tensile load, holes can nucleate in the polymer, allowing a rapid increase in strain without catastrophic failure of the part. The stress conditions leading to crazing are shown relative to the Tresca and von Mises yield criteria in Figure 3.15. The crazes form a porous web structure, illustrated in Figure 3.25.

One possible effect of composites is to limit or enhance the number of alternative deformations within a solid, leading to increased strength. The addition of carbon to iron gives rise to non-cubic structures in steel. The non-cubic structures have fewer slip modes, making the metal stiffer and stronger. The addition of fibers or additives to ceramics creates a location for new surface area to form, making the ceramic less brittle. In polymers, variations in primary, secondary and tertiary structure can all modify the mechanical properties. An example of primary structural changes affecting the modulus is copolymerization. Many additives and reinforcements alter the ability of chains to fold, or modify the proximity of interchain neighbors, affecting the secondary structure will also modify the modulus. Increased crystallinity, or tertiary ordering, will result in an increased modulus, as the number of deformation modes is reduced in

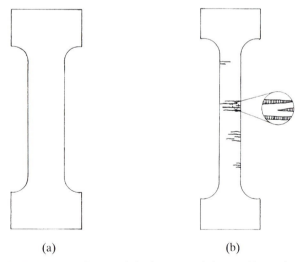

(a)   (b)

FIGURE 3.25 Polymer crazing: (a) A dogbone sample is tested in tension. (b) Prior to the predicted yield failure, microscopic cracks form in the gauge section. The cracks are bridged by fibrils.

transforming from a disordered amorphous to an ordered crystalline state.

FRACTURE

The failure of a material depends on design criteria. In many applications, a part will exhibit failure if yield occurs. At this point, the solid body no longer bears the applied weight while retaining the initial shape. If the material is used in a bridge, however, the bridge does not fail if it displaces by a few inches over 100s of years. Therefore, some plastic deformation does not necessarily indicate failure. Failure by yielding may not represent catastrophic failure. The extreme limit of material failure is fracture. Fracture, or breaking one part into two pieces, represents complete failure.

Different materials fracture in different ways. Ceramics, many metals and some brittle polymers experience brittle fracture, with sudden disruption of the structure and catastrophic failure. Some metals and most polymers exhibit ductile failure, with gradual tearing leading to slow, but complete, failure. Some useful models can help understand the fracture mode of materials.

Fracture can be visualized as the growth of pre-existing cracks. The source of the original cracks could be scratches, cracks, molecular imperfections or any other sharp feature that can focus the stress. The specific origin of the crack is unimportant for developing a model of fracture, because failure will always occur at some initiating flaw. At the crack tip, the stress concentration caused by a flaw will increase the most rapidly. Failure will start at the crack tip and propagate through the solid.

Consider a solid containing an elliptical crack under a uniform stress, perpendicular to the crack, as shown in Figure 3.26. Note that if the crack is parallel to the applied stress, there is no drive to enlarge the flaw. The stress is concentrated by the crack, with the highest stress being at the crack tips. The stress at the crack tip is given by:

$$\sigma_t = \sigma_0 \left(1 + 2\sqrt{\frac{a}{r}} \right) \qquad (3.26)$$

σ_t is the stress at the crack tip, σ_0 is the applied stress, a is the crack length and r is the radius of the crack tip. The crack-tip stress increases as the crack lengthens and as the radius of the tip decreases. As the crack grows without a commensurate increase in crack radius, the stress concentration will increase. Therefore, crack propagation is generally spontaneous. This is why drilling a hole at the tip of the crack to blunt the stress

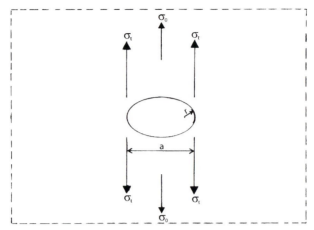

FIGURE 3.26 An elliptical crack, with a width of *a* and a crack tip radius of *r*. A general tensile force, σ_0, has a stress concentration σ_t at the crack tip.

concentration can sometimes arrest crack growth. Crack blunting, as the crack-tip intersects a reinforcing particle, is one proposed mechanism by which composites improve fracture. A nanoreinforcement will be less effective at blunting a crack in this manner.

A problem with the geometric model of a sharp elliptical crack, however, is that as the tip radius is decreased the stress concentration approaches infinity. An approach to solving the problem of infinite stress concentration uses an energy balance. In this approach, developed by Griffith, the energy required to create a sharp crack is related to the energy required to create new free surface. If one assumes that in forming the crack the stress is relieved in a zone surrounding the crack, an amount of energy is dissipated. An amount of energy is also required to create new free surface. When the energy required to create the free surface is lower than the amount of energy released by the crack, the crack will grow, as shown by:

$$\sigma_f = \left(\frac{2E\gamma}{\pi a}\right)^{\frac{1}{2}} \tag{3.27}$$

σ_f is the failure stress, E is the tensile modulus, and γ is the surface energy of the material. The form of this equation holds true for many brittle materials. However, by determining the failure stress as a function of the crack length, calculated values of surface energy are far greater than the true value. This indicates that even where linear elasticity holds, some plastic deformation is contributing to the total energy consumption. The energy dissipated by plastic deformation swamps the surface energy. Therefore, the surface energy is usually replaced by a Griffith fracture energy, G_c. In plane stress, the equation then has the same form as the Griffith criteria:

$$\sigma_f = \left(\frac{EG_c}{\pi a} \right)^{\frac{1}{2}} \tag{3.28}$$

In the more common plane strain criteria, and for linear elastic deformation, the expression is more complex:

$$\sigma_f = \left(\frac{EG_c}{\pi(1 - v^2)a} \right)^{\frac{1}{2}} \tag{3.29}$$

From a processing point of view, eliminating flaws will improve the resistance to fracture. E and G_c can be treated as materials properties for a given set of testing conditions. The fracture strength of a given material is governed by the maximum size of the flaws in the material. If the flaw size can be reduced, the fracture strength can be improved. If the flaw size is known, the fracture strength can be predicted. However, several brittle polymers behave as if they contain flaws of a given size even if they are carefully prepared and polished to remove any damage. This leads to the concept of "inherent flaw size", and may be related to the formation of crazes. In polystyrene, the inherent flaw size is about 2 mm.

Fracture depends on the general mechanical behavior of the bulk material, the crack size and the part geometry. For a general crack, a more general form of the above equation is more helpful:

$$\frac{1}{b} \frac{\partial}{\partial a} (F - U) \geq G_c \tag{3.30}$$

where F is the external work, U is the internal energy, and b is the part thickness. Linear elastic fracture mechanics apply specifically where Hooke's law is obeyed by the bulk of the material. In this case, the force-displacement curve is as shown in Figure 3.27. P is a generalized load, and D is a generalized displacement. The internal energy before the crack grows is based on the load and displacement:

$$U_1 = \frac{1}{2} PD \tag{3.31}$$

The final state includes the changes in displacement as the load changes:

$$U_2 = \frac{1}{2}(P + \partial P)(D + \partial D) \tag{3.32}$$

The work done by displacement during crack formation is a function of the change in energy:

$$\partial F = P\partial D + \frac{1}{2}\partial P\partial D \qquad (3.33)$$

For a crack to occur, a critical energy must be surpassed:

$$G_c = \frac{1}{2b}\left(\frac{P\partial D}{\partial a} - \frac{D\partial P}{\partial a}\right) \qquad (3.34)$$

The compliance modulus, C, the reciprocal of Young's modulus, provides a useful simplification, as follows:

$$C = \frac{D}{P} = \frac{1}{E} \qquad (3.35)$$

$$\partial D = P\partial C + C\partial P \qquad (3.36)$$

Combining these expressions provides a relationship for the Griffith fracture energy, expressed as:

$$G_c = \frac{F_c^2}{2b}\frac{\partial C}{\partial a} \qquad (3.37)$$

F_c is the critical fracture force for the part. C as a function of a may be

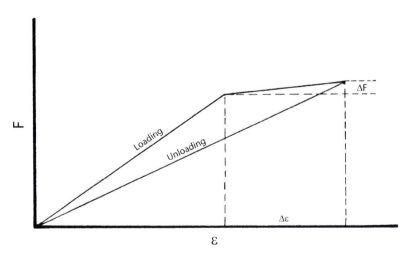

FIGURE 3.27 The force-displacement plot for crack opening. Loading increases until cracking causes a precipitous increase in sample strain, $\Delta\varepsilon$, for a relatively small increase in force, ΔF. On unloading, the force is completely recovered.

determined analytically or experimentally. Thus, a measure of F_c can provide a value for G_c. This indicates that a greater energy is required for failure in compliant materials with smaller flaws.

Another approach to understanding fracture mechanics is the stress-intensity factor. Here, the stress at the tip of the crack can be determined from the applied stress, as:

$$\sigma_{ij} = \frac{K}{(2\pi r)^{1/2}} f_{ij}(\theta) \tag{3.38}$$

σ_{ij} are the stress tensor components at a point, r and θ are the coordinates of the point and K is the stress intensity factor relating the applied stress to the local stress as a function of sample and crack geometry. There are three modes of crack loading, shown in Figure 3.28. Of these, Mode I loading is the most commonly encountered and most likely to result in fracture failure.

Returning to the elliptical crack, the crack tip stresses are:

$$\begin{Bmatrix} \sigma_{11} \\ \sigma_{12} \\ \sigma_{22} \end{Bmatrix} = \frac{K_I}{(2\pi r)^{1/2}} \cos\left(\frac{\theta}{2}\right) \begin{Bmatrix} 1 + \sin\left(\frac{\theta}{2}\right)\sin\left(\frac{3\theta}{2}\right) \\ \sin\left(\frac{\theta}{2}\right)\cos\left(\frac{3\theta}{2}\right) \\ 1 - \sin\left(\frac{\theta}{2}\right)\sin\left(\frac{3\theta}{2}\right) \end{Bmatrix} \tag{3.39}$$

$\sigma_{23} = \sigma_{13} = 0$, and, for plane stress $\sigma_{33} = 0$, or for plane strain $\sigma_{33} = v(\sigma_{11} + \sigma_{22})$. As $r \to 0$, the stress $\sigma_{ij} \to \infty$, so the stress is not an appropriate criterion. Irwin postulated that since K_I defines the stress around the crack, then $K_I \geq K_{Ic}$ represented a fracture criterion, where K_{Ic} is the critical value for crack growth. The advantage of the stress-intensity approach is that for any Mode I problem:

$$K_{Ic} = Q\sigma_c a^{\frac{1}{2}} \tag{3.40}$$

σ_c is the stress at the onset of growth and Q is a parameter determined by the geometry. The form of Q is a non-dimensional series expansion dependent on crack size and specimen geometry. Table 3.4 shows the values of K_{Ic} for various specimen geometries.

Microscale yielding contributes to the total fracture energy. Small-scale, non-affine deformations can absorb significantly more energy than is predicted from forming new free surface. If the zone of yield-

TABLE 3.4 Test geometry and K_{Ic}.

Test	Geometry	Written in terms of the engineering values
Single Edge Crack		$$K_I = \left[1.99 - 0.41\left(\frac{a}{w}\right) + 18.7\left(\frac{a}{w}\right)^2 - 38.48\left(\frac{a}{w}\right)^3 + 53.85\left(\frac{a}{w}\right)^4\right]\sigma_0 a^{1/2}$$ w = width of specimen
Three Point Bend		$$K_I = \frac{Fs}{bw^{3/2}}\left[2.9\left(\frac{a}{w}\right)^{1/2} - 4.6\left(\frac{a}{w}\right)^{3/2} + 21.8\left(\frac{a}{w}\right)^{5/2} - 37.6\left(\frac{a}{w}\right)^{7/2} + 38.7\left(\frac{a}{w}\right)^{9/2}\right]$$ F = Applied load, s = Span between supports, b = Sheet thickness
Double Torsion		$$K_I = Fl_m\left[\frac{3}{lb^3 b_n(1-\nu)}\right]^{1/2}$$ Plane Strain (thick sample, $b < l/2$) l_m = length of moment arm, b_n = sheet thickness in plane of crack, ν = Poisson Ratio
Double Cantilever		$$K_I = \frac{2F}{bb_n^{1/2}}\left(\frac{3a^2}{h^3} + \frac{1}{h}\right)^{1/2}$$
Tapered Double Cantilever		$$K_I = 2F\left[\frac{K}{bb_n}\right]^{1/2}$$ Tapered so that $\frac{3a^2}{h^3} + \frac{1}{h}$ is constant
Compact Tension		$$K_I = \frac{F}{bw^{1/2}}\left[29.6\left(\frac{a}{w}\right)^{1/2} - 185.5\left(\frac{a}{w}\right)^{3/2} + 655.7\left(\frac{a}{w}\right)^{5/2} - 1017\left(\frac{a}{w}\right)^{7/2} + 638.9\left(\frac{a}{w}\right)^{9/2}\right]$$

Table after Kinloch and Young

ing is small, however, the elastic strain field may not be disturbed significantly. By making this assumption, the plastic zone may be defined by the elastic stress.

One model proposed by Irwin is that in order for the crack to advance, the yield stress in a zone preceding the crack must be deformed sufficiently. This circumstance is illustrated in Figure 3.29. This type of zone is observed in highly plastic materials. In order for the crack to proceed some distance r_y, a roughly spherical plastic zone is created, centered on the future tip of the crack. If $r_y \ll a$, the size of the plastic zone is given by:

$$r_y = \frac{1}{2\pi}\left(\frac{K_{Ic}}{m_p \sigma_y}\right)^2 \tag{3.41}$$

m_p is the plastic constraint factor, m_p = effective yield stress/σ_y. This accounts for the constraint on the plastic zone created by the surrounding elastic material. This constraint partially compensates for the uniaxial deformation, so the fracture stress is greater than for uniaxial tension. For plane stress, $m_p = 1$, whereas for plane strain, $m_p \approx \sqrt{3}$.

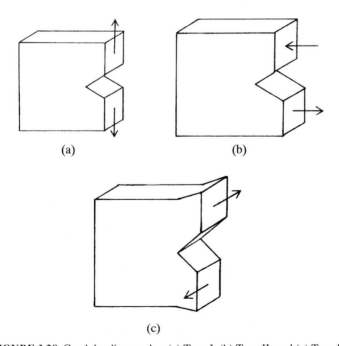

(a)   (b)

(c)

FIGURE 3.28 Crack loading modes, (a) Type I, (b) Type II, and (c) Type III.

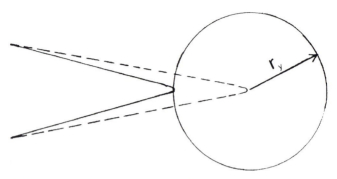

FIGURE 3.29 Irwin model of the plastic zone. For the crack to extend, a plastic zone in front of the crack of radius r_y must fail through yielding.

In many brittle polymers, the plastic zone is more closely approximated by a line. Dugdale modeled this situation, as shown in Figure 3.30. Yielding at the crack tip makes the crack longer by R. The force applied to opening the crack is cancelled by internal stresses acting on the boundary of the plastic zone, until the crack advances. The length of this plastic zone is given by:

$$R = \frac{\pi}{8}\left(\frac{K_k}{\sigma_p}\right) \qquad (3.42)$$

σ_p is usually taken to be the yield stress, as the yield stress defines the point where permanent damage begins in a material.

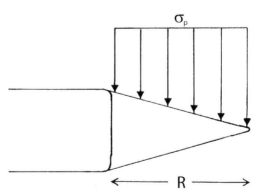

FIGURE 3.30 Dugdale model of the plastic zone. For the crack to extend, the crack must open sufficiently to cause yield along a length R. Crack opening is resisted by a restoring force σ_p.

For linear elastic fracture mechanics, G_{Ic} and K_{Ic} are related. Since polymers and composites seldom follow linear elastic mechanics, the relationship in them is more complex.

RUBBERY ELASTICITY AND VISCOELASTICITY

In a polymer, deformation will occur first in the amorphous regions. The stress results in local chain conformation changes. The chain connectivity applies a restoring force, especially the resistance-by-proxy along the chain. Over time, the force applied by proxy along the chain can be relaxed as the chain reorganizes. The extent of this reorganization is affected by the time a material is held under force, by the temperature and any mechanical interconnections between chains via cross-links, by crystallinity and so forth. Thus, the mechanical properties of a polymer are time and temperature dependent. This phenomenon is viscoelasticity.

The continuum approach to viscoelasticity describes any material as simple combinations of mechanical elements. The base mechanical elements are the spring and the dashpot. A typical dashpot is the cylindrical device attached to a screen door that keeps it from slamming closed once it has been opened. The spring is an adequate description of metals and ceramics in the linear elastic region. A dashpot is a representation of the result of applying a force to a liquid, where the mechanical response is a delayed conformation to the applied force, resisted by the viscosity of the liquid. Linear polymer viscoelasticity may be described by a series or parallel combination of these elements.

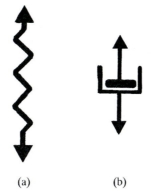

(a)   (b)

FIGURE 3.31 (a) spring and (b) dashpot elements used in continuum viscoelasticity models.

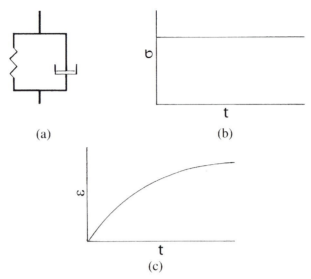

(a)

(b)

(c)

FIGURE 3.32 Creep behavior. (a) The Kelvin element, with a spring and a dashpot in parallel. (b) Constant stress. (c) Time-dependent strain.

Creep

A polymer under a constant load will slowly elongate, as shown in Figure 3.32. This is different from metal creep in that (1) the creep may introduce no defects, and (2) the deformation may be partially, or completely, recoverable with time or temperature. This sort of response is typical of the tire of a parked car, a pipe containing flowing water, or a loaded beam.

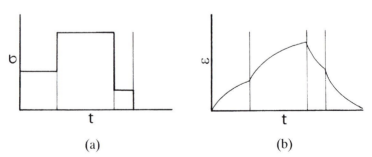

(a)

(b)

FIGURE 3.33 Boltzmann superposition. (a) A series of constant stresses are applied over time. (b) The resulting aggregate strain response is a linear superposition of the individual time-dependent strain responses to each stress.

The response can be modeled by a Kelvin element. This parallel combination of spring and dashpot resists the initial deformation. With time, the dashpot will allow displacement, modeling the time dependence of creep:

$$\varepsilon(t) = \frac{\sigma_0}{E}[1 - e^{(-t/x)}]$$ (3.43)

$\varepsilon(t)$ is the time-dependent strain, σ_0 is the constant force, and t is time.

τ is a relaxation constant, derived from the elastic modulus and the internal viscosity:

$$\tau = \frac{E}{\eta}$$ (3.44)

The modulus, E, is the instantaneous modulus. The internal viscosity is a more curious idea: the viscosity of the solid. The internal viscosity is the resistance to bond reorganization.

Since this equation is being used to describe linear viscoelasticity, the stresses are superimposable, and the strain responses are additive.

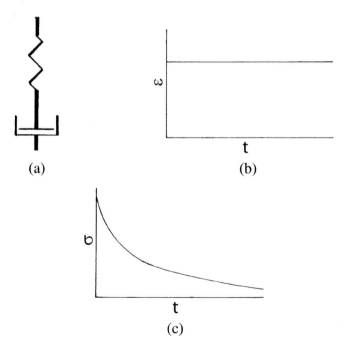

(a)   (b)

(c)

FIGURE 3.34 Stress relaxation behavior. (a) A Maxwell element, with a spring and a dashpot in series. (b) Constant strain. (c) Time-dependent stress.

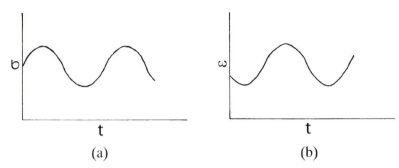

FIGURE 3.35 Dynamic load response. The stimulating stress (a) is followed by (b) a phase-shifted strain response.

Stress Relaxation

A polymer under a constant strain will slowly release the stress response, as shown in Figure 3.34. This type of response is typical in rubber bands, the residual stresses in bent pipes, and so on. The stress as a function of time is described by a Maxwell element. The material exhibits an initial deformation resisted by the spring. Over time, the dashpot will displace, allowing the force to relax.

This behavior is defined by the time-dependent response of a Maxwell element. This is a series combination of the spring and dashpot:

$$\sigma = \sigma_0 e^{(-t/x)} \tag{3.45}$$

where $\sigma(t)$ is the time-dependent stress, σ_0 is the initial stress, t is time, and again τ is relaxation time. Superposition also applies to stress relaxation.

Damping

If a cyclical load is applied to a polymer, the material will exhibit a cyclic strain. Viscoelasticity implies that there will be some lag between the applied stress and the resulting strain. This is accounted for by an angle of phase shift:

$$\varepsilon = \varepsilon_0 \sin \omega t \tag{3.46}$$

$$\sigma = \sigma_0 \sin(\omega t + \delta) \tag{3.47}$$

ω is the period of the oscillating stress or strain. δ is the phase shift in the response between the stress and strain.

Real Materials

A real viscoelastic material exhibits creep, stress relaxation, and damping. Each of the elements above provides a governing equation for one type of viscoelastic response. However, a single element is insufficient to describe the full range of properties. One approach is to combine multiple elements. As shown in Figure 3.36, this gives a combination of properties that can empirically recover something similar to real behavior. A combination of dashpots can be used in a single element to give a complex relaxation time profile. All of these exhibit some utility in generalizing the properties of a single polymer system from a limited number of measurements. The empirical nature of these models makes the value of predictions somewhat limited.

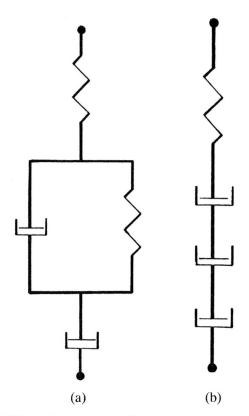

(a)   (b)

FIGURE 3.36 (a) The standard element model, with springs and dashpots in both series and parallel. (b) Dashpots in series in a Maxwell element give a relaxation spectrum.

Box 3.3 Is there any basis for using springs and dashpots?

An answer can be found by looking at the polymer chain according to the Rouse model (which has had many contributors). The Rouse model is a semi-molecular model in which chain elements are reduced to a center of mass with connectivity and a resistive environment (Figure 3.37). A chain element is usually considered to be on the order of a monomer. The center of mass is related to the molecular weight of the monomer. The connectivity is normally ascribed to a spring constant related to the modulus of the polymer. The resistive environment is related to the energies of interaction between elements of the chain.

In effect, the Rouse model provides a picture where each chain element gives rise to a spring and dashpot. The overall chain connectivity gives rise to the upper limit of the number of elements in series. The viscoelastic response, then, is a superposition of scales of interacting units. Vibration-displacement of the individual units is rapid. Vibration of combinations of elements becomes more complex. The longest combination is vibration of the length of the entire chain, which requires significant energy, and would likely be related to the glass or melting transitions.

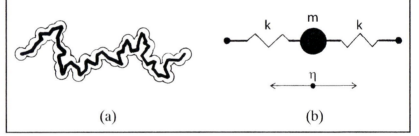

(a) (b)

FIGURE 3.37 The Rouse model. A polymer chain (a) is resolved as a simplified mechanical system with (b) a center of mass, spring inter-connectivity between the centers of mass, and a flow-resistive environment impeding motion of the center of mass.

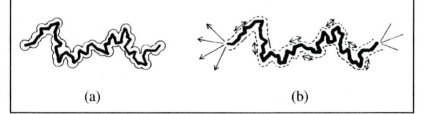
FIGURE 3.38 The reptation model. Here, the polymer chain (a) is constrained by a tube formed by neighboring chains. Therefore, (b) motion is restricted to sliding along the chain, gradually releasing the stress at the ends of the chain.

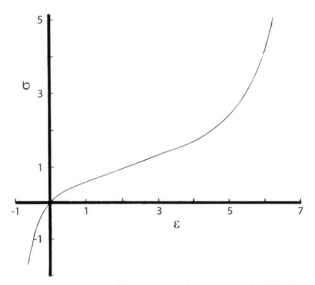

FIGURE 3.39 A representation of the stress-strain response of rubber, from compressive to tensile deformations.

Rubbery Elasticity

Important features of rubbery elasticity were explained long before significant theoretical tools became available to describe viscoelasticity. Rubbery elasticity describes the ability of certain polymers with small quantities of cross-links, at temperatures well above the T_g, and containing no significant interchain or intrachain interactions, to experience and completely recover large deformations. Rubber elasticity is a thermodynamic, not a kinetic effect: rubber deforms nearly instantaneously, holds the deformation indefinitely, and recovers reversibly.

Rubber elasticity, like elasticity and viscoelasticity, starts with the free energy expression. Recall that for traditional elasticity, with strong interactions between atoms, the basic force equation can be expressed by:

$$F = \frac{\Delta E}{\Delta x} \tag{3.48}$$

where F is the elastic force, E is the internal energy, and x is the displacement from the equilibrium position. For rubbery elasticity, we must determine the energy. To do so, we revisit our previous discussion of energy, enthalpy and energy (Box 3.1). We begin with the free energy:

$$\Delta G = \Delta H - T\Delta S \tag{3.49}$$

G is the Gibbs free energy, H is the Helmholtz energy, T is the absolute temperature, and S is the entropy. By elaboration of the expression for traditional elasticity:

$$F = \left(\frac{\partial G}{\partial \ell}\right) = \left(\frac{\partial H}{\partial \ell}\right) - T\left(\frac{\partial S}{\partial \ell}\right) \qquad (3.50)$$

The Helmholtz energy is the internal energy stored in the system, minus that energy expended in work done by the system, which is:

$$\Delta H = \Delta U - \Delta PV \qquad (3.51)$$

In the case of a linear, one-dimensional deformation, ΔH is ΔPL.

On the other hand, the entropy must be accounted for by counting all of the accessible states for the chains:

$$S = k \ln \Omega \qquad (3.52)$$

Since we are interested only in the change in entropy, we can limit the counting only to the change in number of states, which will often make the evaluation more tractable. In order to evaluate the change in the number of states, we must consider the nature of the deformation of atomic elasticity vs. rubbery elasticity. In atomic elasticity, the atoms are displaced small amounts, in the context of the energy well of the atomic bonds. The positions of the atoms are displaced, but the way in which space can be filled is not changed. On the other hand, in a rubber, the situation is very different. Relatively large deformations can be accomplished by rotating about a relatively small number of bonds. Restating the requirements for rubbery elasticity, the rotations are resisted only by weak interactions between chains and atoms normal to the backbone of the chain.

For traditional crystalline elasticity, the entropy can be neglected. On the other hand, for rubber the ΔH term is neglected. Therefore, the change in energy depends only on properly formulating the entropy. How can entropy cause elasticity? Visualize gas sealed in a cylinder, with a movable piston at one end. Initially, the gas is in equilibrium. If the piston is displaced, thereby decreasing the volume available to the gas, there will be a restoring force exerted on the piston. The volume describes the number of states available initially, which can be expressed as:

$$\Omega_1 = \left(\frac{V_1}{a^3}\right)^N \qquad (3.53)$$

where Ω_1 is the number of states, V_1 is the initial volume, a is the diameter

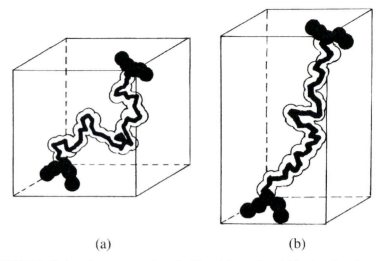

(a)   (b)

FIGURE 3.40 Atomic representation of rubber deformation. (a) In the relaxed state, a chain stretches from one corner to another of a unit volume in the rubber. There are many ways for the chain to cross from corner to corner. (b) After deformation, the unit volume has been deformed, and there are fewer chain conformations that can successfully cross from corner to corner across the volume element.

of a gas molecule, and N is the total number of gas molecules, 6.02×10^{23} for one liter of gas. Taking the ratio of the two states:

$$\frac{\Omega_1}{\Omega_2} = \left(\frac{V_1}{V_2}\right)^N \tag{3.54}$$

The change in energy driven by this entropic change is:

$$T\Delta S = T[\ln \Omega_1 - \ln \Omega_2] = T \ln\left(\frac{V_1}{V_2}\right)^N \tag{3.55}$$

The entropy can be a very large contribution even for a small change in deformation. How does the number of states change for a rubber chain? For an individual chain, the relevant counting of states can be reformulated as the probability for a given chain to cross between one cross-link and the next, as shown in Figure 3.40.

If the distance between any two cross-links is small, there are many ways for the chain to cross the space. Under tension, the distance between these points increases, and the number of paths that will cross this distance is reduced. In order to make the explanation of the deformation tractable, the sample is assumed to deform affinely, so that the increase in separation of the cross-links is equivalent to the applied strain. It is further assumed that neighboring chains will have no effect on the reconfiguration, or rather the accessible conformations for this calculation. This is variously referred to as the Gaussian chain, ideal chain or phantom chain assumption. The probability P_N for a given end-to-end separation of an undeformed chain, $\bar{R}$, is given by:

$$P_N(\bar{R}) = \left[\frac{3}{2\pi N\ell^2}\right]^{3/2} e^{\left(-\frac{3\bar{R}^2}{N\ell^2}\right)} \tag{3.56}$$

where N is the number of chain segments, and ℓ, is the segment length. Thus, the expression for S for a chain is given in terms of the probability of the chain conformation of the first box in Figure 3.40:

$$S = k\ln[P_N(\bar{R})] \tag{3.57}$$

If the chain is deformed by λ_x, λ_y and λ_z along the x, y and z directions, respectively, the change in entropy is the change in the probable conformations:

$$\Delta S = k\{\ln[P_N(\bar{R}_1)] - \ln[P_N(\bar{R}_0)]\}$$

$$= -\frac{3k}{2N\ell}[(\lambda_x^2 - 1)R_{0x}^2 + (\lambda_y^2 - 1)R_{0y}^2 + (\lambda_z^2 - 1)R_{0z}^2] + K - K \tag{3.58}$$

K is the natural logarithm of the normalization front factor, appearing in both the deformed and undeformed state:

$$K = \frac{3}{2}\ln\left[\frac{3}{2\pi N\ell^2}\right] \tag{3.59}$$

The contributions of the individual chains must be summed over all chains. An approximation of the number of chains is given by the volume, V, and the concentration of subchains, v, expected per unit volume:

$$N = vV \tag{3.60}$$

Furthermore, the root-mean-square end-to-end distance of a polymer chain in three dimensions is a function of subchain length:

$$\langle\bar{R}_0^2\rangle = N\ell^2 \approx \langle\bar{R}_x^2\rangle + \langle\bar{R}_y^2\rangle + \langle\bar{R}_z^2\rangle \tag{3.61}$$

This yields an effective measurement of the entropy change for tensile elongation:

$$\Delta S = -\frac{kvV(\lambda_x^2 + \lambda_y^2 + \lambda_z^2 - 3)}{2} \tag{3.62}$$

The force exerted by the conformation change is given by the change in entropy as a function of the change in length:

$$F = -T\frac{\partial S}{\partial V} \tag{3.63}$$

For a uniaxial deformation in an incompressible liquid, the component

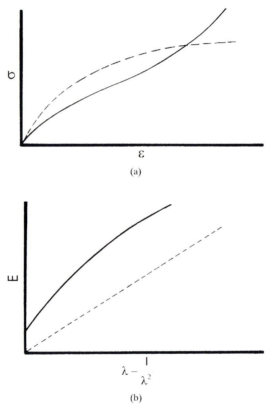

(a)

(b)

FIGURE 3.41 (a) Representation of the tensile response of (——) a real rubber and (- - - -) the stress-strain response for a Gaussian chain calculation. The Gaussian chain deviates from real behavior, both by overshooting the predicted stress at small strains and under-predicting the response at larger strains. (b) According to the Gaussian chain calculation (- - - -), a plot of modulus vs. $\lambda - 1/\lambda^2$ should be a straight line with an intercept at zero. Real rubber (——) exhibits neither a constant slope nor a zero intercept.

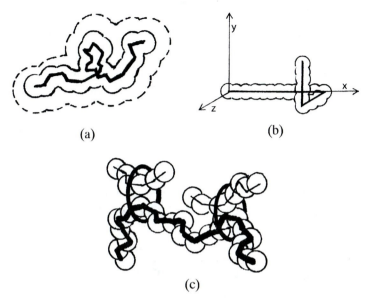

FIGURE 3.42 Alternative models of chain interactions. (a) The chain may be constrained to a tube, as in the Doi-Edwards approach. (b) The tube may be resolved into independent x, y and z components as in the Gaylord-Douglas random tube model. (c) The chain may be constrained by virtual hoops, as in the Higgs-Ball model.

deformations along each principle axis are defined with the help of the Poisson ratio:

$$\lambda_x = \lambda, \lambda_y = \lambda_z = \lambda^{-1/2} \tag{3.64}$$

This results in an expression for stress as a function of strain:

$$\sigma = \frac{f}{A} = -T\left(\frac{1}{lA}\right)\frac{\partial S}{\partial \lambda} = kTv\left(\lambda - \frac{1}{\lambda^2}\right) \tag{3.65}$$

A is the cross-sectional area, and l is the length. This leads to some valuable conclusions that correspond with measurements of real rubbery behavior. First, at small strains λ^2 is very small, and the equation reduces to Hooke's law. The modulus can be estimated from:

$$E = 3kTv \tag{3.66}$$

This would be explicitly calculable if the local concentration of chains could be known. Secondly, the modulus is linearly proportional with T. As temperature increases, the material will shrink, rather than extend, as occurs with crystalline elastic solids. The stress is also not a linear function of strain. For a simple rubber, a plot of σ vs. λ should give the curve shown in Figure 3.41. The deviation from the ideal rubber curve at high strains is understandable: the statistics used are only valid around the relaxed chain coil. Also, at very large extensions, the chains will reach a fully extended state where the statistics used no longer apply. The deviations at small extensions are more troubling.

Rubbery elasticity models are available to give more physically meaningful expressions to account for the resistance to the reconfiguration. For example, one model describes the polymer chain as constrained completely to the diameter of the chain, with the chain slipping out of this confinement in a generalized, non-affine fashion.

Rubbery matrices can provide a composite with good impact resistance. The addition of a ceramic reinforcement phase can increase the modulus.

COMPOSITES AND NANOCOMPOSITES

Bulk property modifications are introduced by compositing. Traditional composites consist of strong, stiff, micron-sized, ceramic particles embedded in an impact resistant, impermeable polymer matrix. This is the structure in materials ranging from graphite reinforced tennis rackets to ABS refrigerator liners, up and down the performance scale. For most composites, some properties will improve, while other properties will not. The ceramic will usually improve the barrier properties and the compressive modulus of a polymer matrix. The polymer improves the fracture resistance of the ceramic. However, the tensile and shear modulus of the polymer usually are not improved, even when large amounts of ceramic, as high as 50% by volume, are used. When, instead, the ceramic modifier is nanoscopic, the tensile and shear moduli can be doubled or better even at relatively small loads, such as 10% by volume. There are also improvements in fracture resistance and barrier properties.

Nanoreinforcement can be more effective than traditional fillers. The increase in previously unimproved properties lies at the core of the interest in, and promise of, nanocomposites. To understand why the improvement happens, one may use the atomistic view.

To begin with, there are two general classes of interaction between the reinforcement and the matrix. If the interactions between particle and surroundings are strong, any force exerted on one phase can be transmitted across the boundary into the other phase. In this case, the particles are reinforcing. If the interactions are weak, force in one phase cannot be transmitted across the boundary between components. Here the particles

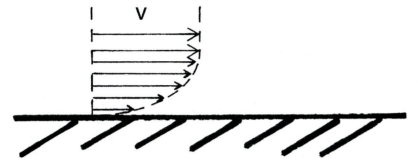

FIGURE 3.43 The velocity of a liquid is *v* in the bulk, but vanishes to zero at the wall.

are fillers, occupying space in the material. Useful composites can be made from materials of either type. Polymers can be filled with cheap ceramic, such as clay or talc. The result is a polymer with less permeability and reduced coefficient of thermal expansion (CTE). Filling a polymer with a ceramic can result in an improved compressive modulus, decreased permeability, and increased apparent surface hardness. Ceramics, such as concrete, can be filled with an aggregate to increase the damage tolerance and decrease the shrinkage of the material, allowing much thicker parts to be cast. Filling a ceramic with rubbery particles can block crack growth and create a fracture resistant material.

Restricted Mobility

One of the most common explanations for the reinforcing properties in traditional composites, and traditional composites literature, is a local restriction on mobility created by proximity to a surface, or chain pinning. The polymer chains near a surface cannot move in the same way chains in bulk polymer do. As the size of the particles decreases, proceeds the argument, the surface area for a given mass of reinforcement increases. This leads to better reinforcement as particle size decreases, because more of the polymer chains are restricted in movement. The basis for this explanation is a mental picture that when in contact with a stationary surface, the molecules of a fluid will have zero velocity. This is a continuum approximation commonly used for viscosity and flow modeling.

There is little experimental evidence for stationary chains on the atomic level. The properties specific to a polymer result from the local movement of the chain, confined by the presence of neighboring chains. A polymer chain, or the elements constituting the chain, will move at a velocity defined by the temperature of the atoms. The mobility of individual chain elements is restricted by connectivity to the next element along the chain and the confining presence of other, neighboring chains. The presence of an adjacent surface can exclude some specific conformations for a chain, possibly affecting the mobility of the chain, and consequently the incremental properties contributed by that chain to the whole material.

Computer models do show a change in the conformations available to an individual chain at a stationary boundary. Some computer models use an *ab initio* calculation where the positions, velocities, and rules of individual atoms are inserted into a model and allowed to move according to set rules. Other models use the Smolochovski or Langevin formulations

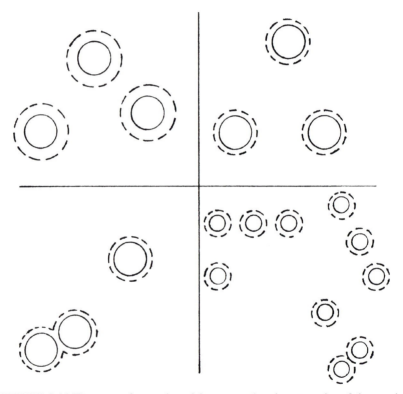

FIGURE 3.44 The zone of control model suggests that the properties of the matrix within some radius r of the particle are influenced by the particle. The zone of control picture works relatively well for isolated particles. The amount of material converted by the particles depends on the value of *r*. However, for the same value of *r*, as particles become smaller, less material is influenced. Also, if the particles are near to each other, the nature of *r* becomes uncertain.

FIGURE 3.45 Foreign particles can initiate crystal growth. This transcrystalline layer is highly organized, but usually of limited range. The rest of the material then goes through more standard crystal growth.

for atomic diffusion, combined with rules on atomic connectivity and force fields exerted by neighboring atoms. The computer model predictions show changes in the mobility of chains near surfaces, which can have limited persistence into the matrix.

The simple continuum description is usually formulated in one of two ways. First, the ceramic particle can be regarded as transforming a larger volume into ceramic phase. This volume can be defined by half the average distance required to change from completely ceramic properties to completely polymer properties within the matrix. Such distances are sometimes measured using microindenters or nanoindenters that probe the hardness of a cross-section of the composite. The hardness of the ceramic is measured at the center of an isolated ceramic particle. The tip is moved away from the center, and, ideally, the hardness will begin to decrease at the edge of the ceramic particle. As the indenter moves farther from the center, the hardness will be observed to decrease, until the hardness of the bulk polymer is measured. One half of this distance, then, can be regarded as the ceramic volume. The difficulty lies in deconvoluting the response of the tip from the response of the matrix.

Alternatively, the restricted local motion concept can be formulated as a boundary layer, with properties neither polymer nor ceramic. This can be especially true when the particle can nucleate a transcrystalline layer, where the polymer boundary layer has a significantly different structure than the bulk of the matrix.

Either formulation, while based on an observable and model-consistent premise, is simplistic. According to the restricted-motion scenario, any stationary boundary should induce the same property changes, with the possible exception of boundaries with compressible materials such as rubber. Therefore, any ceramic particulate should provide the same predictable and measurable property changes to a polymer matrix. Unfortunately, experiments do not fully bear this out. The models can be rescued with fitting parameters, but the physical meaning of these parameters is often unclear.

Load Transfer

Early tests of mechanical property changes as a function of interaction strength showed no difference in properties in materials with predictions of good matrix/particle interactions and with poor observed matrix/particle interactions. This seemed counterintuitive and unsatisfying, but was consistent between many studies. A possible explanation lies in load transfer across the interface.

The load in a composite is borne by both components, some by the matrix and some by the reinforcement. In order for the load to be borne by the reinforcement, the force must be transferred between the matrix and the reinforcement. The load in the matrix builds up along the interface with the reinforcement. A portion of the load is then transferred to, and borne by, the reinforcement. In this way, load is shared between the phases of the composite. The load transfer to a small particle is less efficient than to a long fiber, and so fibers generally improve the mechanical properties better than particles, at least when the force is parallel to the fibers. The load transfer to a nanoparticle should be even less efficient. However, experiments instead have observed improvements in the properties consistent with good load transfer.

Applying an excess load can result in partial or complete failure in the composite. The failure can occur in the matrix, in the reinforcement, or at the interface between the two. Failure in the matrix occurs due to a weak matrix and is straightforward to solve by increasing the matrix strength.

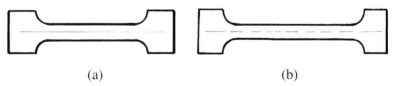

(a)   (b)

FIGURE 3.46 A single fiber test of adhesion between matrix and reinforcement. (a) A single fiber is embedded in a transparent matrix. Under tension, the load transfers from the matrix to the fiber through the interface. When the load on the fiber exceeds the fiber strength, the fiber breaks into a critical length. The critical length is characteristic of the strength of the interface between fiber and matrix.

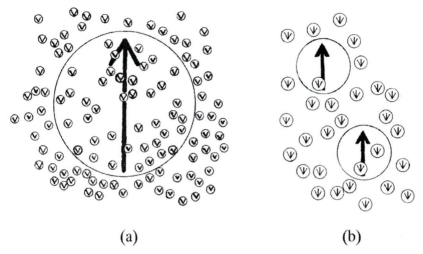

FIGURE 3.47 An illustration of load transfer between smaller and larger particles. (a) If the particles are too small, there is little connection between the load in the small particles and the large particles. (b) If the particles are closer to each other in size, the load on the large particles can be transferred to the smaller particles more efficiently.

If the interface strength is overwhelmed, the reinforcements will tear free from the matrix. No load transfer is possible after this type of failure. Failure can be observed in the reinforcement, as more and more load is borne by the reinforcement. This behavior is typically observed in fibrous composites.

In fiber composites, the load transferred to the fiber can result in breaks in the individual fibers, usually at some consistent, critical length. Measuring the length of the fiber between breaks provides an estimate of the effectiveness of load transfer across the interface. Particles are smaller than this critical length, and so bear less than the optimal load. Transmitting the load across a curved boundary also can result in a local increase in stress, particularly shear stress. This can result in overcoming the interactions at the surface of the particle and failure of the interface. After the interface fails, there is no longer any load transfer. This is the point where nanoparticles can show distinct improvements above larger particles. While the curvature at the surface is even sharper, there is much less surface per particle to transfer load. However, there is more surface area for load transfer per added reinforcement volume element. The net result is decreased amount of load borne by each individual particle. However, since every particle contributes, and the load is insufficient to overload the interfacial adhesion, load transfer can be more effective.

The load transfer approach recognizes both a concentration transition and a shape factor for load transfer. The concentration transition occurs when particles are close enough together that the building stress field around individual particles interacts, as the load is transferred from the matrix to the reinforcement. Even for zero-dimensional, spherical particles, there is a concentration, varying with particle size and shape, where this interaction occurs, usually around 10%. If there is an eccentricity to the particles, or one-dimensional particles, fibers, the load transfer will be better parallel to the long axis than to the short axis, leading to anisotropic reinforcing behavior. In addition, interaction occurs at smaller particle fractions. If the particles are flakes, or two-dimensional particles, there are two dimensions of increased load transfer and yet lower interaction concentrations.

Classic Reinforcement

Mechanical damping is also affected by the presence of fillers and by filler/matrix interface conditions. Damping behavior is particularly interesting around the glass transition temperature. The expression relating damping behavior to polymer volume fraction is usually reported as:

$$\tan \delta_c = \tan \delta_p (1 - \phi_f) \qquad (3.67)$$

where ϕ_f is the volume fraction of the reinforcement, and $\tan\delta_c$ and $\tan\delta_p$ are the loss moduli of the composite and polymer, respectively. This equation is valid for composites filled with non-agglomerated, rigid, solid particles that are not interacting with the matrix.

Significant physicochemical interactions or bonds between the polymer and the reinforcement have a different response. The interface theory has been used to include a modification based on an immobile polymer interface:

$$\tan \delta_c = \tan \delta_p (1 - B\phi_f) \qquad (3.68)$$

The correction parameter B re-introduces the effective thickness of the particle-matrix interface layer:

$$B = (1 + \Delta R / R^3) \qquad (3.69)$$

where R is the mean radius of the dispersed particles, and ΔR is the thickness of the interface.

The rule of mixtures provides one bound to the modulus, reflecting strong adhesion between matrix and reinforcement, and a high filler aspect ratio:

$$M_c = M_p \phi_p + M_f \phi_f \qquad (3.70)$$

The harmonic mean is found to be more applicable to rigid spherical particles with poor adhesion:

$$M_c = \frac{M_p M_f}{M_p \phi_f + M_f \phi_p} \qquad (3.71)$$

M_c is the modulus, shear, elastic or bulk, of the composite, M_p and M_f are the moduli of the polymer matrix and filler, and ϕ_p and ϕ_f are the volume fractions of the polymer and filler, respectively. The moduli of most reinforced polymers lie somewhere between these two limits. These expressions are also sometimes used to evaluate the modulus for semicrystalline polymers, where the crystals are regarded as reinforcements.

Particle geometry, size, degree of dispersion (distribution through the matrix) and of agglomeration (number of particles in particle clumps), and interfacial interactions strongly affect the final properties.

So far, the argument has focused on mechanical properties. This approach also predicts changes in barrier properties. With the addition of filler, there is less diffusion volume available for vapor transport. At the critical particle concentration, the path between particles becomes suddenly more tortuous. Thus, the rate of change in barrier properties will be greater beyond this concentration.

SURFACE MECHANICAL PROPERTIES

The surface properties are different from those of the bulk material for several reasons. First, one primary assumption in predicting bulk responses is the continuity of the material. There should be no surfaces nearby to distort any force fields. Second, the surface of a material is often chemically different from the bulk. Metals form oxides on the surface, impurities and lower quality chains tend to segregate to the surface of a crystalline polymer. Small differences in surface energy, compatibility, or density can result in a reinforcement accumulating at a free surface.

Surface properties include scratch and wear resistance, hardness and friction, or non-stick properties. Wear resistance is the durability of a surface to abrasion by a dynamically applied force. Hardness is a quantitative measure of the resistance of a material surface to displacement from an applied, static, perpendicular force applied over a small area. Scratch resistance describes a response to a highly localized force applied in a

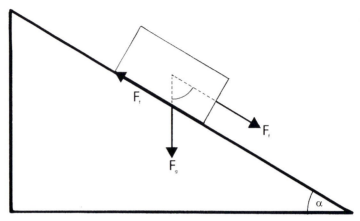

FIGURE 3.48 Friction on an inclined plane. At the critical angle α, the resolution of the gravitational force, F_g, along the incline overcomes the friction at the interface between the block and the plane, F_f.

dynamic test. Unsurprisingly, the surface properties are somewhat interconnected. A hard material generally wears less easily, a scratch-resistant material is more likely to prove hard.

Friction is a primary force by which load is transferred to the surface of a material. Any discussion of surface properties begins with an understanding of friction.

Friction

The study of friction is the province of tribology. Any externally applied force, whether dynamic or static, interacts with the surface. For any

TABLE 3.5 Coefficient of friction between pairs of materials.

Material Pair	COF
Gold and Gold	2.5
Copper and Copper	1.2
Mild Steel and Mild Steel	0.6
Brass and Steel	0.3
Polyethylene and Steel	0.6
Teflon and Steel	0.15
Tungsten Carbide and Tungsten Carbide	0.35

Table after Bowden and Tabor

Box 3.5 *The apparent surface area of interaction.*

In most circumstances, contact between two flat surfaces is limited to three points. If the block is highly polished, the amount of contact can increase. If the two surfaces are extremely rough, some extra energy may be required to push asperities past each other. The apparent surface area of the body also can change as the body remains at rest. Consider a table with a plastic coaster resting on it. The coaster alone, initially after placement, will slip easily. If a heavy drink is placed on it, the friction increases. However, if the coaster is slightly compliant, then the longer it remains on the surface, the more it will conform to the table. This will increase the amount of real surface area, and the resultant coefficient of friction.

The dynamic friction coefficient is generally determined by the strength of rapidly formed interactions, or bonds. Metallic and secondary bonds are rapidly formed, and so metals and polymers will generally show higher friction than ceramics. Chemical compatibility, for example, between ions, exposed permanent dipoles, and so forth will increase surface interactions and friction between a material pairing. Chemical inertness, such as occurs in Fluorine that has filled the electronic shell in a covalent bond, leads to poor interactions for materials such as Teflon™.

Softer materials, with better compliance with the rest surface, will generally exhibit higher friction through larger regions of contact with the substrate. Forming a barrier between two surfaces with a liquid will bar the formation of interactions and drop the friction dramatically, a factor that allows ice-skating. Thin films of soft metals or polymers on hard substrates can allow high friction within a hard, strong part.

In effect, friction is a type of temporary adhesion between two surfaces, with the effect maximized by increasing the number and strength of interactions. Static friction, then, can be time dependent, because over time, two surfaces may conform through material deformation. This is a temporary effect, because once the surfaces are in motion relative to each other, only the three highest points will be in contact, as was initially the case, and limited to the fast forming interactions at these points.

---

force to be transferred to the surface, the force must be transmitted across the boundary between the surface of the material and the object exerting the force. The friction coefficient, μ, is a dimensionless constant of proportionality between the force to move an object and the normal force exerted by the object on the surface plane of rest, which is expressed as:

$$F = \mu N \qquad (3.72)$$

One way to measure the coefficient of friction is to incline the surface of the plane on which a block of the material rests. The tangent of the steepest angle that the block can experience before sliding, as in Figure 3.48, or alternatively the shallowest angle that causes the block to stop, is the friction coefficient.

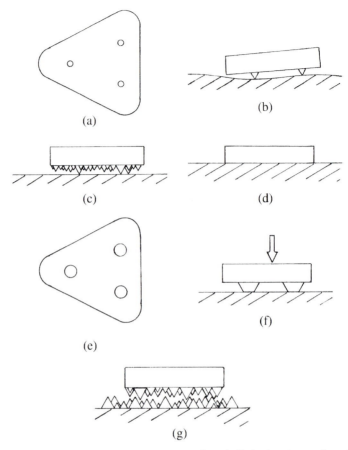

FIGURE 3.49 Sliding friction between two surfaces is limited to three points. (a) simplified three-point surface, a planchette, (b) the sliding friction arising from three contact points. (c) A real surface also exhibits three random contact points. (d) The interactions between two atomically smooth surfaces will exhibit anomalously high friction. (e) Over time, the contact points may deform. (f) This causes a temporary increase in friction, but once the object begins moving again the interactions return to three points. (g) A very rough surface exhibits higher friction through the energy required to move the contact points up and over asperities.

$$\mu = \tan\theta \qquad\qquad (3.73)$$

The coefficient of friction has been observed to be independent of the surface area of the block, but seems to depend on the composition of the block and the roughness of both the block and the rest plane. Table 3.5 shows coefficients of friction for several material pairs.

The friction of composite surfaces can be complicated by several factors. First, the reinforcement and matrix often have different surface energy, and so an improvement in free energy is to be had if the high-energy material segregates to the surface of the part. Second, even if there is no segregation, the highest points of the material tend to be reinforcement. During sliding, any wear and removal of the reinforcements at the high points may cause the resting points to shift. Alternatively, loose reinforcement particles can interfere with the sliding behavior, creating a rolling surface and lowering the friction. Third, the continuum approach, in which the entire, heterogeneous solid is replaced by a homogeneous solid with average properties, may not provide clear property performance predictions.

Hardness

Hardness is the resistance to deformation by an impinging, perpendicular, indenting force. An indenting tip is pressed into the surface, usually to cause permanent deformation. The depth of the damage is related to the hardness. There are many types of hardness measurement, varying primarily according to shape of the indenter. Shapes include pyramidal, narrow pyramidal, micro and ball shaped, and there are, unsurprisingly, different measures from each. The indention resistance is partly controlled by the compression modulus and compressive yield strength. However, the local expansion created by the Poisson effect is resisted by surrounding material, so shear modulus and plastic yield will play a further role in the hardness response. If the material is heterogeneous, the size of the indenter is particularly important. If the indenter is large, an average measurement of the hardness is reasonable. If the indenter is small, the hardness measurement is local, and may reflect the separate hardness of the matrix, reinforcement, or some convolution of the two. Microindenters have been used near particles in composites to provide evidence for a matrix zone surrounding the particle with properties that gradually change to the bulk behavior. These measurements are very difficult to make unambiguously, as the tip shape, and the shape of the particle that may include mass just beneath the surface of the polymer, must be deconvoluted to provide accurate data. With proper attention to the potential complications, the hardness generally predicts surface mechanical properties, including wear and scratch resistance. Comparisons between tests with the same tip type, on homogeneous materials can give reliable results. Comparisons between tests using different tip types or on heterogeneous or composite materials are less reliable.

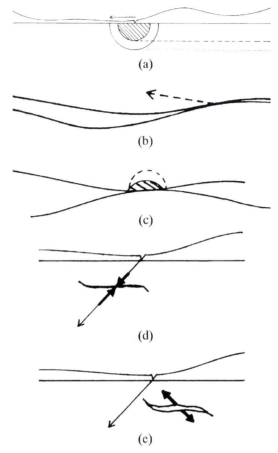

FIGURE 3.50 (a) An asperity between two moving bodies concentrates force beneath the surface of a solid in a plastic region. (b) Even without an asperity, contact transmits force into the substrate. (c) If the plastic damage exceeds the yield stress, a crack can form at the yield boundary. (d) Cracks close before an advancing contact point. (e) However, behind the contact point, there is an opening force.

Wear

When two bodies are moved in relation to one another, the friction, or adhesive, interaction between them will lead to wear damage. For interactions between hard surfaces under small loads, the damage can be minimal. For greater applied force or softer materials, the wear will be more significant.

Two bodies in contact will form temporary bonds with varying strength. When the bodies are at rest, the strength and number of these bonds can increase with time of contact. When the two bodies are moving relative to each other, the temporary bonds must be broken. If the new bonds between the two separate materials are easily broken (Figure 3.50) the wear is slight, perhaps the transfer of a few atoms, and nanoscopic roughening. If more material is removed, the wear becomes more noticeable.

The classic wear configuration, as described, is between two bodies. Some adhesion between surfaces occurs at the interface between the bodies. For hard, rigid, flat bodies with little interaction, the adhesion is easily overcome, causing little damage. For softer bodies, the interaction occurs over a greater area, requiring more applied force to initiate relative motion between the surfaces. Some energy is dissipated between the two bodies due to motion against friction. For surfaces with stronger interactions, the force required to overcome the interactions can be dissipated instead through subsurface fracture and wear. In this case, some energy will be transformed into plastic deformation of the softer body, or fracture in a brittle material.

If sufficient damage is done below the surface to cause a fracture plane, large pieces of the material can be torn free. This results in rapid wear. If there are small, poorly bound particles, such as in some composites, the particles can be dislodged, again resulting in rapid wear, with extensive material loss. For extremely rough bodies, the force is transferred between bodies at the tips of protruding roughness. This can involve localized stress, and non-normal shear force that can cause these protrusions to fracture. Once wear begins, particles of significant size can be introduced between the two bodies, creating a three-body condition.

Three-body wear is far from the base, two-body wear case. Primarily, the particles will engage in rolling, thus serving as a lubricant between surfaces and potentially reducing wear. Even irregular particles will align so that the most symmetrical axis is perpendicular to the shear force, minimizing the accumulation of shear strain between the bodies. However, if the loose particles are very irregular, or if they are chemically different and have better adhesion with one or more stationary surfaces, they can enhance wear by forming stress concentration points.

This stress concentration is also the source of another form of wear, abrasion. A surface with asperities can cause a local force concentration resulting in large amounts of yield, fracture, and other damage in the counter body. The damage quickly removes material. This is the basis of sanding and grinding.

Corrosion is another source of wear. Chemical action on the surface of a substrate can result directly in loss of material through ionization and solubilization, or through conversion to a poorly bound oxide layer that flakes loose from the surface.

Three types of specialized wear, which are really combinations of adhesion, abrasion, and or extra mechanically applied force, are also commonly observed. Fretting occurs in situations such as drive shafts, where the natural adhesive wear is enhanced by the presence of particles in a lubricant layer, resulting in rapid material damage. Particles impinging from a fluid stream cause erosion. Since the particles can strike unevenly across the surface, erosion as a wear process can be uneven, creating an uneven wear surface in consequence. Finally, a surface in contact with a liquid, such as a pump impeller, can suffer from unevenly applied forces if the liquid cavitates, that is, if bubbles of gas are suddenly introduced. These bubbles are collapsed, usually rapidly and abruptly, causing transient, often high, loads at the surface. This can cause fracture and enhanced wear.

Scratch

In a scratch test, a point is dragged across the surface of a material body. The scratch needle is generally steel, making reports of scratch data uniformly part of an x/steel material pair. The scratch is a combination of hardness and asperity-based wear test. The scratch test is something like a dynamic hardness test, where the indenter is used to apply both a normal and a shear force. The scratch test is also something like applying a single asperity from a very rough surface to a substrate. The tip can slide across the surface at low applied normal component, causing only surface, adhesive damage. Greater applied normal force transmits force deeper into the substrate, perhaps accumulating plastic damage that can be measured. Usually, sufficient force is applied to plow a furrow in the surface. The depth and character of this furrow reflects the hardness convoluted with the resistance to shear failure. The mechanics of the scratch force, especially in a heterogeneous body, are complex. The scratch test is usually interpreted empirically.

DIFFUSION AND PERMEABILITY

Diffusion and permeability are barrier properties. Diffusion in this context is the transport of small molecules through the composite. Permeability is the case of diffusion through the thickness of the composite. Together, these properties describe the mobility of environmental substances in and through the material. The improvement in barrier properties of composites is a primary advantage in composite materials. Thus, some discussion of diffusion, permeability, and barriers to corrosion may facilitate future exploration. The basic principle is that as more of the permeable polymer is converted to impermeable ceramic or metal, less liquid or gas may pass through.

Diffusion

In the broadest sense, diffusion describes the motion of particles through space. The rate of particle movement is described by the mass flux, Q:

$$Q = \frac{M}{At} \tag{3.74}$$

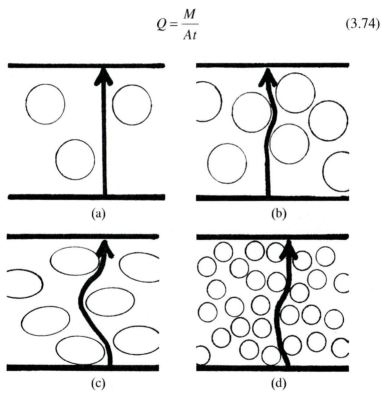

(a) (b)

(c) (d)

FIGURE 3.51 The particles in a composite change the free path for diffusion between surfaces. (a) For a dilute system, the diffusion path is affected only by the volume fraction of impermeable phase. (b) If the system is semidilute, the diffusion path is significantly interrupted, becoming tortuous. (c) For flake reinforced systems, the occlusion in a semidilute system occurs at a lower volume fraction. (d) For a concentrated system the diffusion path becomes very restricted.

M is the mass passing through a cross-sectional area A over a time t. Fick observed that the flux along a single dimension could be characterized by a single diffusion constant:

$$Q = -D\frac{dC}{dx} \qquad (3.75)$$

This, Fick's first law, observes that mass will diffuse away from a high concentration, C, toward a low concentration, resulting in a diffusion gradient in concentration. The time change in concentration resulting from diffusion is described by Fick's second law:

$$\frac{dC}{dx} = D\frac{d^2C}{dx^2} \qquad (3.76)$$

In a material, the particles may be as small as atoms or as large as high polymer chains. The space in any solid is occupied partially by the atoms or molecules of a material. The occupied space hinders the mobility of any species by occlusion, and may additionally provide an attractive or repulsive energy barrier beyond the traditional view of hardcore repulsive atomic volumes. The diffusion rate in a crystalline solid is so low compared to an amorphous polymer that diffusion is usually regarded as occurring exclusively in the amorphous portions of the polymer. Diffusion in this region is governed by polymer chain segmental mobility, or the rate of exchange between a polymer segment and a hole representing a solute molecule.

The diffusion coefficient is related to both the diffusing species and the material. At small loads of an impenetrable particulate, up to about 10% but depending strongly on the shape of the particulate, the flux will be reduced by the decrease in cross-sectional area caused by an increasing volume fraction of the solid phase. Above the concentration limit the diffusion path becomes torturous, though this is not expected to be a sharp transition. The tortuosity contributes to the diffusion rate by increasing

---

Do It at Home 3.1

Fill a piece of pipe with gelatin, made the traditional way. Put water above the gelatin and a piece of paper towel under the gelatin. Measure the amount of water that passes through the gelatin over 1 hour and 1 day. Make another pipe-full of gelatin containing sand. Repeat the measurements.

---

the effective length of the path length of diffusion in addition to decreasing the cross-sectional area. This has been shown to occur where the particles overlap and the shortest path is no longer a direct line through the material.

Permeability

Permeation can be visualized as a steady state diffusion system, with a reservoir of liquid or vapor on one side of the film. Therefore, Fick's laws also govern the amount of a diffusing species passing through a material to a sink on the other side.

Predictions of permeation processes in heterogeneous systems are as complex as the mechanical responses. In addition to the segmental mobility through the polymer, and the tortuosity of the diffusion path length, there are also defect structures, such as micro-cracks, pores, and variations in density caused by the reinforcement that may contribute to increases in diffusion rate. Regions within a polymer inaccessible to the diffusing species include both the occluding reinforcing particles and any crystallinity. In the case of plate or flake-shaped particles, diffusion depends strongly on orientation. The permeability of polymers filled with flaky or lamellar particles is usually lower than those filled with equivalent volumes of spherical fillers. Regions in a composite that can enhance diffusion include holes and weak interfacial boundaries.

The quantification of changes in the permeation resistance associated with the heterogeneities in a continuous polymer medium is not straightforward, but several simplified approaches have been developed. Permeability of polymers containing an impermeable phase such as fillers or impermeable polymer crystals in the absence of porosity can be estimated from:

$$P_i = P_0 (1 - V_i)^2 \qquad (3.77)$$

where P_i is the permeability of polymer containing an impermeable phase, P_0 is the permeability of an amorphous or unfilled polymer, and V_i is the volume fraction of the impermeable phase.

REFERENCES

McCalister, W., Materials Science and Engineering, 6th ed., John Wiley & Sons, New York, 2003.

McCrum, N., B. Read and G. Williams, Anelastic and Dielectric Effects in Polymeric Solids, Dover Publications, New York, 1991 (reprinted from Wiley, 1967).

Young, R. and P. Lovell, Introduction to Polymers, 2nd ed., Chapman Hall, London, 1991.

Hiemenz, P., Polymer Chemistry, Marcel Dekker, New York, 1984.

Kinloch, A. and R. Young, Fracture Behavior of Polymers, Elsevier, London, 1983.

Bowden, F. and D. Tabor, Friction, an Introduction to Tribology, Doubleday and Company, Garden City, 1973.

CHAPTER 4

Features of Nanocomposites

A nanocomposite is composed of two materials: nanoscopic reinforcement and a binding matrix. In Chapters 2 and 3, the structure and properties of the components of the materials constituting the phases were discussed, as were the interactions between the matrix and reinforcement. This chapter will focus on the particulates used to produce nanocomposites.

NANOREINFORCEMENTS

The reinforcing phase in a nanocomposite must be, in one or more dimensions, nanoscopic. This is, in some ways, an unusual criterion. Chemical additives are on an angstrom scale, and many fillers, and granular, abrasive, and coating materials are microscopic to macroscopic in size. Nanoparticulates are intermediate to molecular and microscopic. Many materials, both naturally occurring and synthetic, are available in this size range.

The dimensionality of a particle is somewhat unclear. The definition used here is drawn in analogy with crystal shapes. In other sources, the number of nanoscopic dimensions is described, as an alternative. In this view, zero-dimensional particles are nanoscopic in all dimensions, and have an aspect ratio close to 1. In the other classification, spherical particles have three dimensions at the nanoscale. While an atom may be realistically depicted as a mathematical point defining materials space, a silica sphere of 100 nm cannot. However, a spherical nanoparticle is not unique in being a three-dimensional reinforcement, either.

There are various ways to classify the particles useful for nanocomposites. The majority are metals and ceramics, but there are organic particles such as graphite, nanotubes, and nanofibers. The particles may be classified according to source, natural or synthetic. They may be classified according to use, such as conductive, mechanical or barrier property improvement additives. Here, classification will be based first on the dimensionality, or pseudo-dimensionality, of the particle, and second on the technique used to produce the particle. Initially the size and shape of the particles will be considered.

Reinforcement Size and Shape

Particle Size

The size of a single sphere is an easily measured quantity, well defined by the radius or diameter. Knowing the radius provides information including largest chord length, volume, and surface area. When dealing with real particles, however, a characteristic measure is more difficult to determine. The average diameter of the particle is a useful number for spheres with a normal distribution of particles. However, there are many possible averages to choose from, and the distribution may affect the accuracy of average calculations. The problems increase if the particles are not spherical, but irregular or ellipsoidal.

There are several common techniques for measuring the diameter of irregular particles. Martin's diameter is measured as the distance between opposing sides of a particle, measured crosswise to the particle, on a line bisecting the projected area. A photograph edge is used as the reference line, and all of the diameter lines are measured parallel to this edge. Feret's diameter is obtained by measuring the distance between two tangents on opposing sides of the particle. Where Martin's diameter measures a chord of the particle, Feret's diameter measures the gap required to fit a particle in an imaginary set of calipers. Again, a reference edge must be chosen so that all measurements are consistent. The final common technique uses a set of quantified circles, and the shape is matched to the circle with the closest match in area. The diameter of the particle is taken as the diameter of this circle.

Martin's diameter can be used to estimate the specific surface area, or area per unit volume. This is:

$$M.D. = \frac{4}{DS} \qquad (4.1)$$

M.D. is Martin's diameter, *D* is the packing density, and *S* is the specific

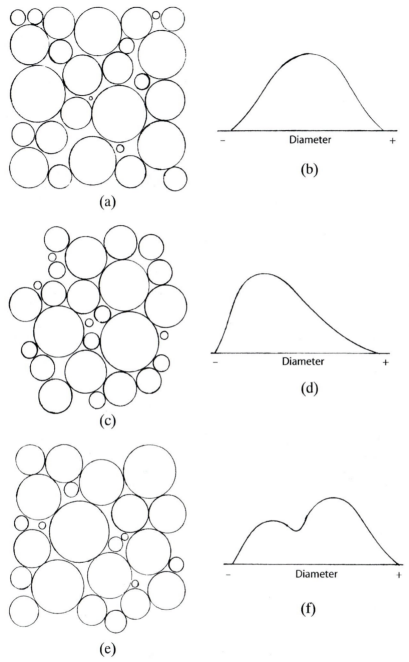

FIGURE 4.1 (a) A collection of particles described by (b) the normal distribution. (c) A second collection of particles, described by a distribution (d) skewed to smaller particles. (e) a third collection of particles, described by (f) a bimodal distribution.

127

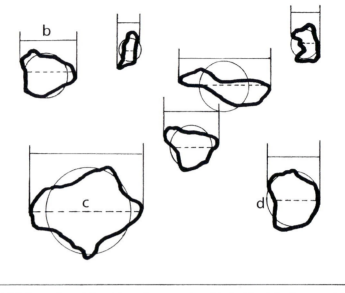

FIGURE 4.2 Several particles. (a) a reference edge. (b) the Martin's diameter. (c) the Feret's diameter. (d) the equivalent circle area.

surface area. The packing density is the number of particles in a unit area. More on analysis of particles is given in Chapter 5.

The distribution of particles will play a role in the average diameter obtained for an evaluation of the size. For a normal distribution, any calculation of the average size will provide a similar estimate. However, for irregular distributions, this is no longer true. The variations become quite important depending on the technique used to estimate particle size and the size average that determines particle behavior. As with molecular weight, the size measurement may have several moments that describe the distribution of particle sizes.

The arithmetic mean of particle size is calculated using the summation:

$$\bar{d} = \frac{1}{n} \sum_{i=1}^{p} d_i f_i \qquad (4.2)$$

The summation is across all particles, but can be simplified by dividing observations into groups, for example 1 – 10 nm, 10 – 30 nm, and so on. p then becomes the number of counting intervals. f is the number of observations in a given interval, and d is the median diameter of the interval. n is the total number of measurements collected.

In addition to the size of the particles, the distribution is also important. Powder size is often assumed to have a normal distribution for most samples. A log normal distribution is also quite common, where the distribution has a tail at larger particle diameters. Care must be taken to ensure that the missing tail of the distribution is a real feature of the particles, rather than an artifact from a technique that is insensitive to, and therefore excluding of, the smallest particles.

The arithmetic mean is the first moment of the distribution about the mean particle size. The width of the distribution can be characterized by the variance:

$$m_2 = \frac{1}{n} \sum_{i=1}^{p} (d_i - \bar{d})^2 f_i \qquad (4.3)$$

The variance is in units of l^2. The standard deviation provides a characteristic of the distribution in units of length:

$$s = \sqrt{\frac{1}{n} \sum_{i=1}^{p} d_i^2 f_i - \bar{d}^2} \qquad (4.4)$$

An estimate of the skewness, or asymmetry, of the distribution is provided by the third moment of the average. The third moment has units of l^3. Dividing by the cube of the standard deviation reduces the skewness to a dimensionless number:

$$\frac{m^3}{s^3} = sk = \frac{\frac{1}{n} \sum_{i=1}^{p} (d_i - \bar{d})^3 f_i}{\left(\frac{1}{n} \sum_{i=1}^{p} d_i^2 f_i - \bar{d}^2 \right)^{\frac{3}{2}}} \qquad (4.5)$$

The sharpness of the data, or the tendency to form a peak, can be estimated by the fourth moment. Again, in order to remove the sensitivity to units, especially with a dependence on l^4, the moment may be divided by the standard deviation to the fourth power:

$$\frac{m_4}{s^4} = pk = \frac{\frac{1}{n} \sum_{i=1}^{p} (d_i - \bar{d})^4 f_i}{\left(\frac{1}{n} \sum_{i=1}^{p} d_i^2 f_i - \bar{d}^2 \right)^{2}} \qquad (4.6)$$

Other traditional definitions of the central value, or average, diameter, such as the mode, may be less useful than the arithmetic mean. The relation of these definitions to characteristic properties of the particles or to specific measurement techniques is uncertain.

Some other techniques for calculating the average bias the calculated diameter in specific ways. The distortion can result in a diameter differing from the arithmetic, traditional mean, but also one more directly applicable to specific measurement techniques or properties. The geometric mean may be an alternative appropriate for evaluating some properties:

$$\bar{d}_g = (d_1^{f_1} d_2^{f_2} \ldots d_n^{f_n})^{\frac{1}{n}} \tag{4.7}$$

This mean emphasizes the frequency of various counts. The geometric mean is a better estimate of the mean for log normal distributions. More frequently observed diameters would have a larger contribution than single observations. The geometric mean will therefore apply more for properties that are dominated by the most frequent contributors. The harmonic mean may also have some utility:

$$\bar{d}_h = \left(\frac{1}{n} \sum_{i=1}^{p} \frac{f_i}{d_i} \right)^{-1} \tag{4.8}$$

The harmonic mean de-emphasizes the frequency of the observation. Using a different approach based on the moment of the distribution, values of the mean may be determined from the diameter providing an equivalent surface area:

$$\bar{d}_s = \left(\frac{1}{n} \sum_{i=1}^{p} f_i d_i^2 \right)^{\frac{1}{2}} \tag{4.9}$$

The mean surface diameter is based on the particle with a surface area that is the mean of all surfaces. Likewise, a diameter based on an equivalent particle volume can be calculated:

$$\bar{d}_v = \left(\frac{1}{n} \sum_{i=1}^{p} f_i d_i^3 \right)^{\frac{1}{3}} \tag{4.10}$$

The mean volume diameter is based on the particle with a volume that is the mean of the volumes of all particles.

As the geometric mean emphasizes the frequency of counts, the linear mean diameter emphasizes the particles with the largest diameter. The linear mean diameter is obtained in a manner analogous to the moments of the molecular weight, where $\overline{M}_n$ is the arithmetic mean of the molecular weight and $\overline{M}_w$ emphasizes the longer chains:

$$\overline{d}_l = \frac{\sum f_i d_i^2}{\sum f_i d_i} \qquad (4.11)$$

If instead of multiplying by the length, then dividing by the average diameter, a surface mean diameter can be derived by multiplying by the area and dividing by the diameter average:

$$\overline{d}_a = \frac{\sum f_i d_i^3}{\sum f_i d_i^2} \qquad (4.12)$$

Multiplying by the density and normalizing appropriately provides the weight mean average diameter:

$$\overline{d}_w = \frac{\sum f_i d_i^4}{\sum f_i d_i^3} \qquad (4.13)$$

This moment is sometimes called the volume-averaged diameter, but the relationship between the volumetric term in the density calls attention to the importance in measurement techniques and properties that depend primarily on the mass of the particles. The presence of only a few large particles will account for the majority of the mass of a powder sample.

Particle Shapes

Particles with only one or two dimensions at the nanoscale exhibit an aspect, and can align in flow fields, interact with force fields according to orientation, and so on. The primary characteristic of particles with shape is the aspect ratio.

Aspect Ratio

The aspect ratio, p, can be assigned by describing the particle as an ellipsoid. The ratio of the major to the minor axes is the aspect ratio. So prolate ellipsoids, or rods, have an aspect ratio greater than one, and oblate ellipsoids, or disks, have an aspect ratio less than one. Another approach gives the aspect ratio as the ratio of the longest dimension to the shortest dimension. The proper aspect ratio depends on the definition used for a given piece of work.

B, L and T

If a particle were allowed to come to rest on a surface, the position of greatest stability would be assumed. The breadth, *B*, is defined as the smallest possible Ferret's diameter projected by the profile of this resting particle. The length, *L*, is the Ferret's diameter perpendicular to the breadth. The thickness, *T*, is the Ferret's diameter measuring perpendicular to the plane of greatest stability. The flakiness is *B/T* and the elongation is *L/B*.

Shape Factors

Even irregular particles tend to have consistent values of $\bar{S} / \bar{d}^{2}$ and $\bar{V} / \bar{d}^{3}$ for a given $\bar{d}$ for the same material. *S* is the surface area, and *V* is the volume. Therefore, these ratios make reliable shape factors for powder types.

Particle Measurement

The particle size reported by the manufacturer is based on one of a limited number of available, repeatable characterization techniques.

Mass Transport

Diffusion is a method for measuring particle sizes. The motion of liquid molecules displaces suspended particles, and the smaller the particle, the faster diffusion will occur. Diffusion is measured using the flux of particles, *Q*, using Fick's first law:

$$Q = D \frac{\Delta C}{\Delta x} \qquad (4.14)$$

ΔC is the change in particle concentration with respect to Δx, the distance traveled. *D* is the diffusion constant:

$$D = \frac{kT}{3d\pi\eta} \qquad (4.15)$$

k is Boltzmann's constant, *T* is the absolute temperature, *d* is the particle diameter, and η is the viscosity of the suspending liquid. The diffusion rate is usually very small for large particles, but can be successful in measuring nanoparticle diameters. The diffusion velocity is proportional to the density, and so is generally more descriptive of a mass-average diameter.

A similar technique is provided by sedimentation. Again, the involvement of density is indicative of a mass average. Larger particles will sediment more rapidly than smaller particles. Using the pipette method, the suspension is agitated and aliquots of the solution are collected to confirm the initial concentration of particles. The solution is allowed to sit quiescent and aliquots are collected at 2:1 geometric time intervals. The aliquot is dried and weighed to determine the mass of particles remaining. Then, using Stoke's law, the diameter of the particles that have left the solution at a given time, based on the velocity of the particles that have settled out, v, is calculated, according to the following:

$$d = \sqrt{\frac{18v\eta}{g(\rho_p - \rho_L)}} \qquad (4.16)$$

where ρ_p is the particle density, ρ_L is the liquid density, and g is the gravitational constant. A correction must be added for the change in water column height when the sample is taken. The Stoke's equation was derived for spherical particles. In sedimentation techniques, elliptical particles will behave in terms of an average diameter.

FIGURE 4.3 The pipette technique for measuring particle size. A well-mixed solution of particles is placed in a column. A sample is collected from the bottom of the column at regular intervals. The mass of particles in each sample is measured. The particle settling rate is a function of particle size, liquid viscosity, and the difference in density between the particle and the liquid.

Using a flowing mixture of particles, sedimentation or sedimentation accelerated by centrifugation may also be used to classify and separate particles. In a fluidized bed, for example, the upward flow of gas balances the force of gravity on the particle, again according to Stokes law. The particles are thereby separated by mass. This is the basis for air classification.

Nanoparticles are subject to Brownian motion, and generally leave solution far too slowly for practical measurement by sedimentation or flow techniques.

Finally, inertial effects may be used to separate particles, and allow measurement of the particle size. The particles suspended in a gas will possess a momentum different from that of the surrounding liquid. As the gas passes around a bend, as in a cyclone separator, the particles will be removed from the gas.

All of the mass transport phenomena provide mass-average-based information on the particles. Mass transport phenomena are also useful for separating particles based on size. The various facets of mass transport and particle behavior are discussed more thoroughly in Chapter 6.

Filtering

Filtration is the process of removing particles from a liquid or gas stream by introducing a porous media. Sieves will generally separate large particles. Particles smaller than the holes in the sieve will pass through, while larger particles will be caught. A stack of sieves can separate a sample of particles into a series of particle size intervals. Particles with aspect, such as fibers, may be able to pass through a sieve lengthwise, and so only a portion of the particles will be collected. Nanoparticles are smaller than available sieves, but can sometimes be passed through fabric filters. Fabric filters may be woven or random web. A filter catches particles both by physically blocking large particles and adsorbing particles electrostatically. Filtration can be used as a particle collection and separation technique, but is generally poor for use in particle sizing.

Turbidity and Light Scatter

When light shines through a dilute suspension of particles, part of the light will be scattered by the particles. Some of the light is transmitted, and measuring the ratio of intensity of the transmitted light to the initial intensity provides a measure of turbidity. Turbidity is one technique for measuring the particle size, as the intensity is proportional to r^6. For very large particles, the forward scattering adds to the intensity and complicates the measurement.

Measuring the angle and intensities of scattered light provides another assessment of particle size. Particles that are small relative to the wavelength of the light, scatter according to Raleigh scattering, while larger particles obey Mei scattering. The angle of scattered light provides information on particle size, again as r^6. Nanoparticles are usually much smaller than the wavelength of the light, and may interact strongly with the light, rather than weakly, as is typical of larger particles. This happens because the light can be absorbed and transformed into other vibratory modes. Therefore, light scattering is often unsatisfactory for measuring the diameters of nanoparticles. Light scattering is dependent on the index of refraction difference, the concentration of the particles, and the reflectivity of the particles. Provided these factors are known, light scattering can provide a complete distribution of particle diameters. Light scattering assumes spherical particles. Irregular particles will rotate and tumble in solution, so that all diameters will be averaged.

Surface Area and Permeability

The quantity of gas absorbed on a surface is proportional to the surface area. Non-porous particles with an easily interpreted shape are typical assumptions. Gas porosimetry techniques are discussed more in Chapter 10. This technique measures the surface area and calculates a diameter based on this value. Therefore, surface area provides an estimate of the surface average diameter.

Alternatively, the permeability of a liquid through a packed bed of the particles is characteristic of the particle size. The flow velocity through the bed can be described by the Carman Kozeny expression:

$$v = \frac{p^3}{(1-p)^2} \cdot \frac{g\Delta P}{k\eta L S_w^2 \rho^2} \qquad (4.17)$$

v is the liquid velocity through the particle bed, p is the bed porosity, ΔP is applied pressure differential, g is the gravitation constant, L is the bed thickness, η is the liquid viscosity, and ρ is the particle density. S_w is the specific surface, in area/unit mass, which may be calculated or determined by gas porosimetry. The specific surface area may depend on the liquid composition, and of course temperature. k is a dimensionless constant that depends on the orientation of flow relative to the perpendicular to the orientation of the particles:

$$k = \frac{3}{\sin^2 \phi} \tag{4.18}$$

There are several ways to measure the porosity of the bed. A calculation may be based on the packing of spheres. However, for non-spherical particles, the bed porosity may be better estimated from the difference in volume between a continuous bulk specimen and the equivalent volume of the particle bed. The volume difference may include closed porosity, however. Closed porosity consists of closed pores inside the material, whereas open porosity consists of interconnected pores, often with an open path to the surface of the particle. Measuring the volume of a liquid penetrant is an appropriate physical technique, assuming all open porosity is penetrated by the liquid, and contributes to flow through the bed. As with porosimetry, permeability provides a measure of the particle size measured by the specific surface area, and therefore dependent on the area averaged diameter.

Microscopy

The best technique to measure an accurate, arithmetic average is to measure and count using a microscope. Optical microscopy is insufficient for micron-sized particles and smaller. Most scanning electron microscopy has a resolution limit of about 100 nm, but some field emission microscopes can achieve resolution below 10 nm, depending on sample preparation and scope conditions. Getting a count for samples with a large distribution of particle sizes may require measurements made at several magnifications.

Particles

The simplest nanoparticles are spherical or otherwise particulate. Pseudo-zero-dimensional reinforcements are naturally occurring or synthetic. Fly ash and fumed particles, carbon black, jewelers rouge, and other pigment materials are examples of finely divided particles suitable for use in fashioning nanocomposites. Synthetic materials may be created by burning, grinding, spraying or chemical reactions. Table 4.1 provides a list of spherical nanoparticulates.

Grinding

Naturally occurring and synthetic bulk materials must be reduced to nanoparticles. One technique to accomplish the reduction is grinding.

TABLE 4.1 Spherical Nanocomposites.

Metals	Ceramics	Flakes
Iron	Alumina	Smectic clays
Gold	Silicon Carbide	
Silver	Silica	**Fibers**
Copper	Graphite	Carbon nanotubes
Nickel	Copper sulfate	Cellulose whiskers
	Iron Oxide	
Polymers	Titania	
Polypropylene	Tungsten Carbide	

From Utracki, Vol. 1

The grindability of a material is qualitatively described based on hardness. Sufficient damage must be applied to cause a crack to penetrate the diameter of the particle. The breaking process is dominated by the critical yield stress of the material.

The most common techniques for grinding are the various forms of milling. Typical mills are ball mills, disk mills, jaw mills, hammer mills, roller crushers, and so on. The cost of milling is about an equal combination of the mill material costs and the energy consumed. A rough rule for the energy required to reduce the particle size is provided by Kick's law:

$$E = C \log\left(\frac{X_f}{X_p}\right) \tag{4.19}$$

where E is the energy, C is a constant dependent on the milling material, X_f is the particle size of the feedstock, and X_p is the desired size. Breaking the particles into ever smaller pieces is an ever more expensive proposition.

Particle fracture is easier if the material is brittle. Size reduction in ceramics is often a crushing and grinding process, with filtering or centrifugation used to remove particles of sizes in excess of the target value. Grinding to produce large quantities of nanoparticles smaller than 100 nm is difficult, verging on impossible.

Atomization and Spraying

Atomization is similar to the process used to deliver paint, perfumes, or drugs. A pressurized gas passes an orifice. The pressure drop across

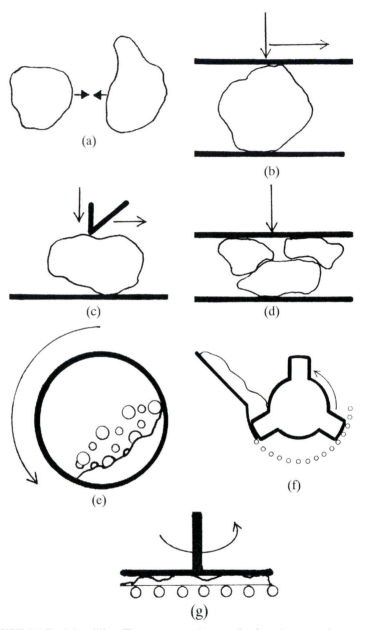

FIGURE 4.4 Particle milling. There are several types of actions that can reduce particle size. Examples include: (a) particle-particle impact, (b) crushing and shearing, (c) pinching and shearing, (d) crushing with multiple particle interactions. Examples of mills: (e) In a ball mill large balls apply impact force due to the powder in a rolling jar. (f) In a hammer mill, rotating hammers scrape particles against an attrition screen. (g) In a disk mill, a spinning disk is used to apply force against an attrition screen.

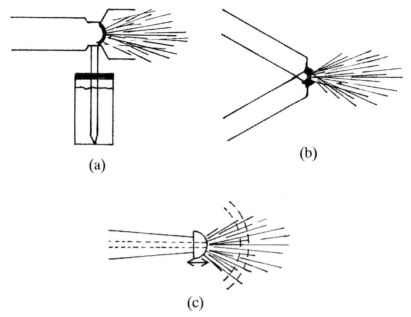

FIGURE 4.5 Small particles can be made by droplet formation. (a) In a pressure drop nebulizer, the moving gas draws liquid into the gas stream, where the liquid is blown into small droplets. (b) In an impinging jet nebulizer, a liquid stream is injected into a moving gas stream. (c) Liquid is fed through an ultrasonic nozzle. The vibrating nozzle breaks up the liquid into small drops.

the orifice draws a molten or dissolved material through the orifice. The high velocity gas separates the liquid into very fine drops that cool to form particles if the material is molten, or that evaporate to leave a particulate residue if the material is in solution. There are several alternative enhancements to the basic approach. The molten material may be heated and pressurized sufficiently to flow through the orifice. The outlet liquid will then be impinged on a cold, high-pressure jet of cooling liquid. This jet does additional work to disintegrate the forming droplets of liquid, simultaneously providing cooling to solidify the particles. Another refinement adds an ultrasonic horn with the liquid orifice passing through the center. The vibrating nozzle adds considerable energy to the liquid passing through the orifice, and an impinging gas jet can further break the liquid up. Atomization generally is restricted to stable materials, those that can be melted and have low viscosity, and melts that do not oxidize readily in gas streams. Using an inert gas stream increases the utility of the technique but increases the cost.

Fuming

Fumed particles are part of the natural output of combustion processes. There are important commercial sources of nanoparticles available from fume sources. Fumed silica is generated by burning $SiCl_4$ in a hydrogen flame. The resultant products include water, hydrochloric acid, and nanoscopic silica powders that find use in cosmetics, paints and pigments, and ceramic processes. Fly ash is a mixture of silica, alumina, and iron oxide, with sundry other atomic components resulting naturally from combustion, especially of coal. Some fly ashes contain a large quantity of residual carbon after burning. The composition depends strongly on the composition of the fuel. Fly ash undergoes the same chemical changes as cement, and is similar in composition to clay and shale. Fly ash is a waste product of energy production, recently finding use in brick and concrete production. Carbon black is the product of burning an hydrocarbon liquid in a reducing atmosphere. Carbon black finds application in such diverse applications as rubber reinforcement, photocopier toner, and other pigments, leather treatments, dry cell batteries, and ceramic production of high-temperature insulators.

Fumed materials are airborne, and do not settle by gravity or typical sedimentation techniques. The particles are generally separated by electrostatic precipitation. Very small particles are often more highly electrostatic, picking up charge quickly. The suspended particles are typically exposed to a corona discharge. The particles pick up the charge and are then attracted to a collection plate. The charge depends on the size, and so the charge on the plate can be used to control both the size of the particles collected and the rate of particle collection. The electrostatic precipitator is an important component of pollution control systems. Filtering is particularly ineffective with nanoscopic particles, unless the filter membrane is charged.

Chemistry

Precipitation and Sol-gels

Amorphous silicas with particle sizes greater than those obtainable by fuming often are prepared by precipitation. Sodium silicate primarily is precipitated from solution with sulfuric acid, with silica and sodium sulfate as the reaction products. The size of the particles formed can be controlled by temperature, mixing, and other parameters. Acids other than sulfuric acid may be combined in part to improve control of particle size.

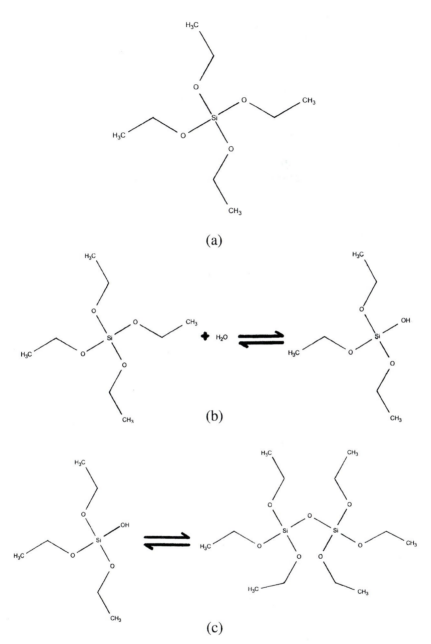

FIGURE 4.6 (a) Tetraethoxysilane (TEOS), an alcoxysilane is a chemical modification. (b) In the presence of water, one of the protecting ethoxy groups is hydrolyzed. (c) Two hydrolyzed silanes will spontaneously condense. A series of such reactions can create a polymer chain.

141

Other materials can likewise be prepared by precipitation of soluble metallic salts, precipitated using strong oxidizing agents.

Sol-gel processing is similar to precipitation. Inorganic precursors used include silicon or other metal alcoxides. The alcoxy group is hydrolyzed in water. Silicon is the most common metal, but aluminates and titanates are available. The hydrolyzed material is precipitated in an alcohol condensation reaction to create the sol. The sol is polymerized in a water condensation to drive gel formation. The gel is more malleable than many ceramic or pre-ceramic forms. The gel can be dried to create a material with controlled porosity, with size controlled by reaction conditions. pH, catalysis, temperature and ratio of metal alcoxide to water are all variables that contribute to the structure of the gel formed. Under reaction conditions favoring chain growth over chain branching, primarily acidic, the gel can be spun into fibers. Basic conditions tend to create monodisperse particles, with changes to the reaction rate leading to either spherical or disk-shaped particles.

Silsesquioxanes

Polycondensation reactions such as those used in the sol-gel process result in a particularly unusual nanoparticle if the precursor silane mate-

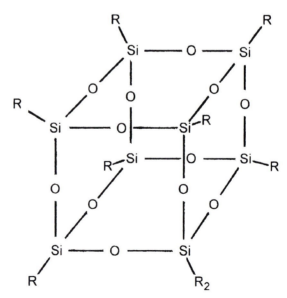

FIGURE 4.7 Silsesquioxanes are a class of organic silane assuming a cubic configuration.

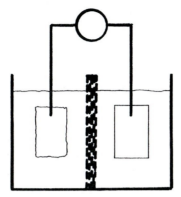

FIGURE 4.8 A battery cell. Pure metal precipitates at the cathode. The precipitate is usually friable, and easily broken up into smaller particles.

rials are trifunctional. These silanes assemble under condensation to create a cube, with an organic group pendant at the corners of the cube. The cubes are about 1 nm on a side. By virtue of the pendant organic group, diverse chemistries are possible centered on the cube. For example, depending a double bond containing chain allow the silsesquioxane to participate in a free radical reaction.

Reduction

Metal particles can be produced chemically by reducing metal oxides. The brittle oxide can be reduced by grinding to an appropriate size. The particles are heated in a reducing environment, such as hydrogen, carbon monoxide, ammonia, or methane. The temperature must be kept relatively low, to prevent coarsening or sintering the particles. Both the initial particle size and the reaction conditions control the final metal particle sizes. Low temperatures prevent coarsening, higher temperatures and water vapor in the reducing gas can result in larger particles.

An alternative to crushed metal oxides is treating waste metal scrap with strong acids or bases to create water-soluble salts. pH adjustment will usually cause these salts to nucleate precipitates, which can then be treated in a reducing atmosphere.

Electrolysis

Electrolysis of an aqueous solution of metal salts is another source of metallic nanoparticles. A high acidity and current density is maintained, and so hydrogen and metal are released at the anode. A spongy material that is easily broken apart is formed, harvested at frequent intervals. Organic additives to the solution can help reduce the particle size obtained from physical reduction of the accumulated metal harvested from the cathode.

Oxalates

Another approach to producing fine metallic nanoparticles is dissolving metal sulfates or chlorides in hot, distilled water. Ammonium oxalate is added, causing insoluble metal oxalates to precipitate from solution. Heating at about 400 °C causes decomposition, with the elimination of carbon monoxide and carbon dioxide.

Latex

Due to high fracture toughness, polymers are somewhat difficult to produce as nanoparticles from the bulk. A typical class of

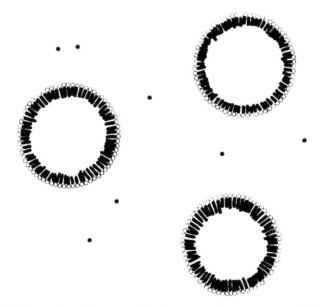

FIGURE 4.9 Latex formation. A surfactant creates uniform spheres containing monomer. The initiator is typically in the water. When an initiator molecule diffuses through the surfactant layer into a monomer particle, the entire particle becomes a single molecular chain. This technique creates uniform, monodisperse particles.

nanoparticulate polymers is available in latex formulations. Latex particles are created by mixing a monomer in water, creating a suspension by adding a surfactant. The surfactant allows micelles containing the monomer to form. The micelles have a normal, mono-modal, narrow distribution. The size of the micelles is controlled by the concentration and composition of the surfactant. An initiator is added to the water phase. From the solution, a single initiator diffuses through the micelle boundary to start polymerization. The resulting polymer particles are thus narrowly distributed in size, containing chains with nearly uniform molecular weight. Latexes composed of polymers sufficiently below the glass transition can be dried to a fine powder. Latexes above the glass transition will tend to fuse into a continuous film during drying.

Polymer Nanocrystals

There have been experiments conducted on semicrystalline polymeric materials where bulk material is treated with a strong acid, such as fuming nitric acid. The amorphous phase is destroyed by the acid. After washing away the acid and drying, the powder left behind is composed of uniform organic crystals. These crystals have usually experienced surface scission of the fold surface, and the chains are dramatically reduced in length. These organic nanocrystals will have a shape factor consistent with the growth habit of the parent material and dependent on the degradation time. Therefore, slowly formed crystals from which all traces of amorphous component are removed will have the plate structure characteristic of lamellae. Crystals that begin as hedrites or spherulites, and are less thoroughly etched, may appear as sheaves or spheres. Cellulose is an example of material produced using chemical reduction of a natural, polymeric bulk source.

Fibers

One-dimensional nanoparticulates consist of fibers, lathes, whiskers, and tubes. Asbestos and other minerals naturally form nanoparticulate lathe-like crystals. Additional contributions from the ceramic world are whiskers and tubes grown step-by-step starting at one end. Nanofibers are created from non-conductive organic materials by electrospinning or ultradrawing.

Whiskers and Nanotubes

Carbon nanotubes are currently notorious tubular reinforcements.

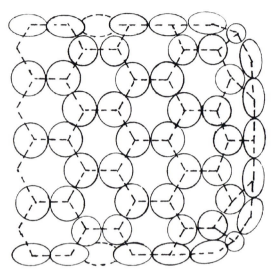

FIGURE 4.10 Carbon nanotubes are composed of rolled sheets. The sheets are formed of resonant, hexagonal, six-carbon rings. The ends of the tubes can be closed if five-member and six-member rings form to create the proper geometry.

While entering commercial availability, the materials are still rather expensive, primarily due to difficulties in purification. Carbon nanotubes are a common component of soots, carbon blacks and other combustion sources, albeit at generally poor purity. A more productive synthesis route deposits the carbon on a nucleating surface, and tubes grow by cold or hot vapor deposition. By careful practice, plates with dense arrays of single walled nanotubes can be grown. Multi-walled tubes are typically co-formed with single-wall tubes in pyrolysis techniques.

A naturally occurring form of nanotube is found in weathered pumice, called imogolite. The material is a paracrystalline material with a mixed composition of alumina, silica, and waters of hydration. The material is paracrystalline, rather than crystalline, due to a pronounced defect structure and its chemical composition. Imogolite has roughly two silicon atoms per aluminum atom. Recall that silicon forms four bonds while aluminum forms three. This mixture of 4-point and 3-point intersections leads to the characteristic curvature, and creates a reactive, functional surface that allows raft-like groups to assemble. A synthetic analog can be created by precipitation from a solution of $AlCl_3$ and Na_4SiO_4.

Whiskers have long been available as a source of sub-micron reinforcements with high aspect ratio. Whiskers are typically grown using epitaxy. Epitaxy is assembly initiated at a surface that provides a surface feature that matches one or more crystallographic growth planes. Whisker growth is more common in ceramics with a fibrillar crystalline growth habit, such as boron nitride and tungsten disulfide. The material is heated sufficiently to allow rapid diffusion of the source material. A feed supply of gaseous boron nitride can be created using a heated hollow tungsten electrode with boron nitride in the center. Long, single crystals of the ceramic will then grow away from the nucleating surface. The whiskers are harvested from the surface for application.

Narrow vanadium pentoxide nanosheets are produced by condensation from solution, using chemistry similar to the sol gel procedure employed for silica gels. The structural features, including ribbon width and thickness, can be modified using magnetic fields, organic templates, or thermal treatment. Nanotubes can also be created.

Electrospinning

Electrospinning is a process for creating polymeric nanofibers. A polymer is dissolved in solution. The solution is ejected very slowly via a syringe pump. A voltage is applied across the gap from the extrusion orifice to a deposition surface (Figure 4.11). The voltage causes the surface energy of the liquid to break down in a small surface area, and a jet results. The jet has a charge on the head that is attracted to the opposing electrode. The solvent in the jet evaporates as the particle travels, condensing to form a fiber. A random bed of fiber is deposited at the collection electrode. The jet processes are random and chaotic. The jet

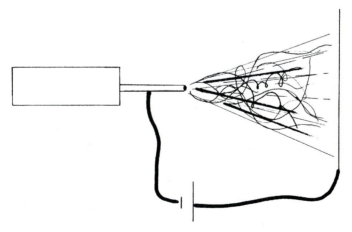

FIGURE 4.11 Electrospinning.

direction cannot be controlled, and so alignment of the fiber bed has yet to be accomplished. The fiber diameter is controlled by the concentration of the polymer in the solvent and by the voltage.

Electrospinning from a melt has also proven tractable, but the fibers tend to have a larger diameter. Despite the relatively smaller diameter of electrospun fibers, crystallites in the fiber are not directed preferentially along the fiber axis. The fibers are therefore no stronger than bulk polymers, albeit with very small diameters. Electrospun fibers are therefore quite weak.

Ultradrawing

Polymer melts and concentrated solutions can be drawn to form fibers. Typical polymer fibers are on the order of microns and larger in diameter. Some work has been done to "ultradraw" fibers, in order to achieve diameters in the nanometer range. The expected improvement over electrospun fibers is improved alignment of the chains along the fiber. The improvements in crystallinity should result in fibers approaching the maximum predicted strength based on covalent bonds and bond angle stretch.

Plates

Finally, two-dimensional, planar particulates are found as plates and sheets. Natural clays such as Montmorillonite are currently the biggest area of study within nanocomposites. Synthetic, flake nanoparticles are available as well.

Flakes

Sources of two-dimensional particles are again both natural and synthetic. The most common materials are naturally occurring, and synthetic materials are often modifications of natural materials. Naturally occurring two-dimensional materials include clay. That is, the material behavior recognized as belonging to clay arises characteristically from the sheet-like layers that comprise the inorganic phase of clay. Clays have a fine crystal structure, usually plate shaped. They have a large specific surface area, so are physically sorptive and chemically active. They hydrate well, generally by absorbing water between plate layers, allowing the crystals to displace readily. Natural materials are usually named after the geographical location of the first mines to produce the material, such as montmorillonite (Montmorillon, France) and bentonite (Fort

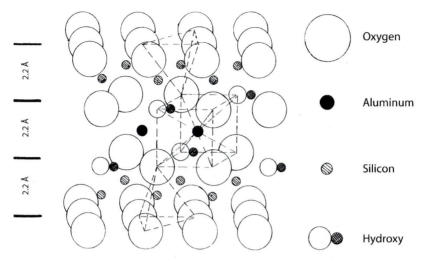

FIGURE 4.12 A layer of a montmorillonite unit cell: The oxygen is in four layers, with regular substitutions of OH$^-$ in the two inner layers to maintain the charge balance. Silicon atoms are located in tetrahedral sites, and aluminum is located in central octahedral sites between the layers.

Benton, USA). The natural flake materials often have inorganic contaminants: quartz, mica, feldspar, sand, opal, volcanic ash, carbonates, sulfates, sulfides, fossils, and more. These contaminants contribute to inconsistency from batch to batch of clay raw materials supplies. Synthetic clays have been produced by chemical and thermal treatments to remove impurities and optimize the chemical functionality of natural clay.

The basic unit cell for clay is illustrated in Figure 4.12. There are two tetrahedral layers of covalently bonded silica, SiO_2, and one layer of ionic, octahedral aluminum, Al. The aluminum atoms are in a +3 oxidation state. The aluminum has six neighboring oxygen atoms, two shared from each of the two neighboring tetrahedral sheets and two from an OH group that occupies the largest interstitial site between the tetrahedral sheet and the octahedral sheet. The positive charge on the aluminum is balanced by the negative charge on the atoms in the shared oxygen layers. This unit cell is charge balanced, or stable, and has little edge charge that would make the material more susceptible to chemical attack. The triclinic crystalline sheets stack to form larger crystals.

The majority of variations between clay types are caused by cation substitutions for the aluminum, for example, by Mg^{2+}, Fe^{2+}, Fe^{3+}, Ti^{4+}, and Mn^{2+}. The silicon atoms in the tetrahedral sites also may be substituted by aluminum. Silicon substitutions create a surface charge requiring an interlayer cation to balance, and can lead to natural curvature. The balancing cations introduce space in the interlayer, serving as a route for rapid water infiltration and rendering the sheets, and the clay, expansible. The cations also may exchange with the environment provided the charge balance is maintained.

There is a host of commercially available clays. Kaolins, serpentines, micas, vermiculites, smectites are all classes of ceramics with flake structures and properties that may be modified. The most common types for nanocomposite research are the smectites. Smectite clays include montmorillonite and bentonite. The aluminum is partially substituted by Mg, Fe, Cr, Mn, or Li. The surface charge is usually balanced by Na, K, or Ca. The smectites expand in water and alcohol. The thickness of the tri-layer sheet is about 6.60 Å, while the minimum spacing between sheets corresponds to a monolayer of water. Montmorillonite is sometimes used as a name for the entire smectite clay class. Bentonite is rich in montmorillonite. Montmorillonite is the most commonly used smectite. In montmorillonite, one out of every six aluminum ions is replaced by an Mg^{2+} ion. This puts an average charge of -0.67 on each unit cell, balanced by sodium ions.

Monolayers

Monolayer self-assembly is another interesting approach to creating planar nanoparticles. Surfactants and lipids are examples of self-assembling materials. These molecules have both hydrophobic and hydrophilic tails. In systems where one hydrophilic phase is present, the

TABLE 4.2 Properties of montmorillonite.

Elemental Composition	%	Property	Flakes
SiO_2	51.14	Density	2.3 – 3.0 g/ml
Al_2O_3	19.76	Unit Cell	Monoclinic
Fe_2O_3	0.83	Unit Cell MW	540.46 g/mole
ZnO_2	0.1	Hardness	1.5 – 2.0
MgO	3.22	Endothermic Transitions	140°C
CaO	1.62		700°C
K_2O	0.11		875°C
Na_2O	0.42	Exothermic Transition	920°C
H_2O	22.8	Swelling in H_2O	up to 30-fold

From Utracki, Vol. 1.

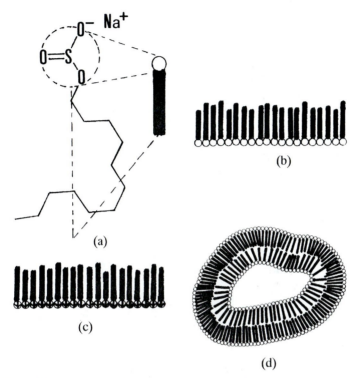

FIGURE 4.13 Surfactant modification: (a) an example of a surfactant, with a polar, hydrophilic head group and a long, hydrophobic body. (b) The surfactant will form a monolayer to create a barrier between a hydrophilic surface and a hydrophobic liquid, or vice versa. (c) In cases where the hydrophobic head can polymerize, such as TEOS described above, the surfactant can form a permanent layer. (d) In a water solution, the surfactant can create a double layer, with water both inside and outside, but the interior is protected by a double layer composed of the hydrophobic tails. Transport from outside to inside this boundary requires diffusion across the hydrophobic double layer.

hydrophobic end must be excluded from the solvent. If small quantities of the lipid are present, a monolayer will be formed. If the surface is too small, multiple layers can form. If a portion of the chain contains functional groups, such as double bonds, then the layered structure can be polymerized to create a two-dimensional film.

Treatments

The interaction between matrix and reinforcement was discussed in Chapter 3. The performance of a composite will depend on the load transfer between the particles and the matrix. The natural surface of the particles may present properties that do not support the desired interactions. The performance of the particles as reinforcement can be enhanced through specific treatments. All particles can be surface modified to control interactions. In addition, the nanosheets of clays often require separation to improve matrix interactions. In this section, some of the processes by which the reinforcement may be modified for use in composites are discussed.

Surface Functionality

The surface of the particle is the locus of interactions between the particle and matrix. The reinforcement may be highly, moderately, or poorly

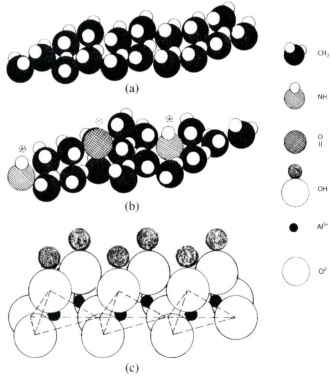

FIGURE 4.14 Surface functionality: (a) Polyethylene presents a uniform carbon-hydrogen, hydrophobic surface. (b) Nylon 11 presents a mixed surface, with polar amine and oxygen groups and hydrophobic carbon-hydrogen functionality. (c) Alumina presents a uniform hydroxyl surface.

interactive with the surrounding matrix. In the absence of specific interactions, the behavior follows the general rule that hydrophobic materials
will interact well with hydrophobic surfaces, while hydrophilic materials
are more compatible with hydrophilic surfaces. Specific interactions are
those caused by specific chemical susceptibilities. Two exposed sulfur
atoms, on separate molecules, will rapidly form an S-S dimer, a feature
exploited by proteins containing cystine residues in biological systems.
Fluorine-containing materials present a particularly non-reactive,
non-interacting surface.

There are additional advantages to modifying the surface of particles.
During physical transport, even within a process stream, the particulates
will interact with external surfaces; air, CO_2 and H_2O; and with each
other. Any of these interactions may result in adsorption or chemical reactions that interfere with the proper and desired function of the reinforcement. For example, exposure to oxygen may cause oxidation, with
resulting, detrimental changes to the surface properties of the particle.
Nanoparticles are particularly sensitive to interactions causing agglomeration. Modifying the particle surface function, for example by adding a
surface charge, may prevent flocculation or oxidation.

The functionality of the surface may be modified to improve the interactions between particle and matrix. Figure 4.14 shows examples of surface functionality. Reinforcement particles are composed of atoms and
molecules, arranged in a crystalline or an amorphous structure. At the
boundary of the particle are located functional groups: some are part of
the native structure in the material, and some result from reactions with
the environment, such as the oxidation of a metal particle exposed to air.
The surface functionality is a characteristic of a material. The surface energy of the material is specific to the nature of the surface functionality,
as well as the specific surface area of contact.

The challenges in modifying particles include: dispersing the particles
in a solvent suitable to conduct the modifications; accomplishing the desired modification as completely as possible without damage to the particles; and collecting the modified particles from solution. These
challenges are similar to the general challenges of producing a
nanocomposite using dissolution or *in situ* polymerization, discussed in
Section II.

Surfactants

The surface energy can be temporarily changed with a surfactant. Surfactants have one or more hydrophilic head groups, sometimes large,
charged species such as that in sodium lauryl sulfate, Figure 4.15, or oth-

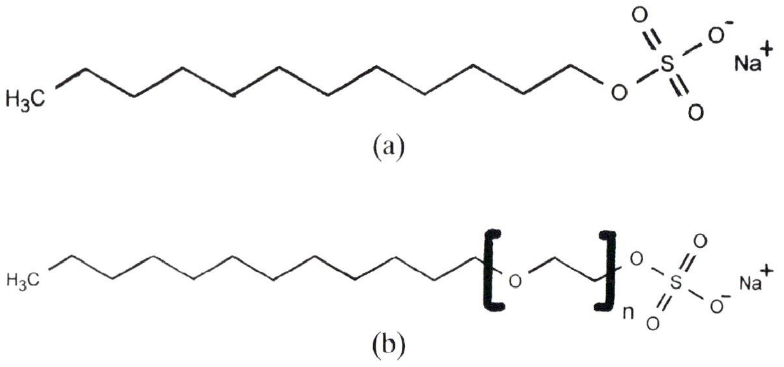

FIGURE 4.15 (a) sodium lauryl sulfate and (b) sodium laureth sulfate, common surfactants.

erwise acidic or basic. Another end or ends are hydrophobic, with some extensive hydrocarbon or glycol chain. The length of the chain may vary, but must be sufficiently long to allow enough mixing or interaction with both the hydrophobic and hydrophilic substances, while at the same time allowing a stable separation between the two substances.

The surfactant will form a stable structure in suspension based on the

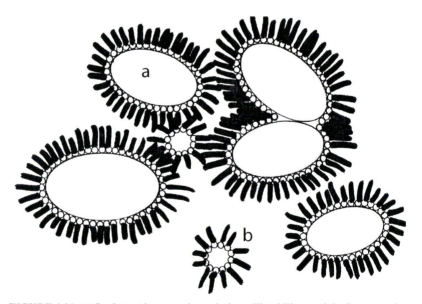

FIGURE 4.16 (a) Surfactant in excess in a solution will stabilize particles in suspension. (b) The excess surfactant can form empty, free micelles.

balance of concentration between surfactant and particle. If the concentration is smaller than sufficient to cover the surface of the particles, then surface modification will be incomplete. By particle agglomeration, the surface area is reduced and the incomplete surface coverage problem disappears, to the detriment of the stability of the particle size in solution. If the surfactant is present in excess, a monolayer of surfactant will form on available surfaces. The excess surfactant will tend to form free micelles. The empty micelles are chemically indistinguishable from micelles containing particles.

The surfactant, and surface modification caused by the surfactant, can be removed almost completely by washing with an excess of almost any solvent. However, surfactants are hydrophilic and hydrophobic. The combination makes complete surfactant removal from any system particularly difficult, since by nature they are partially soluble in any solvent.

Reactions

The simplest techniques for permanently modifying the functionality of a surface are through chemical reactions. A molecule with an acid functional group can react with a surface with a basic group. A molecule with a basic functional group can react with a surface acid group. Pendant double bonds can participate in chain growth reactions with ethylene or styrene, or can participate in oxidation reactions. The same chemistry is available at surfaces as is available in solution. The primary difference is that surface chemistry is much slower. This should come as no surprise: chemical species tethered to a surface are partially constrained, and have less energy available in translation, rotation, and certain vibrations. Therefore, the surface group has fewer degrees of freedom, that is, it is sterically hindered.

While any changes in the functional group will modify the surface interactions, one particular type of modifying group is dominant in modification philosophy. Modifying the surface with short molecules will result in a simple change to the surface functionality. For example, a hydrophilic basic group on the surface is replaced by a hydrophobic propyl group. However, by reacting with one end of a long chain with the sur-

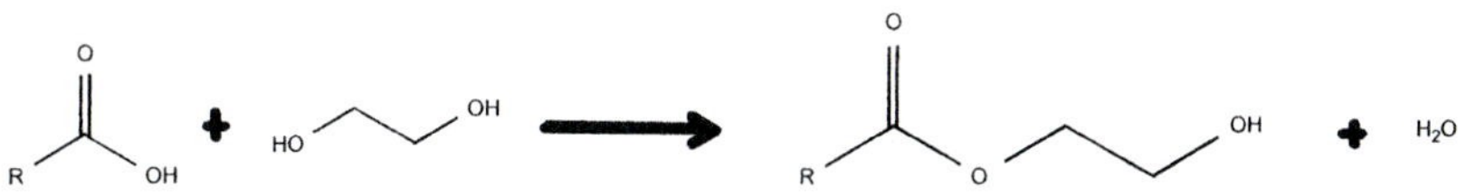

FIGURE 4.17 Ethylene glycol can be added to an acidic surface. The acid-base reaction transforms the acid into a surface with basic chemistry.

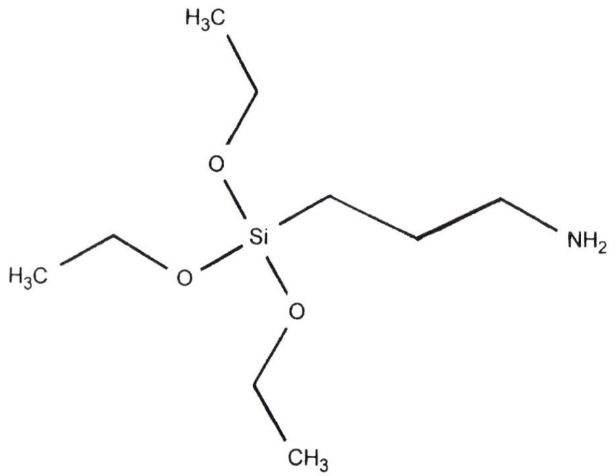

FIGURE 4.18 (γ-aminopropyl) tri(ethoxy)silane, Silicone A-1100. This chemical is also suited for use in a sol-gel reaction. The ethoxy moieties make good leaving groups during the condensation phase, leaving a pendant, functional surface group.

face, the new surface will have some mobility. Any specific interactions with a matrix material are less hindered by the surface.

Silanization

Silanization merits specific mention as an almost ubiquitous modification for silica materials used as reinforcements. Silanization is the common name for a hybrid, specialized surfactant and reaction combination. The name derives from the functional silane chemicals used in the reaction. An example is given in Figure 4.18. The silane will act as a surfactant. Added to a suspension of particles in excess, the silane forms a monolayer coating on the particles through interaction between surface hydroxyl groups and the polar amine group of the silane. The amine group reacts with any hydroxyl functionality on the surface irreversibly. The result is a new and hydrophobic surface. The hydrophobicity can be controlled by the structure of the side groups on the particular silane used.

Oxidation/Reduction

The surface of a particle has a characteristic behavior. For example, a hydrocarbon material has a hydrophobic, non-reactive surface. The behavior is a function of the oxidation state of the surface. If the polymer has polar groups, such as N or O available at the surface, the material will be more hydrophilic and reactive. By contact with a strong acid, an etching group such as potassium permanganate or osmium tetroxide, ozone combined with *uv* light, or oxygen plasma, functional groups can be added to the surface of a particle. The additional functional groups will make the surface both more hydrophobic and more reactive.

By comparison, bare metal is very reactive. Metal will tend to react with oxygen in the environment to form an oxide. The oxide is generally stable and non-reactive, but hydrophilic. Reducing the oxidation state of the oxide, reduction, by introducing a proton, hydrogen, will create a hydroxyl functional group at the surface of the oxide. This will cause the particles to be more reactive. The surfaces of oxide ceramics are also susceptible to reduction, and the surfaces of non-oxide ceramics, such as carbides and nitrides, are susceptible to oxidation.

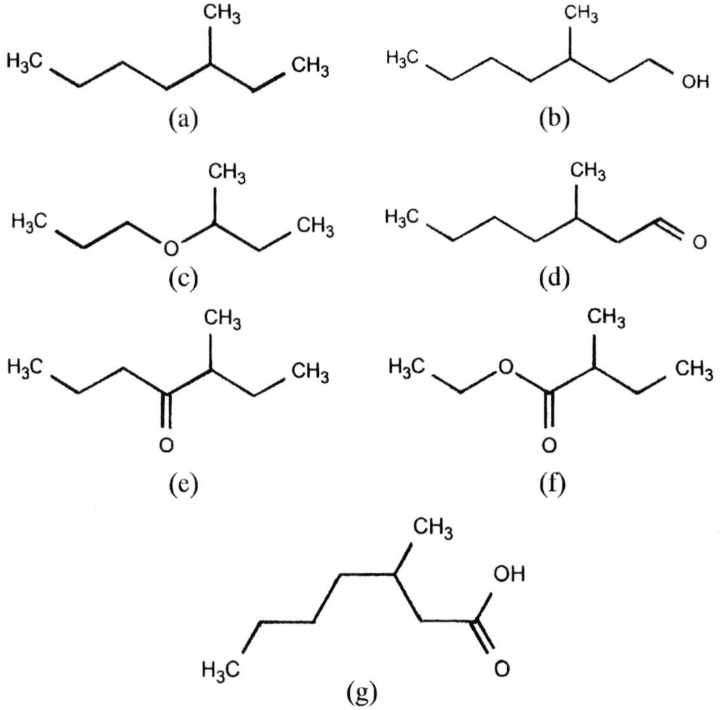

FIGURE 4.19 Surface groups with increasing oxidation. (a) 3-methylheptane, (b) 3-methylheptan-1-ol, (c) sec-butyl propyl ether, (d) 3-methylheptanal, (e) 3-methylheptan-4-one, (f) ethyl 2-methylbutanoate, (g) 3-methylheptanoic acid.

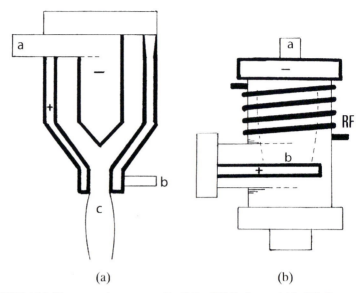

(a)   (b)

FIGURE 4.20 Plasma can be generated by DC or RF discharge. (a) In DC plasma, gas is fed into a cathode-anode gun at a. The gas is converted to plasma. Particles are fed downstream of the plasma chamber, b. The gas expands and exits at high velocity, c. (b) In RF plasma, gas is bled into a moderate vacuum, a. The rarified gas is converted to charged plasma by radio frequency energy and accelerated to the target, b, by an applied voltage differential.

Plasma

The surface of a material exposed to a highly charged gas, plasma, can also cause modifications to the surface of the material. Plasma fields can be created in two general fashions. The first is by direct current discharge. A current is applied to an anode in a high voltage field. The gas atoms pick up electrons from the surface of the anode. The charged gas molecules will be accelerated in the potential field. These particles will be picked up at the cathode, but the particles may pass through a porous electrode to be deposited on a surface beyond the electrode. The second technique for generating plasma uses radio frequency energy passing through a coil. Molecules within the space inside the coil are ionized, and can be discharged toward a target.

There are several possible effects on surfaces exposed to the plasma. The plasma can behave like hard particles. On striking the surface, the ion can remove atoms, causing surface erosion. This can create a rougher surface. The plasma instead may behave like a soft particle. On striking the surface, the ion may deform and create a coating. Finally, on striking the surface the charged species may react with the surface to add a functional group. Applying plasma to particles is problematic. The particles must be suspended in a gas to assure that they are evenly exposed to the plasma. However, the carrier gas will dilute the plasma quickly. One technique is to drop the particles into the gas stream downstream of the ionization zone, thus carrying them into the plasma for treatment.

Coated Particles

Coated particles are important in many technologies. Liquid delivery is often accomplished by encapsulating the liquid in a polymeric shell. Inks and dyes have been encapsulated for delivery under pressure activation. Drugs are encapsulated for delivery to specific environments or release over time. Coatings may provide antistatic or lubrication. Coated particles are usually large, but newer technologies are decreasing particle size.

Coacervation

Coacervation is the process of separating a single phase containing two materials into two immiscible components by adding a third component. The different phases contain different amounts of all three substances. The technique starts with a polymer solution. Adding an aqueous emulsion with a hydrophobic solvent and a third component causes the liquid to separate into a polymer-rich coacervate phase that coats the particles and an aqueous phase containing the mediating third substance. The hydrophobic solvent separates with the polymer-rich coacervate, creating coated particles of a size determined by the size of the hydrophobic domains present in the original emulsion. The particle size of these hydrophobic domains is controlled by the agitation rate.

Aerosols

An aerosol is a suspension of a solvent or a solid particle in a gas stream. Aerosol particles are typically of a size or density such that settling is very slow. Where the solvent is water, the aerosol may be called a mist. Aerosols can be used to create capsules. A solid, suspended aerosol particle can nucleate a solvent condensing out of the vapor phase. The vapor phase can be a metal, or a liquid monomer, which can be polymerized in a secondary step.

Alternatively, two different types of aerosol particle, substrate and coating material, can be mixed in a highly turbulent stream. The result is coalescence of the particles. When a liquid and a solid particle coalesce, a coated or partly coated particle is obtained. The liquid particles may also coalesce into larger drops, which then coalesce with a solid particle, resulting in some uncertainty in coating thickness. Also, uncharged solid particles, or coated solid particles with tacky surfaces, may flocculate into undesirable agglomerates. The impinging aerosol approach results in more variation within the coated particle product.

Reactions

Reactive materials can be coated by replacement at the surface layer. For example, zinc can be coated with lead or copper by starting with zinc particles, then replacing the outer layer with a material higher in galvanic potential. The outer surface can be exposed to oxygen to cause the formation of an oxide layer.

Milling

A ball mill can be used to create coated particles, also. The two particles to be combined must be very different in both size and hardness. For example, a 100 nm alumina particle can be coated with 10 nm zinc particles in a ball mill. Alternatively, a 100 nm styrene latex particle can be coated with 7 nm fumed silica in a ball mill. The force applied by the mill causes the hard particle to deform and penetrate into the softer particle, thereby creating the coating.

Intercalation

While the plates in clay are nanoscopically thin, they tend to form lamellar stacks. In order to take advantage of the thin dimension of the plates, the layers must be split, or exfoliated. Flake nanocomposites have three states. The first is the native, lamellar structure. The particles in this form are not nanoscopic and will reinforce as a unit, perhaps allowing for some interlaminar shear inside the particle under stress. The second state is an intercalated structure, wherein through chemical or physicochemical means, the layers are separated beyond the equilibrium state. By inserting a monomolecular or multilayer of molecules between the lamina, the process of separating the layers is begun. The packed lamina may incorporate as a unit, but processing or later deformation may be accommodated by layer displacement. Finally, through chemical compatibility and or shear, the lamina will be pulled away from each

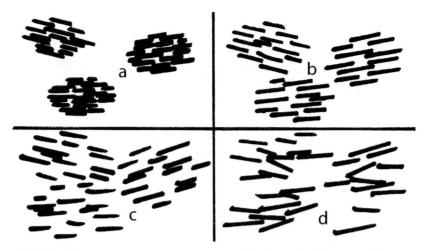

FIGURE 4.21 Some geometries assumed by clay (after Utracki). (a) In unmodified clay, the layers remain tightly packed. (b) In a compatible material, the layers of clay may separate slightly. (c) If a small species is used to intercalate the layers, they can separate effectively during processing. (d) If the layers can be made to be easily separated through modification, and the clay is compatible with the matrix, thorough dispersion is possible.

other completely. All of the property modifications expected based on particle size will manifest only after complete separation of the layers.

The process of exfoliation proceeds through three phases. First, the gallery separation is increased. Second, the attractive forces between the layers are disrupted. Finally, the interaction between the plates and the matrix is increased. In practice, these steps may be subdivided or partially integrated.

The layers in the stack of clays containing a surface charge, such as montmorillonite, are separated by a cation monolayer. Dried, the gallery contains only the charged ions and a monolayer of water. These materials are known to swell in water and alcohol, as molecules from the liquid phase form multilayers. The cation in the interlaminar gallery layer also is susceptible to exchange with larger cations. After swelling with solvent, the cation may be exchanged with quaternary onium ions, such as ammonium or other large cations. At this point, phase one of the exfoliation is complete, and phase two is beginning. Further swelling is now possible, as the charges are balanced without surface proximity. The further-increased gallery spacing may at this point be large enough to support exchange with a surfactant or functional silane or other surface-sizing agents, thereby temporarily or permanently neutralizing the surface charge. This carries into phase three of exfoliation.

Montmorillonite has both surface and edge charges. The charges on the edges are easily accessible to modification, but do not accomplish much improvement in interlaminar separation. These sites do represent an opportunity to attach functional groups. Candidates include sizing agents, or perhaps longer species such as end-tethered polyethylene glycol or polyvinyl alcohol oligomers. A reactive intermediate may also be chosen that will support later *in situ* copolymerization with a monomer. Examples include acid, hydroxyl, glycidyl, maleic, or isocyanate groups. Polyester-compatible chains will support incorporation in addition polymerization. *In situ* polymerization is discussed in Chapter 6.

Melts

Polymer melt intercalation between the layers of clay is a difficult idea to entertain, even though many sources make such a claim. The situation envisioned, depicted in Figure 4.22, has polymer chains snaking through

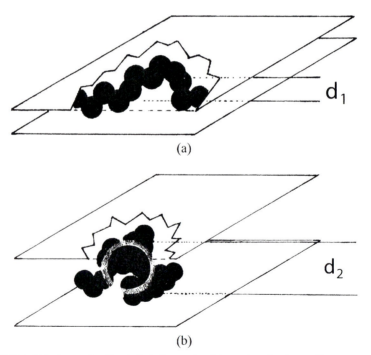

(a)

(b)

FIGURE 4.22 (a) An idealized chain intercalated into a interlaminar gallery must take on a chain-extended conformation. (b) A chain typically takes on a random coil configuration, characterized by the radius of gyration.

the gallery of a stack of clay nanoplates. This is a statistically improbable situation. A flexible chain would have to become highly ordered in order to fit into the interlaminar gallery. The increase in ordering represents a large entropy loss. The entropy loss must be completely overcome by an enthalpy gain if the process is to occur spontaneously. Interactions between the polymer chain and the clay platelet are the source of any enthalpy gains. Intercalation may be possible if a monomer can be used in one of the gallery widening and charge neutralization steps. The monomer can then be polymerized in a second step to create a trapped chain, without incurring the loss in free energy caused by the reduction in entropy. However, the subsequent polymerization step may not proceed even if adequate monomer functionality is available. It may not proceed if surface binding protects one of the necessary functional groups, or restricts the orientation of the monomer, such that ineffective reactions, for example, self-quenching, become competitive with chain growth.

The gallery spacing is sometimes observed to increase slightly in the presence of polymer melts, however. The increase may be a hydrostatic effect originating from the distribution of the clay particles with the melt. Characterization of structure is considered in Chapter 9. Melt exfoliation begins when the interlaminar gallery becomes large enough to allow significant diffusion of the molecular chains into the opening between plates.

The diffusion has multiple contributing forces. The polymer is diffusing along a diffusion gradient. Further, through proper choice of the preceding intercalation steps, there is an energetic gain in entering the gallery. However, the space constraints are also significant. On entering the gallery even partially, however, the gallery spacing will be disrupted, which is probably strongly dependent on both the length of chain having entered and on the temperature.

Once the interlaminar spacing has been increased, and especially after neutralizing attractive forces between the platelets, shearing force will improve exfoliation. Low viscosity materials do not transfer much force to the crystals, but the melt can transfer shear if the interactions between melt and silica surface are reasonable. As the lamina break apart and the interaction increases, the flow properties will change. Melt processes are discussed in Chapter 7.

Challenges

The primary enemies of nanoparticles are particle growth and chemical transformations. Particle growth may occur due to agglomeration, sintering, and ripening. The increased surface area available for oxidation and other reactions enhances potentially damaging chemical transformations.

Agglomeration

Due to their ready susceptibility to acquiring charge, and their low mass, nanoparticles tend to agglomerate readily. Agglomerated particles will not reinforce in the same manner as smaller, individual particles. Commercial spherical particulates, carbon black and fumed silica, are usually supplied as aggregates. The aggregates are stable for two reasons. The aggregate reduces the surface area, thus reducing both surface available to collect charge and the energy stored in the free surface. The aggregates are thermodynamically favorable. The small particle size makes subdividing the particles, even along relatively weak boundaries, difficult. Impact will seldom provide enough energy to separate the particles, and can instead provide the energy to fuse particles already in contact.

Collections of particles, particle beds, are created simply to transport the material. The bed is susceptible to further aggregation due to compression. Bed compression depends on the particle size and the porosity of a compact of the powder. The pressure required to compress particles in a bed is explained by the logarithmic expression:

$$\log P = aV_0 + b \tag{4.20}$$

where V_0 is the bulk volume of a substance, while a and b are material-dependent constants. The smaller the particles, the lower the pressure required to compact the powder. Further, the increase in V_0 due to the porosity of the bed is smaller as the particle size decreases.

Ultrasonic vibrational energy may accumulate at weak interfaces, where impedance mismatches occur. Likewise, electromagnetic energy may accumulate for the same reasons. Such energy sources may improve deagglomeration. Coating the particles with a surface that prevents interactions between moieties may also help, but agglomeration often occurs even during particle production.

Sintering

Large areas of free surface are thermodynamically unfavorable. The energy gain through agglomeration is limited. One way to further improve the stability of the particles is through sintering. If there is suffi-

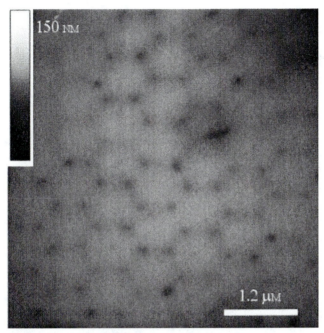

FIGURE 4.23 Sintering of an acrylic latex, nominal 300 nm particles. The particles begin in a closest-packed arrangement. The particles spread, deform, and merge with neighbors.

cient atomic or molecular mobility, the surfaces of the particles at the contact point will start to combine. The first stage of sintering is the formation of a narrow bridge, or neck, at the contact point, where the particles become fused. A new grain boundary is formed if the particles are crystalline. The neck thickens until the spaces between particles become pores. The pores decrease in size and change shape as the trapped gas diffuses out of the new, larger particle.

Ripening

If there is a broad distribution of particle sizes, then the large particles will grow larger still by cannibalizing from the smaller particles. The material from a neighboring smaller particle diffuses across the contact point to add to the growing surface of the particle. The difference in size required to allow ripening to become competitive is controlled by the molecular diffusion rate. Ripening is also more likely for crystalline materials, or in other systems with an additional energy advantage for growth beyond just surface area reduction. In these systems, growth may not be even in all directions. The ripened particles may change from the initial shape.

Oxidation

The increased surface area of a nanoparticle also increases the number of exposed reaction sites. The oxidation of metal surfaces is spontaneous and rapid. As with any bulk metal surface, an oxide layer will form. With nanoparticles, this oxide layer will form a significant fraction of the composition, soon after the metal particle is created. Creating pure, stable metallic nanoparticles is difficult.

Organic molecules are also more susceptible to chemical reactions. Bear in mind that nanoparticles may be only 10 times larger in diameter than the molecule in solution. If a chemical reaction occurs in solution,

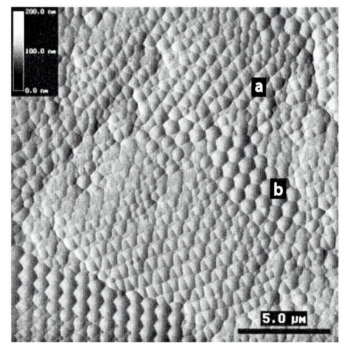

FIGURE 4.24 Ripening of an acrylic latex, nominal 300 nm particles, with prior agglomeration to an average of 900 nm average diameter. (a) The original packing is irregular. (b) Some particles get larger at the expense of neighboring particles.

the same reaction may cause a significant effect if a nanoparticle is subject to the same conditions.

MATRIX MATERIALS

Almost any material can be used as a matrix for a nanocomposite. The majority of the rest of the text describes specific features of processing in Section II and characterization in Section III. A brief discussion of various types of materials and the techniques and issues pertaining to incorporating particulate reinforcements follows.

Metals

Metals fail through deformation. The modulus and ultimate strength of a metal are enhanced by processes that prevent defects from traveling through the part. Nanoparticles have the potential for trapping defects, thereby reducing the susceptibility to deformation. Incorporating nanoparticles into metallic parts is the challenge.

Metallic items are created by various casting and forging processes. Casting traditionally involved melting the metal, removing impurities in the form of slag, and pouring the melt into a sand mold. Incorporating nanoparticles in a molten metal faces several potential obstacles. First, many substances are dissolved in metal liquids, rather than remaining dispersed. Liquid titanium could be considered the universal solvent. Second, the surface of the particulates must be wet by the metal, and most particle surfaces are not. Third, small particles are inherently unstable, and heating them increases the chances of aggregation, sintering, and ripening.

Powder metal processes, including powder compacts and powder thermal spray, are a much more promising technique for creating metal-matrix nanocomposites. Powder compacts take mixtures of finely distributed metal powders, and create a fully dense part through sintering below the melting temperature. The high temperatures can still lead to particle growth, and diffusion from the reinforcement into the metal is a real concern.

Forging shapes a material while it is well below its melting point. Nanoparticles could be incorporated by folding the particles into the metal during successive thinning, folding, and fusion steps. The same concerns regarding particle coarsening and diffusion hold for forging as well as for powder processing.

A different approach to incorporating nanoparticles is through solid phase precipitation. Many substances are only partially soluble in a metal. Under the proper conditions, nanocrystals of a second phase can be created spontaneously. This, however, is not a new technology.

Ceramics

Ceramics fail through fracture. Improvements in ceramic properties arise through arresting crack growth, and reducing the susceptibility to surface flaws. Nanoparticles can provide zones that allow local deformation to prevent cracking. The nanoparticles may improve the scratch resistance, preventing surface cracks that will initiate fracture. Nanoparticles may also apply internal compressive stress, which will prevent cracks from forming or opening. Ceramic processing often occurs at very high temperatures. The high-temperature processes create a challenge in suppressing chemical reactions, diffusion, and particle growth in the reinforcing nanoparticulates.

Traditional ceramic processing techniques include firing and slip casting. The ceramic may be produced by packing dry powders into the desired form or by mixing the powders in water to form clay that can be shaped into the desired shape. The piece is then subject to a high-temperature firing step. Depending on the temperature, the particles will either melt or sinter during firing. Both powder and clay techniques are suitable for adding nanoparticulates.

Some ceramics are pozzolans, capable of behaving identically to natural cement. The cement, when mixed with water, exhibits a complex set of hydration reactions. As the mixture dries, a dense ceramic material results. Through this type of chemistry, extremely high temperatures are avoided. This process is similar to *in situ* processes described in this book. Ceramic glasses are inorganic polymers and, therefore, can be processed using many of the same techniques detailed in the following chapters.

Polymers

Improvement in polymer properties occurs through increases in modulus, ultimate strength, and many other properties. The basic premise is that transforming more of the material from a polymer to a metal or ceramic with more desirable properties will provide greater improvements. Of course, the interaction between polymer and reinforcement must be sufficient to transfer the forces across the interface. The processes for producing and characterizing polymer nanocomposites are explored in subsequent chapters.

HAZARDS OF PARTICLES

Handling particulates has hazards different from handling bulk materials or liquids.

Explosion

Particles can be an explosion hazard. As discussed above from a chemical point of view, the smaller the particles are, the larger the number of available surface groups available to participate in chemical reactions. Finely divided aluminum is the primary component of the highly explosive gray powder used in many fireworks and pyrotechnics. Grain dust is highly explosive, and has been the root of several agricultural disasters. Organic and metallic powders, both of which can oxidize, must never be exposed to open flames or sparks.

Inhalation

Airborne particles are a potential health hazard. Particles breathed into the lungs can accumulate in various places. Large particles, down to about 5 microns, interact early with the bronchial tubes. There, cilia move the particles safely out of the lungs. Particles between 5 microns and about 100 nm are mostly exhaled. However, a significant fraction interacts with and sticks to the alveolar sacks, from which particles leave very slowly. At smaller particle diameters, about half of the particles are caught in the bronchial tubes and about half are trapped in the alveoli. The smallest particles may also be transmitted through the lungs into the bloodstream.

Crystalline particulates have been shown to cause fibrosis in the lung. Crystalline minerals with high aspect ratios are particularly damaging, but any lathe-like particle seems to trigger the formation of fibers in the lungs.

Irritants

Particulates are naturally foreign material that can cause irritation. The surface chemistry may enhance the irritation. Nanoparticles have a very high specific surface area, increasing this effect further. Further, nanoparticles can move under physiological conditions. Proper safety equipment should always be used. Filter masks are expected to be ineffective for the very smallest particles. Positive airflow should be effective, however.

REFERENCES

Cadle, R., Particle Size, Reinhold Publishing, New York, 1965.

Perry, R. and D. Green, Perry's Chemical Engineers' Handbook, Chapter 20, McGraw-Hill, 1999.

Utracki, L., Clay-Containing Polymeric Nanocomposites, vol. 1, Rapra Technology, UK, 2004.

Nanocomposites Processing

There are two reasons to discuss nanocomposites processing. First, the processing conditions determine many of the important structural features of a material, and thus the performance. Secondly, in order to successfully enter the marketplace, composites and nanocomposite products must become profitable to fabricate. Many of the same practices for plastic part manufacture are useful in composites manufacturing. The particles play a significant part in the processing behavior, and nanoparticles even more so.

This section contains three chapters introducing facets of processing. Chapter 5 introduces the concept of viscosity, or resistance to flow, adding in the influence of particles on the process. Viscosity is common to all processing techniques. Chapters 6 and 7 cover two broad approaches to fabricating composite pieces. One approach, covered in Chapter 6, covers mixtures of particles in low viscosity solutions. This approach is typical of *in situ* and solution processing. This chapter introduces dissolution, particle settling and *in situ* polymerization. Chapter 7 discusses melt processes, traditionally used for thermoplastics. This chapter introduces mixing, melting, extrusion, injection molding, compression molding and other important process techniques. Nanoparticles have particular impact in the viscosity, particle settling velocity, and melt mixing.

CHAPTER 5

Viscosity

Material properties are the driving force for material selection. However, once the resin and reinforcements have been chosen, a final product must be manufactured in a practical manner. A number of concepts directly affect all processing techniques. The first subject to be considered is viscosity. The viscosity of a liquid describes the resistance to flow under applied force. A high-viscosity liquid will flow less under the same force as a low-viscosity liquid. Viscosity is a fundamental property of the liquids and solids to be formed into composite parts, and may be better understood separate from the influence of the viscosity on specific processing techniques.

Two classes of nanocomposites processing techniques have been drawn, based on the viscosity regime used. When a monomer is combined with a reinforcement, together with subsequent polymerization, the viscosity of the solution of mer and particle is generally low. Likewise, if the polymer is dissolved in a solvent, whether by means of organic dissolution, emulsion, or supercritical gas injection, the viscosity of the resulting solution is low. On the other hand, direct attempts to mix reinforcement into molten polymer generally occur in a high-viscosity regime. Traditional polymer processing techniques mostly fall in this category. Compression molding and thermoforming, extrusion, cold drawing, injection molding and blow molding are applied primarily to polymers in the molten state.

Viscosity is particularly significant in melt processing. As will be discussed, even small amounts of added reinforcement can dramatically increase the viscosity. As the particle volume concentration increases, or as the particle diameter decreases at constant volume, the viscosity increases more sharply still. The purpose of this section is to introduce conceptual frameworks for understanding viscosity in processing. There are many mathematical descriptions of viscosity provided, predominantly without derivation.

TYPES OF FLOW

Processing involves displacement of matter, or flow. Flow patterns fall into three general categories. A smooth, steady-state flow pattern is characteristic of laminar flow. Discontinuities in the flow field are characteristic of turbulent flow. In some circumstances, the direction of flow is uniformly downstream, but the pattern may have time-dependent variations in direction vectors or velocities. This is characteristic of streaming flow.

In laminar flow, a steady-state flow pattern is established in which all elements of the liquid have a downstream direction vector. The flow field is known to be parabolic, with the velocity a maximum in the center of the flow and decreasing to 0 at the wall under ideal circumstances. Laminar flow is held to apply where the Reynolds number, Re, is less than

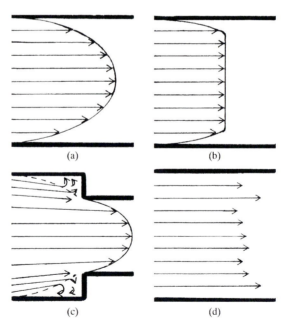

FIGURE 5.1 Flow fields: (a) laminar, (b) turbulent, (c) turbulence caused by narrowing, (d) streaming flow.

2000. The Reynolds number is a unit-less constant relating inertial to viscous responses in a liquid:

$$\text{Re} = \frac{\rho v d}{\eta} \tag{5.1}$$

where ρ is the density, v is the velocity of the liquid, d is a characteristic length, such as the diameter of a pipe, the size of a particle, and so forth, and η is the viscosity, also given the symbol μ. For polymers, the viscosity is typically very large and flow rates are generally low, so the conditions for laminar flow are more likely to be met for polymers than in any other type of liquid.

Turbulent flow is characterized by a non-uniform flow. All parts of the liquid may or may not have a downstream velocity vector. Streaming flow is a fast flow that behaves, on average, as laminar flow, with slip at the boundaries. Solutions applying to turbulent and streaming flow tend to be less transparent and therefore less instructive than laminar flow calculations. The concepts taught through laminar flow arguments are also present in turbulent flow: the loss of energy through viscous flow, as well as the factors leading to increased viscosity.

Another dimensionless number important to viscosity and rheology is the Deborah, or Weissenberg, numbers:

$$\text{De} = \frac{\lambda}{t_{flow}} \tag{5.2}$$

$$\text{We} = \frac{\lambda_{steady}}{t_{flow}} \tag{5.3}$$

where λ is the relaxation or retardation time, the same as that for the viscoelastic response, and t_{flow} is the characteristic flow time for the processing method. t_{flow} is usually taken to be l/v where l is the characteristic length scale of the flow and v is the characteristic fluid velocity. λ_{steady} is measured instead from steady-state flow measurements. The Deborah number represents the balance between the natural response time of the polymer and the deformation rate applied during the process. A large Deborah number describes a system where the material will behave in a more viscoelastic manner; a small Deborah number is characteristic of liquid flow. If the Deborah number is large, the Newtonian description of flow is insufficient, and a time-dependent viscosity description will be required.

VISCOSITY

Viscosity and rheology may be unfamiliar fields of study, but are a familiar feature of daily life. Honey and oil have higher viscosity than milk or water. Jelly and ketchup are common liquids with complicated viscosity behavior. Viscosity, as the resistance to flow, is the dominant property in understanding processing. Incorporating particles also requires mixing, which can be a particular challenge for polymer melts.

Observations show that the relation governing simple flow operations, such as stirring, reflect a linear, proportional relationship between the force required, using, for example, a spinning rod and the resulting velocity of the liquid. The faster a liquid such as water is stirred, the harder the water is to stir. Near the stirring rod, the liquid is moving most rapidly, while the velocity of the liquid drops linearly with the distance from the stirring rod. This can be analyzed by considering a small element of the water, near to a moving plate, as illustrated in Figure 5.3. The type of flow created by a moving surface is called drag flow, a common component of many processing techniques. The analysis is general to an object moving through a liquid, the motion of liquid near a moving surface, or a liquid moving near a stationary wall. The only difference between geometries will prove to be the boundary conditions to the flow. The discussion that follows primarily is applicable to laminar flow conditions.

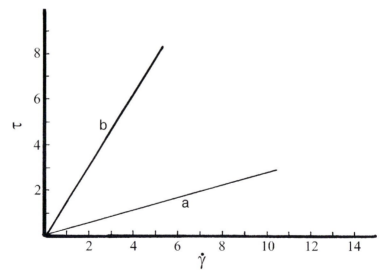

FIGURE 5.2 Newtonian liquids with, a, low or, b, high viscosity.

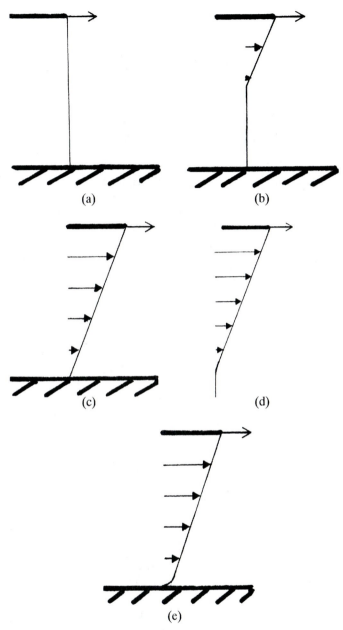

FIGURE 5.3 Plate moving at velocity, *v*, through a liquid. (a) Initially, the fluid is still. (b) After a short time, the surrounding liquid accelerates to match the velocity of the liquid. (c) The liquid near the stationary wall has zero velocity, so a stable velocity gradient is established. (d) In the absence of a stationary wall, the velocity profile will eventually, naturally drop to zero at some distance from the moving plate. (e) Slip at the wall will distort the velocity profile.

Examining the water "element" near to a moving surface, the liquid in contact with the surface should be moving at, or infinitesimally close to, the same velocity as the plate. At a stationary wall, the liquid element must be moving at, or infinitesimally close to, zero velocity. In the absence of a neighboring surface, however, far from the moving plate the liquid will remain undisturbed. A velocity gradient develops through the liquid layer, and a restraining force is exerted on the plate by the neighboring, more slowly moving liquid. If the force is removed, the liquid should stop flowing. In the drawing of Figure 5.3, the force is applied over the surface of the plate, resulting in a corresponding shear stress, τ. Provided the shear is not sufficiently high as to generate turbulent flow, a steady-state velocity gradient, Δv, which varies with distance away from the plate, Δy, will develop as shown in Figure 5.3. The force will thereafter maintain the velocity gradient:

$$F_s \propto \frac{\Delta v}{\Delta y} \tag{5.4}$$

The linear gradient depicted here is observed for many substances, and is the defining characteristic of a Newtonian liquid. Alternatively, the velocity gradient can be rewritten as $\Delta x/\Delta t$:

$$F_s \propto \frac{\left(\dfrac{\Delta x}{\Delta y}\right)}{\Delta t} \tag{5.5}$$

For infinitesimally small elements and displacements, the numerator is an expression of the shear strain, γ. Therefore, the time change in shear strain is the shear strain rate, $\dot{\gamma}$. This yields Newton's law of viscosity:

$$\tau = \eta\dot{\gamma} \tag{5.6}$$

The viscosity, η, is the constant of proportionality between shearing stress and the resulting deformation rate in a liquid. The viscosity is a property that depends on the interactions between molecules. The velocity gradient initially varies with time, but will establish a steady-state over a limited period, which is dependent on the viscosity. An alternative way to look at viscosity is as a contributor to energy loss:

$$\frac{F_s}{A}\frac{\left(\dfrac{dx}{dt}\right)}{dy} \propto \frac{dv}{dy}\left(\frac{dv}{dy}\right) \tag{5.7}$$

Ady is a differential volume and $F_s dx$ is an energy dissipated through shear. Therefore:

$$\frac{dE}{A} = \eta \left(\frac{dv}{dy} \right)^2 \tag{5.8}$$

where E is the energy per unit volume lost to viscosity, and dv/dy is the incremental velocity profile as a function of lateral distance from the plate.

At this point, the Navier-Stokes equation can be used to represent flow for the parallel plates being discussed to this point. In one dimension for an incompressible liquid, the equation is written as:

$$\rho \left(\frac{\partial v}{\partial t} + v \frac{dv}{dx} \right) = -\frac{\partial P}{\partial x} - \eta \frac{\partial^2 v}{\partial x^2} + \rho g \tag{5.9}$$

where ρ is the density, v is the velocity, g is the gravitational constant, and P is the pressure. This equation describes conservation of momentum. The terms on the left side of the equation describe changes to the momentum of the fluid, reflected in changes to the velocity profile with position and time. The right side describes forces applied to the fluid in the region of interest: pressure gradient, viscous loss, and gravitation. From conservation of matter, $\partial v / \partial x = 0$. Gravity does not act along the x direction. The flow is in a steady-state, so $\partial v / \partial t = 0$. Finally, since we are imposing a shear force and not a pressure gradient, the pressure term is also zero. For the displacement of two plates separated by a film of liquid, then, the conservation of momentum yields the expression:

$$0 = \eta \frac{\partial^2 v}{\partial x^2} \tag{5.10}$$

This has the solution:

$$v = C_1 x + C_2 \tag{5.11}$$

The boundary conditions are: at the stationary wall, $x = 0$ and the liquid velocity $v = 0$ to match; at the moving plate $x = h$, the thickness of the liquid layer, the liquid velocity $= V_{plate}$. Therefore, the velocity of the fluid as a function of position is:

$$v = \frac{Vx}{h} \tag{5.12}$$

The velocity profile is linear across the thickness of the liquid film for an incompressible Newtonian liquid. From the velocity as a function of position, the flow rate, Q, can be calculated from $\eta = \tau \dot{\gamma}$:

$$Q = \int_A v dA \qquad (5.13)$$

This integration yields a flow rate for drag flow of:

$$\frac{Q}{w} = \frac{Vh}{2} \qquad (5.14)$$

Here w is the width of the plate. This expression allows evaluation of the amount of liquid that will move as a result of displacing the plates.

Normal Forces

The discussion so far has focused on the shear and strain rate as scalar quantities. Flow may be three dimensional, or have a three-dimensional response. In order to consider three-dimensional flow, the three-dimensional tensor forms of the relevant equations must be considered. This is generally more detail than can be covered here. Nevertheless, there are several important cases for polymers and solutions of particles where non-linear effects may appear in even simple processes, including elongational flow and shear flow.

The stress tensor includes nine terms. The vector notation is given:

$$T = \begin{vmatrix} \tau_{11} & \tau_{12} & \tau_{13} \\ \tau_{21} & \tau_{22} & \tau_{23} \\ \tau_{31} & \tau_{32} & \tau_{33} \end{vmatrix} \qquad (5.15)$$

where:

$$\tau_{ij} = \eta \left(\frac{\partial v_i}{\partial x_j} + \frac{\partial v_j}{\partial x_i} \right) \qquad (5.16)$$

and:

$$\tau_{ii} = -P + 2\eta \frac{\partial v_i}{\partial x_i} \qquad (5.17)$$

Here P is the pressure. The stress tensor is typically symmetric, meaning that $\tau_{ij} = \tau_{ji}$. For a Newtonian liquid in the situation depicted in Figure 5.2, only τ_{12} and τ_{21} are nonzero. That is, flow in a Newtonian liquid is a pure shear state. By symmetry, the cross-shears, τ_{13} and τ_{23}, also must be zero even for non-Newtonian liquids. The normal stresses, τ_{11}, τ_{22} and τ_{33}, however, can be non-zero. The pressure is typically non-zero, as well. For a gas, the pressure can be calculated from the proper constitutive equation, such as:

$$P = \frac{nRT}{V} = \frac{\rho RT}{M} \qquad (5.18)$$

This is the ideal gas law, where ρ is the density and M is the molecular weight. Unfortunately, the density is not a strong function of pressure for incompressible liquids, polymers, or solutions, and does not extrapolate to zero. Therefore, the pressure component cannot be separated from the extra force component for many of the liquids of interest in composites processing. In rheology, a common approach is to consider the normal stress differences instead. There are two normal stress differences, which represent pressure components exerting a restoring force on a liquid in elongational or shear flow:

$$N_1(\dot{\gamma}) = \tau_{11} - \tau_{22} \qquad (5.19)$$

$$N_2(\dot{\gamma}) = \tau_{22} - \tau_{33} \qquad (5.20)$$

This results in three viscosity components defined in terms of the shear stress, and two normal stress components:

$$\eta(\dot{\gamma}) = \frac{\tau}{\dot{\gamma}} \qquad (5.21)$$

$$\Psi_1(\dot{\gamma}) = \frac{N_1}{\dot{\gamma}^2} \qquad (5.22)$$

and

$$\Psi_2(\dot{\gamma}) = \frac{N_2}{\dot{\gamma}^2} \qquad (5.23)$$

The normal stress coefficients serve a similar role in liquids to the Poisson effect in solids. The viscosity and first and second normal stress coefficients (Ψ) are recognized as potential functions of $\dot{\gamma}$. Consider the case of a positive first normal component (stress difference) during shear flow. The stress at the face of the advancing liquid, τ_{11}, is the downstream pressure. The compressive force in the y direction, which is holding the plate down, τ_{22}, must be compressive and larger than the downstream pressure, in order to give the positive component sought. Thus, there is a normal force exerted by the flowing liquid, and downward pressure must be applied to keep the plates from separating.

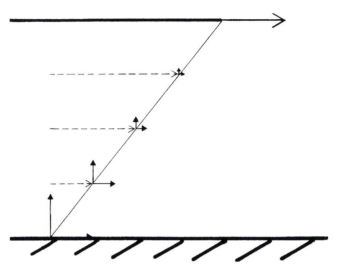

FIGURE 5.4 Shear forces caused by a moving plate also cause normal forces, exerting an upward force on the shearing plate.

Elongational Flow

Shearing flows have non-zero off-axis stresses. Zero shear flows, including elongational and biaxial flow, have no off-axis components. Elongational flow is the type of liquid motion occurring along the center-line in fiber spinning, extrusion with drawing, biaxial stretching flow, or planar flow. The velocity is defined in uniaxial elongational flow similarly to tension in an incompressible solid:

$$\underline{v} = \begin{pmatrix} -\dfrac{\dot{\varepsilon}(t)}{2}x \\[2ex] -\dfrac{\dot{\varepsilon}(t)}{2}y \\[2ex] \dot{\varepsilon}(t)z \end{pmatrix} \tag{5.24}$$

Putting this into the formula for a Newtonian liquid yields:

$$\underline{\tau} = \begin{pmatrix} \eta\dot{\varepsilon}_0 & 0 & 0 \\ 0 & \eta\dot{\varepsilon}_0 & 0 \\ 0 & 0 & -2\eta\dot{\varepsilon}_0 \end{pmatrix} \tag{5.25}$$

Following two particles along the z-axis, there is an expression for the separation between them, l, based on the definition of strain:

$$\frac{l}{l_0} = e^{\dot{\varepsilon}_0 t} \tag{5.26}$$

Like shear, the displacement between two particles will grow continuously. Unlike shear, where the particles separate linearly in time, in elongational flow the displacement is exponential in time. Due to this acceleration, shear-free flows are strongly deformed. Under shear-free flow, the polymer chains and particles in the flow stream may reorganize more readily. Blow molding, being a biaxial extensional flow, also exhibits this character, as does fiber drawing.

Viscosity Dependence on Molecular Weight (Part I)

Recall that viscosity also describes energy dissipation. One way to evaluate the loss of energy to friction is via an analysis of the motion of a molecule in a flow field. For non-spherical molecules, there will be part

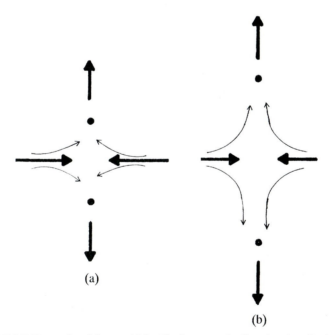

FIGURE 5.5 Elongational flow, exhibited in, for example, fiber drawing. Particle locations (a) initially and (b) some time later.

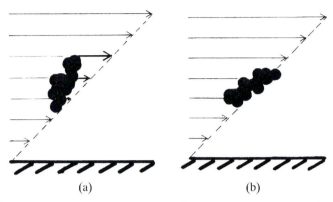

(a)   (b)

FIGURE 5.6 (a) A polymer chain in a flow field will experience different flow rates at different points along the chain. (b) The flow field will induce tumbling in the particle or chain, as different ends of the particle are differently accelerated, causing rotation.

of the molecule moving slower than the motion of the center of mass, and parts moving faster than the center of mass. As a result, some of the energy is put into tumbling instead of translation. The angular velocity under flow of either the polymer molecule or a suspended particle is described by:

$$\omega = \frac{\omega_x + \omega_y}{2} = \frac{1}{2}\frac{dv}{dy} \tag{5.27}$$

The average holds because for elements moving along the x-axis, the angular velocity minimum of zero, whereas along the y-axis, elements experience an angular velocity maximum of dv/dy:

$$\omega = \frac{\dot{\gamma}}{2} \tag{5.28}$$

Resistance to tumbling is supplied by the viscous force:

$$F_{viscous} = fv \tag{5.29}$$

where f is a friction factor and v is the velocity of the molecule, or a representational portion of it. For larger objects, the forces and interactions can be summed piece-wise. For polymers, the piece size is on the order of the mer, and even molecular liquids may be broken into segments. Thus:

$$F_{viscous,i} = \zeta v_i \tag{5.30}$$

Here, ζ is the segmental interaction energy, the energy of attraction between a molecule or mer unit of a chain and the surrounding elements.

The work done against friction per segment corresponds to {force x distance}. The velocity of individual elements can also be replaced with the appropriate form of the angular velocity. These modifications lead to a new equation:

$$F_{viscous,i} \cdot v_i = \zeta (\omega r_i)^2 \tag{5.31}$$

Further substitution, without derivation, leads to:

$$\sum_{i=1}^{n} \left(\frac{\Delta E_m}{\Delta t} \right)_i = \frac{1}{4} \zeta \left(\frac{\Delta v}{\Delta y} \right)^2 \sum_{i=1}^{n} (r_i)^2 \tag{5.32}$$

n is the number of segments in the rotating chain, and E_m is the energy loss per molecule. This mathematical description begins to close in on the meaning of viscosity, that is, the force required to overcome the energy lost to the internal resistance to flow. The summation over the pieces of the object must be completed, and $(r_i)^2$ substituted for the two-dimensional radius of gyration. The 2D radius of gyration for a polymer chain is related to the 3D radius of gyration by:

$$\frac{2}{3} r_{g,3D}^2 = r_{g,2D}^2 \tag{5.33}$$

So, for an individual molecule composed of n segments of length l, allowed free rotation:

$$\frac{\Delta E_m}{\Delta t} = \frac{1}{4} \frac{2}{3} \zeta \frac{n \bar{r}^2}{6} \left(\frac{\Delta v}{\Delta y} \right)^2 \tag{5.34}$$

In order to connect the energy loss per molecule to the energy loss per unit volume, the number of molecules per unit volume is needed. This is given by multiplying the equation by a factor converting the size of the chain to the space occupied by the chain:

$$N = \frac{\rho N_A}{M} \tag{5.35}$$

Here, as usual, ρ is the density, N_A is Avagadro's number, and M is the molecular weight. Considering that the molecule is a freely jointed chain comprised of n segments of length l_0 and mass M_0, $\bar{r}^2$ can also be replaced with nl^2. This gives the Debye molecular weight dependence of viscosity as:

$$\eta = \frac{l_0^2 \zeta \rho N_A}{36 M_0^2} M = KM \tag{5.36}$$

K is dependent on: material density; chain composition, through the monomer length and molecular weight; and the segmental friction factor, which includes interchain and chain-solvent interactions. The viscosity is observed to depend proportionally on the molecular weight of the molecule for small and moderate size objects, even for polymer chains below the "entanglement limit." High polymers behave somewhat differently, which we will get to in a moment. This relation allows determination of segmental friction factors from viscosity measurements. Alternatively, this may be related to the viscosity in viscoelastic behavior. The relation also holds for polymers in solution, below a critical concentration.

Viscosity Dependence on Particles

The addition of particles is expected to perturb the viscous response of a liquid. Particles can be large or small. Even at dilute concentrations of particles in solution, where the particles do not interact with each other, the viscosity is perturbed. The conditions required to be dilute are a function of particle size and shape, but are generally lower than 10 %vol. We may turn again to the energy dissipation aspect of viscosity, the energy lost to create a unit strain:

$$\frac{dv}{dy} = 1 \tag{5.37}$$

is

$$\frac{dE}{dt} = \eta_0 \tag{5.38}$$

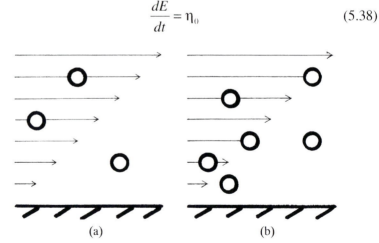

(a)   (b)

FIGURE 5.7 (a) Dilute particles in solution interrupt the flow according to the volume fraction. (b) In concentrated solutions, there may be occlusion between particles, resulting in a dramatic increase in flow disruption.

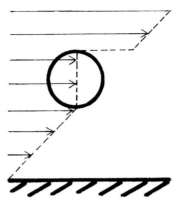

FIGURE 5.8 A particle acts as an interruption in the velocity profile of the surrounding solution.

where η_0 is the viscosity of the pure solvent. As illustrated in Figure 5.8, the particles serve as local holes in the developing velocity profile. Therefore, the mathematics is unchanged if the solution is re-partitioned such that all of the particles are packed together, leaving a clear, but thinner, channel of pure solvent. So, the energy loss to viscosity is modified by the volume fraction of particles present:

$$\frac{dE}{dt} = \eta = \frac{1}{1-\phi}\eta_0 \tag{5.39}$$

Here ϕ is the particle volume fraction, and η is the viscosity of the solution.

A suspension may have a different response to compression than a pure liquid. There may be an additional contribution to the energy loss through interaction of the particles. This is formulated in the Einstein viscosity equation for spheres in a dilute solution:

$$\eta = \frac{\eta_0}{1-\phi} + \frac{1.5\phi\eta_0}{(1-\phi)^2} \tag{5.40}$$

This equation can be rewritten using the series expansion:

$$\eta = \eta_0(1 + 2.5\phi + 4\phi^2 + 5.5\phi^3 \ldots) \tag{5.41}$$

Thus, the viscosity of a dilute solution of particles is perturbed from that of the pure solvent. This equation was derived for spherical particles, and other factors may influence the front factors for the volume fraction terms. This equation may be generalized to:

$$\eta = \eta_0 (A + B\phi + C\phi^2 + D\phi^3 \ldots)\qquad(5.42)$$

Where A, B, C and D are curve-fitting parameters. In the limit of very dilute solutions, that is small ϕ, the second-order and greater terms in ϕ become vanishingly small. Typically, the generalized equation is truncated with B or C, depending on particle volume fraction and shape. Simple mathematical rearrangements of this expression are shown in Table 5.1. Alternative forms of the viscosity have been used to evaluate and predict viscosity and molecular weight dependences in solutions where the second phase is a polymer chain. For a spherical particle, the intrinsic viscosity $[\eta]$ should be 2.5, but this may well be exceeded due to solvation or ellipticity of the particles, such as partially swollen montmorillonite particles or mica flakes.

As the concentration of particles increases above a certain threshold, the simple view of isolated perturbations to flow is no longer valid. In this regime, the particles can interact with each other, and the perturbations to the flow field can overlap. The concentration where this occurs depends on the distribution of particle sizes. Smaller spheres can pack into the spaces between larger spheres: 0.414× smaller will fit in the octahedral interstices; 0.225× smaller will fit in the tetrahedral interstices; 0.125× smaller can pass through even a closest-packed matrix of particles of the larger size. For elliptical particles, the maximum volume fraction is de-

TABLE 5.1 Derived viscosity numbers.

Base equation for solution of particles: $\eta = \eta_0(1 + 2.5\phi + \ldots)$.	
Derivative Expression	Name
$\dfrac{\eta}{\eta_0}$	Relative viscosity, η_r
$\dfrac{\eta}{\eta_0} - 1$	Specific Viscosity, η_{sp}
$\dfrac{1}{C_2}\left(\dfrac{\eta}{\eta_0} - 1\right)$	Reduced Viscosity, η_{red}
$\lim_{c_2 \to 0}\left[\dfrac{1}{C_2}\left(\dfrac{\eta}{\eta_0} - 1\right)\right]$	Intrinsic Viscosity, $[\eta]$
$\dfrac{1}{C_2}\ln\left(\dfrac{\eta}{\eta_0}\right)$	Inherent Viscosity, η_{inh}

Table from Hiemenz.

termined by the maximum that will still allow free rotation of the particles:

$$\phi_{max} \cong \frac{0.62}{p} \tag{5.43}$$

Elliptical particles will affect the viscosity differently than spheres will. Just as with the discussion of molecular weight, the flow around an elliptical particle can cause rotation of the particle, and an additional viscous loss. The loss depends on the axis of rotation: if the axis of rotation is the major axis, prolate, or the minor axis, oblate. The intrinsic viscosity will exceed 2.5 as a result.

Absorption of solvent into the particle will tend to swell the particle. The viscosity depends on the swollen diameter of the particle, not the dry diameter. The total effect of non-ideal contributions to the viscosity can be empirically accounted for by:

$$\frac{[\eta]}{2.5} = A + B\phi \tag{5.44}$$

where A is a combined parameter to account for the effect of solvation, ellipticity, and experimental error, while B accounts for particle overlap and other concentration-dependent effects.

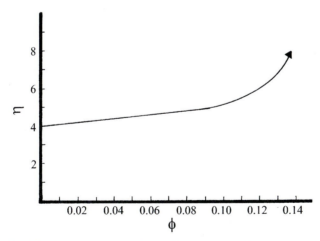

FIGURE 5.9 Viscosity is a function of the volume fraction of particles in solution. The rate of increase rises precipitously at the volume fraction corresponding to particle overlap. The initial viscosity as a function of volume fraction depends on the particle size, as well.

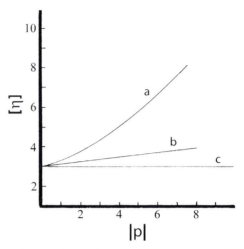

FIGURE 5.10 Intrinsic viscosity as a function of ellipticity, p: (a) rods, $p > 1$. (b) disks, p < 1. (c) Spherical particles have an intrinsic viscosity of 2.5.

EXPERIMENTAL VISCOSITY

Measuring the viscosity of liquids, polymer or solvent, is a matter of constructing well-characterized systems where the geometry of the plate moving through the liquid is known. The study of liquid viscosity is considered in the science of rheology. Some of the salient points of evaluating viscosity using cone and plate, concentric cylinder and capillary rheometers are considered.

Parallel Plates

Parallel plate and cone and plate rheometers (Figure 5.11) are among the simplest to construct and analyze. The shearing plates are similar to the moving plates discussed earlier. In cylindrical coordinates, the z component describes the pressure gradient due to gravity. The r component describes the pressure gradient caused by centrifugal force. The θ component gives:

$$\frac{d}{dr}\left[\frac{1}{r}\frac{d(r\tau)}{dr}\right] = 0 \tag{5.45}$$

The boundary conditions can be identified as: at $z = 0$, $v_\theta = 0$; and at $z = h$, $v_\theta = r\Omega$, where Ω is the rotational velocity of the spinning plate. Solving the velocity profile yields:

$$v_\theta = \frac{r\Omega z}{h} \tag{5.46}$$

where v_θ is the rotational velocity as a function of position, z, and h is the gap width of the rheometer. The velocity has a radial dependence, and so will $\dot{\gamma}$ and τ. The total torque, **T**, applied to the upper disk to maintain a constant $\dot{\gamma}$ is easier to measure than the local shear stress, but is related:

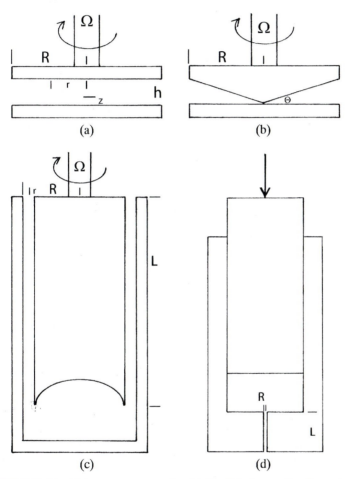

FIGURE 5.11 Liquid rheometer geometries: (a) Parallel plate, radius R, gap height h and rotational velocity ω. The radial distance is measured as r, and the height in the gap is measured as z. (b) Cone and plate, with angle θ, radius R, and rotational velocity Ω. (c) Couette, radius R, gap width measured as r, couette length L and rotational velocity Ω. (d) Capillary, with capillary radius R and capillary length L.

$$\mathbf{T} = \int SxdA \tag{5.47}$$

S is the stress applied at a distance x along the lever arm of the object. For the parallel plate rheometer in cylindrical coordinates this yields:

$$\mathbf{T} = \int_0^R -\tau_{z\theta}(r)(r)(2\pi r dr) \tag{5.48}$$

Using the viscosity relationship for a Newtonian liquid:

$$\mathbf{T} = 2\pi \int_0^R \eta \dot{\gamma} r^2 dr \tag{5.49}$$

And substituting variables:

$$\dot{\gamma} = \frac{\dot{\gamma}_R r}{R} \tag{5.50}$$

Here $\dot{\gamma}_R$ is the rim shear rate. The viscosity, as measured in a parallel plate rheometer of the type shown in Figure 5.11 is:

$$\eta(\dot{\gamma}_R) = \frac{\mathbf{T}}{2\pi R^3 \dot{\gamma}_R} \left[3 + \frac{d \ln\left(\dfrac{\mathbf{T}}{2\pi R^3}\right)}{d \ln \dot{\gamma}_R} \right] \tag{5.51}$$

To measure the viscosity at a single rim shear rate, data for:

$$\dot{\gamma}_R = R\Omega / h \;\; \text{vs.} \;\; \frac{\mathbf{T}}{2\pi R^3} \tag{5.52}$$

must be gathered and plotted as indicated.

This type of rheometer is best for liquids with intermediate viscosities. If the viscosity is too low, the turning plate may cause turbulent flow in the liquid. If the viscosity is too high, it becomes impractical to control the normal force necessary to spin the top plate. The viscosity calculated in this manner is a blend of the shear rates at all values of r. For solutions of particles, the presence of wall effects may change the material response.

Cone and Plate

Putting a small angle on the top plate is a non-intuitive but helpful approach to eliminating the radial dependence of the viscosity. The analysis is conducted in a manner similar to parallel plates, but in spherical coordinates with the origin at the center where the two plates make contact. Using new boundary conditions consistent with the geometry and coordinate system, the velocity profile becomes:

$$v_\phi = \frac{r\Omega}{\Theta}\left(\frac{\pi}{2} - \theta\right) \tag{5.53}$$

Where Θ is the small angle of the upper plate, θ is the angle with respect to vertical, and Ω is again the rotational velocity. The shear surfaces are surfaces of constant θ. The viscosity is:

$$\eta = \frac{3\mathbf{T}\Theta}{2\pi R^3 \Omega} \tag{5.54}$$

The viscosity, shear stress, and shear strain rate are constant at small cone angles, and the strain is homogeneous throughout the sample. This makes the cone and plate geometry effective for measuring the viscosity of suspensions, incompatible blends, and non-Newtonian liquids. The first normal stress coefficient can be shown to be:

$$\Psi_1 = \frac{3\mathbf{F}\Theta^2}{\pi R^2 \Omega^2} \tag{5.55}$$

where $\mathbf{F}$ is the total thrust applied to maintain the gap distance under rotation. The simple measures make cone and plate a popular measurement technique, but the limitations on the measurable viscosity regimes remain.

Concentric Cylinders (Couette)

Measuring very low viscosities benefits from using very large surface areas with a narrow gap to prevent turbulent flow. If instead of plates, the moving surfaces are concentric cylinders of large diameter, the simplicity of calculation is retained. For long cylinders, the end effects are negligible. Where the outer cylinder has a radius of R and the inner a radius of κR (for a gap of $1 - \kappa$), the velocity profile is again a function of radius:

$$v_\theta = \frac{\kappa\Omega(r - R)}{\kappa - 1} \tag{5.56}$$

The viscosity is measured by:

$$\eta = \frac{\mathbf{T}(\kappa - 1)}{2\pi R^2 L \kappa^3 \Omega} \tag{5.57}$$

Do It at Home 5.1

Use a round stick driven by a power drill. Stir various household liquids, particularly water, oil, ketchup, honey and a concentrated mixture of cornstarch in water.

Here, L is the length of the spinning bob. The same answer is obtained if the bob is held stationary and the cup spun with velocity Ω.

A simplified cylindrical rheometer is available in which the outer cylinder is removed. This changes one of the boundary conditions used to derive the equations above. A viscometer of this type is usually calibrated with Newtonian liquids of known viscosity.

Capillary

Higher-viscosity liquids and higher shear rates may be evaluated more easily using a capillary rheometer. Poiseuille considered the flow of a liquid through a capillary. A backpressure is applied to a reservoir of liquid to force flow through a long, narrow capillary. The viscosity is governed by the shear stress and shear strain rate at the wall ($r = R$):

$$\eta = \frac{\tau_R}{\dot{\gamma}_R} \tag{5.58}$$

The shear stress, in cylindrical coordinates, is:

$$\tau_{rz} = \frac{\Delta P r}{2L} \tag{5.59}$$

Therefore, the shear stress at the wall is:

$$\tau_R = \frac{\Delta P R}{2L} \tag{5.60}$$

The shear strain rate can be calculated from the velocity profile. For a Newtonian liquid:

$$v_z = \frac{2Q}{\pi R^2}\left[1 - \left(\frac{r}{R}\right)^2\right] \tag{5.61}$$

The shear strain rate is $-dv_z/dr$:

$$\dot{\gamma}_R = \frac{4Q}{\pi R^3} \tag{5.62}$$

Q is the flow rate. Therefore, the viscosity for a Newtonian liquid is evaluated as:

$$\eta = \frac{\pi \Delta P R^4}{8LQ} \tag{5.63}$$

Capillary rheometers are defined by the R/L parameter. The end effects near the entrance and exit dominate the behavior of flow through a short capillary. At the entrance, there are z variations in the velocity for some length through the die, due to the sudden change in channel diameter. At the exit, die swell or other normal forces can cause further deviations in the z velocity component. In a viscometer, end corrections must be applied to account for these deviations. Flow through a capillary at a constant shear rate, that is, a constant, fixed Q should always give the same wall shear stress. The Bagley correction uses this relation:

$$\Delta P = 2\tau_R \frac{L}{R} \tag{5.64}$$

At constant Q, a plot of ΔP vs. L/R should yield a line with a slope of $2\tau_R$ through the origin. An intercept in y or x offset from the origin, then, can be used to correct for end effects. The y offset can be subtracted from the pressure drop to remove the contribution of end effects at entrance and exit. The x offset can be added to the L/R value used in calculations.

Capillary flow through long channels, where $L \gg r$, can be used to measure the viscosity of incompressible liquids with certain caveats. First, since the shear stress and $\dot{\gamma}$ are evaluated at the wall, any liquid of interest must have representative flow properties at the wall. This may not be true for suspensions, inhomogeneous liquids, or liquids with time-dependent viscosities.

The velocity profile used above is only valid for a Newtonian liquid. The same profile is not generally obtained for more complex liquids, including polymers. For these, the velocity profile generally must be evaluated, and a new relationship for viscosity may be obtained. The Weissenberg-Rabinowitsch correction allows calculation of viscosity using no prior assumption of the velocity profile:

$$\eta(\dot{\gamma}_R) = \frac{4\tau_R}{L\dot{\gamma}_a}\left(3 + \frac{d\ln\dot{\gamma}_a}{d\ln\tau_R}\right) \tag{5.65}$$

$\dot{\gamma}_a$ is the apparent viscosity, the value of the wall shear strain rate if the liquid were Newtonian.

$$\dot{\gamma}_a = \frac{4Q}{\pi R^3} \qquad (5.66)$$

and

$$\tau_R = \frac{\Delta PR}{2L} \qquad (5.67)$$

The viscosity can be evaluated graphically from a plot of $\dot{\gamma}_a$ vs. τ_R. For some liquids, the no-slip boundary conditions defined at the very beginning may not hold. The velocity relationship is:

$$\dot{\gamma}_a = \frac{4v_{ave}}{R} \qquad (5.68)$$

When slip occurs, the standard value for $\dot{\gamma}_a$ over-predicts the shear strain rate:

$$\dot{\gamma}_a = \frac{4v_{ave}}{R} - \frac{4v_{slip}}{R} \qquad (5.69)$$

The postulated slip velocity, v_{slip}, can be evaluated from a plot of $\dot{\gamma}_a$ vs. $1/R$ at constant τ_R, with an intercept giving the slip-corrected velocity at the wall.

Dilute Solutions

Glass capillary rheometers are available to measure the viscosity of dilute solutions of particles or polymers in solution. The chief use of dilute solution viscosity for polymers is to measure the molecular weight of polymers. In the simplest-to-analyze device, the solution is drawn into a reservoir above the capillary. The viscosity is determined by the capillary viscosity equations, where the pressure drop is evaluated from the liquid column height:

$$P = \bar{h}\rho R \qquad (5.70)$$

Here, $\bar{h}$ is the average height of the liquid column. Poiseuille's equation for this situation is:

$$\eta = \frac{\pi R^4 \bar{h}\rho g t}{8Vl} \qquad (5.71)$$

where R is the capillary radius, V is the liquid volume, l is the capillary length, and t is the time to drain. This has the general form:

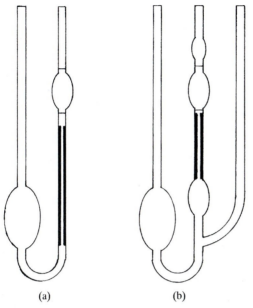

FIGURE 5.12 Dilute solution viscometers: (a) Ostwald and (b) Ubbelohde configurations. The Ubbelohde configuration allows free flow of the liquid through the capillary and simple serial dilutions. The Ostwald configuration has a differential pressure head that must be factored into the evaluation, and must be cleaned between each solution.

$$\eta = A\rho t \qquad (5.72)$$

A is a constant for the viscometer, and so a ratio of two viscosities gives:

$$\eta_1 = \frac{\rho_1}{\rho_2}\frac{\Delta t_1}{\Delta t_2}\eta_2 \qquad (5.73)$$

Solution 1 is usually chosen as the pure solvent, allowing calculation of the reduced viscosity directly from time measurements:

$$\eta_r = \frac{\rho\Delta t}{\rho_0\Delta t_0} \qquad (5.74)$$

The end effects for dilute solution viscometry are usually small.

Melt Flow Index

The melt flow index (MFI) is based on flow through a standardized, simplified capillary. A capillary of $D = 2.095$, an $L/D = 7.637$ is used. A fixed mass is applied to the ram, commonly 2160 kg. The mass extruded over a fixed time, 10 min, is the melt flow index. A high MFI has a low viscosity. The melt flow index is often used as an evaluation of process molecular weight. The melt flow index provides qualitative, comparative information between batches of the same material. Due to all of the limitations, the technique will not provide reliable data for drawing distinctions between different types of materials.

NON-NEWTONIAN FLOW

The viscous behavior of polymer melts and solutions often are observed to be non-linear. That is, many polymers exhibit non-Newtonian flow. Figure 5.13 shows three types of non-linear behavior, pseudoplastic, dilatant and Bingham Plastic. Shear thinning, or pseudoplastic, flow is nearly universal for polymer melts. Dilatancy, or shear thickening, is common for solutions of particles, which may be polymer particles in a theta-solution or describe solutions of nanoparticles in a polymer melt or a monomer/nanoparticle suspension. Bingham plastic behavior is familiar in gels and gums. In Bingham flow, there is no flow below some minimum shear force.

Flow may not be in a steady-state. Figure 5.14 shows some types of non-steady flow experiments. These experiments provide simple repre-

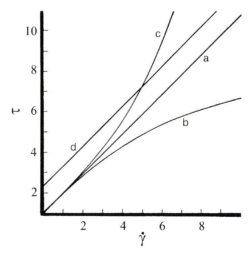

FIGURE 5.13 Different types of flow behavior: (a) Newtonian, (b) shear-thinning, or pseudoplastic, (c) shear thickening, or dilatant, (d) Bingham plastic.

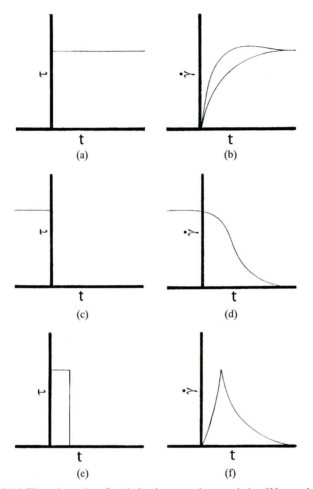

FIGURE 5.14 Time-dependent flow behaviors, not characteristic of Newtonian liquids: (a) under constant shear stress, (b) the shear rate may build slowly, and may even overshoot the steady state shear rate and recover. (c) At the cessation of shear stress, (d) the strain rate may not drop immediately to zero. (e) A single shear event may (f) result in a peak in shear strain rate.

sentations of common transient events, such as initiating or terminating flow, or flow near pipe fixtures. The flow experiences a step change, typically in strain rate, but also, possibly, in shear stress. For a Newtonian liquid, the response to changes in applied variables is instantaneous. For non-Newtonian liquids, the complementary variable can exhibit a time-dependent response that varies from the expected response.

Non-Newtonian liquids may exhibit memory of the previous flow conditions. Figure 5.15 shows some time-dependent viscous responses. Two types of time dependence are rheopexy and thixotropy. For a thixotropic material, the apparent viscosity is observed to decrease with time at constant shear rates. As flow continues, the molecules may become increas-

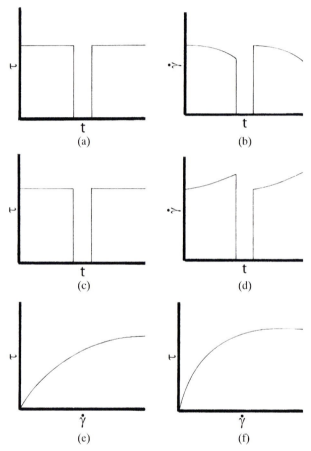

FIGURE 5.15 Memory effects in non-Newtonian liquids: (a) Thixotropic liquids under constant shear will exhibit (b) a steady decrease in shear strain rate. The initial behavior is restored if the shear stress is interrupted and then resumed. (c) Rheopectic liquids under constant shear will exhibit (d) a steady increase in shear strain rate. The initial behavior is again restored by an interruption in stress. The viscous behavior of a Newtonian liquid at two different temperatures typically can be superimposed through a shift in strain rate. A thermorheologically complex liquid will have a stress-strain rate response at two different temperatures (e) and (f) that cannot be superimposed.

ingly aligned and disengaged from neighbors. The resistance to flow is thereby reduced. The change in relation between shear stress and strain rate is reversible with time. Alternatively, a rheopectic fluid shows increasing viscosity with continued steady shear rates. The alignment of molecules during flow may induce order that will increase the interactions between chains, for example, by aligning hydrogen bonding or dipolar interactions, thus increasing the viscosity. Typically, the order is reversible with time. Here, again, temperature can have a significant effect on the viscosity changes shown by these liquids. Typically, temperature will disrupt molecular alignment.

Non-Newtonian liquids can experience unusual behavior, caused by the normal stresses created by time-dependent or position-dependent shear stress responses. When a Newtonian liquid is stirred, the liquid flows away from the rod. Polymer liquids flow toward the rod, resulting in rod climbing. Newtonian liquids tend to retract slightly upon exiting a die. Polymer liquids released from the pressure constraints of the die after extrusion tend to swell, a phenomenon called die swell. Liquids can be drawn through a tube using a vacuum, or siphoned. When the tube is removed from the liquid, siphoning will cease in a Newtonian liquid, but will continue in a polymeric liquid. Fiber spinning exemplifies the use of free surface elongational flow, and can occur in a polymer liquid but not in a Newtonian liquid. When a Newtonian liquid exceeds the conditions required for laminar flow, instabilities will form. Polymeric liquids can exhibit instabilities at conditions far short of Re = 2000. The result may be melt fracture, where the extrudate through a die is disrupted, or spurt flow, where the flow through the die surges, due to slipping at the edges of the die.

Pseudoplasticity

The most common behavior for polymers is shear thinning. For a shear-thinning liquid, the viscosity of the liquid appears to drop with increasing strain rate. That is, as the rate of displacement increases, the incremental shear stress required to increase the strain rate decreases. The decrease in resistance to flow is usually explained by cooperative motion between molecules in the liquid. Due to long range interactions along the chain, goes the reasoning, one chain cannot move without interacting with neighbors. Therefore, small strains require relatively large forces to cause motion, because even small displacements move disproportionately large amounts of the liquid. However, since small displacements have already initiated flow in neighboring molecules, larger displacements have a little head start and require relatively less force to impose.

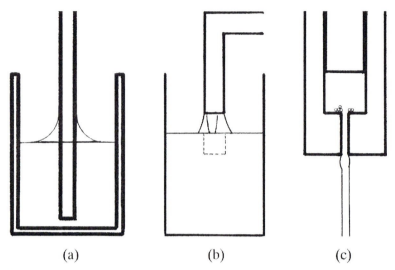

(a)   (b)   (c)

FIGURE 5.16 Non-Newtonian normal stress effects: (a) rod climbing, (b) siphoning, and (c) bubbles accumulate at edges.

Molecular weight is expected to play a significant role in pseudoplasticity. Temperature also will affect the interactions between chains. Pseudoplasticity also is one cause of normal stresses in polymer liquids. There is a non-linear shear strain profile as a function of distance from the agitating surface. This asymmetric profile results in positive normal forces acting toward the shearing surface, for example, a stirring rod.

Fitted Parameter Generalized Newtonian Fluid Models

There are a number of models for pseudoplastic flow. An empirical approach to correcting shear-thinning flow is the power-law fluid. The power-law fluid is an example of a Generalized Newtonian model, since it is designed to correct the steady-state viscosity to Newtonian behavior. The expression for the power-law fluid remains:

$$\tau = -\eta(\dot{\gamma})\dot{\gamma} \tag{5.75}$$

Here:

$$\eta(\dot{\gamma}) = m\dot{\gamma}^{n-1} \tag{5.76}$$

This has two fitted parameters, n and m, the consistency index. m has units of Pa·s^n, in units defined by the value of the exponent. n is dimensionless. These parameters are found by a plot of $\log(\eta)$ vs. $\log(\dot{\gamma})$, which should result in a line with a slope of $n-1$ and an intercept of m. A single set of parameters can usually give adequate predictions of shear, pressure drop, and other process conditions over two to three decades of strain rates. Notice that where $n = 1$ and $m = \lambda$, the Newtonian viscosity is recovered.

For a power-law liquid, the viscosity as measured by capillary viscometer is reflected by:

$$\log \dot{\gamma}_a = \frac{1}{n} \log \tau_R + \log \left(\frac{4m^{-\frac{1}{n}}}{\frac{1}{n} + 3} \right) \qquad (5.77)$$

The flow rate is governed by:

$$Q = \left(\frac{\Delta PR}{2mL} \right)^{\frac{1}{n}} \left(\frac{n\pi R^3}{1 + 3n} \right) \qquad (5.78)$$

The power-law behavior can predict the flow behavior over a range of viscosities.

However, at very small, and at very high, strain rates the viscosity is observed to follow limiting values, as illustrated in Figure 5.17. A fitted-parameter model that more closely recovers this behavior is the

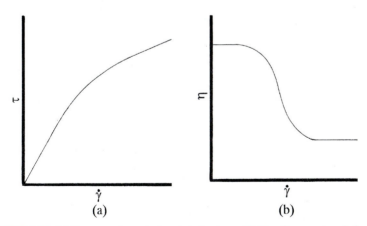

(a)   (b)

FIGURE 5.17 (a) Stress vs. shear strain rate behavior exhibiting Carreau-type behavior, with linear zero-shear and infinite-shear asymptotic behavior. (b) The corresponding viscosity behavior.

Box 5.1 Eyring Theory on Flow of Holes

As mentioned in Chapter 3, flow of material can be envisioned as flow of holes in the opposite direction. We consider tracking the holes because they are likely to be of an average size and resistance to their motion should be uniform, while for the molecules of a fluid, especially a polymer, these conditions may not hold true. In other words, describing the flow of holes is likely to be easier than describing the flow of molecules. The situation is shown in Figure 5.18.

Under equilibrium conditions, with no flow, there is no net energy gain regardless of where the hole is. On the other hand, there is a barrier to the movement of holes, as interactions between neighboring atoms and molecules will be broken and re-formed for the motion to occur. The rate of transport of the holes is governed by a general rate expression:

$$\text{Rate} = h[C] \tag{5.80}$$

Here $[C]$ is the concentration of holes and h is the rate constant. Using the standard form for the energy dependence of the rate constant, the probability of motion of the holes is:

$$h = Ae^{\left(\frac{-\Delta E^*}{kT}\right)} \tag{5.81}$$

This has a front factor A, an activation energy described by ΔE^*, and Boltzmann's constant k. As shown in Figure 5.19, if a pressure gradient is applied, for example using a shear force, then a bias forms for the holes to move upstream and the material downstream.

The applied energy is some amount F_d, where F is the force applied and d is the distance the hole moves. The energy to overcome the barrier to flow downstream lowers by some amount $F_d/2$, while the energy required to return to the original state is increased by an equivalent amount. The new rate expression is:

$$\text{Rate} = A\left[e^{\left(\frac{\Delta E^* - \frac{F_d}{2}}{kT}\right)} - e^{\left(\frac{\Delta E^* + \frac{F_d}{2}}{kT}\right)}\right][C] \tag{5.82}$$

The shear strain rate is $\partial v/\partial y$, which in terms of Figure 5.19 is:

$$\dot{\gamma} = \frac{d\,\text{Rate}}{\lambda} \tag{5.83}$$

λ is the distance away from the first hole required to lose the memory of displacement of the neighboring hole. This concept is the decorrelation distance. The hyperbolic sine can be used to replace the difference in exponentials in the rate expression:

$$\sinh(x) = \frac{1}{2}[\exp(x) - \exp(-x)] \tag{5.84}$$

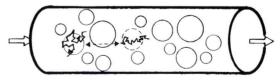

FIGURE 5.18 In flow, the polymer chain moving forward is equivalent to holes in the liquid flowing backward.

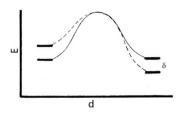

FIGURE 5.19 A pressure differential creates a bias in the energy for moving the hole through the liquid.

Box 5.1 (continued Eyring Theory on Flow of Holes

Since $\tau = F/A$, an expression for strain rate and shear can be obtained using the rate expression:

$$\dot{\gamma} = \frac{2dh}{\lambda} \sinh \frac{dA\tau}{2kT} \tag{5.85}$$

The more familiar arrangement of variables is obtained by rearrangement:

$$\tau = \frac{2kT\beta}{V_h} \left(\frac{\sinh^{-1} \beta\dot{\gamma}}{\beta\dot{\gamma}} \right) \dot{\gamma} \tag{5.86}$$

Here β is a collection of constants, $\beta = \lambda/2dh$ and V_h is the average volume of the holes. As $\dot{\gamma}$ approaches zero, a constant Newtonian viscosity is predicted:

$$\eta_0 = \frac{2kT\beta}{V_h} \tag{5.87}$$

As $\dot{\gamma}$ increases, the viscosity decreases.

The Eyring equation gives a description of a shear-thinning fluid that recovers many important features of the flow of long-chain molecules. The Eyring argument also has been applied to slip of holes at walls. In addition, it can be applied to the movement of dislocations during mechanical deformation. The relaxation time, τ_0, is related to the displacement of holes to relax internal stress within the solid, so $\tau_0 = 1/h \propto \beta$. Again, the viscosity can be connected to the mechanical behavior of the solid polymer. Among the conclusions that may be drawn from this equation are: the hyperbolic, $\sinh(x)$, form is likely to recur in many physical applications; and there may be multiple sources of mobile defects in a liquid, each with a different set of governing parameters.

TABLE 5.2 Hyperbolic identities.

Hyperbolic sine
$y = \sinh x$
$\sinh x = \frac{1}{2}(e^x - e^{-x})$
$x \to 0,\ \sinh x \to 0$
$x \to \infty,\ x \to \dfrac{e^{-x}}{2}$

Table from Hiemenz

205

Carreau-Yasuda model. The equation for the Carreau-Yasuda model has five parameters:

$$\frac{\eta(\dot{\gamma}) - \eta_\infty}{\eta_0 - \eta_\infty} = [1 + (\dot{\gamma}\lambda)^\alpha]^{\frac{n-1}{a}} \qquad (5.79)$$

The zero-shear viscosity, η_0, occurs at small strain rates. The infinite-shear viscosity, η_∞, holds for high $\dot{\gamma}$. The remaining parameters, n, λ and a, govern the behavior of the power law region of the viscous behavior. Not only is the Carreau-Yasuda model strictly empirical, it requires the simultaneous fitting of five parameters. The expression does not yield tractable analytical results, but is feasible for numerical solutions to pressure drop and shear calculations. As flow approaches Newtonian, $\eta_\infty \rightarrow \eta_0$, the right side of the equation becomes indeterminate.

Viscosity Dependence on Molecular Weight (Part II)

Earlier, the viscosity was found to depend on molecular weight. However, as the molecular weight increases, there is a range, at close to a molecular weight of $\approx 2,000$, where the dependence changes to the 3.4 power, as shown in Figure 5.20. The value of the exponent and the molec-

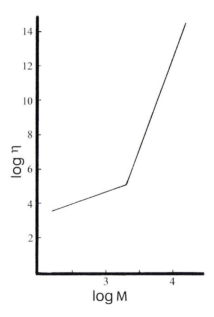

FIGURE 5.20 The viscosity is dependent on molecular weight to the 1st power at lower molecular weights, and on molecular weight to the 3.4^{th} power for longer chains.

Box 5.2 Reptation and Beuche

Entanglements describe a kinetic phenomenon that generally describes the temporary interactions between chains. This includes a physical barrier to translation, as well as potential hydrogen bonding, chain rigidity, and a host of factors that are also difficult to explicitly evaluate. One way of accounting for the interactions is described by the Bueche theory, in which the movement of a chain requires the force be transmitted to connected chains.

This theory accepts predominantly a physical barrier, but the nature of the interactions between chains is not definitive. The Bueche theory eventually leads to a molecular weight dependence of 3.5 power. However, the form of the equation does not yield the Debye viscosity at low molecular weights, and incorrectly predicts other flow-related phenomena.

A more successful theory has been the Reptation theory of deGennes. The fundamental picture here is that a molecule is trapped in a tube created by neighbors. The nature of interactions is included through the segmental friction factor, ζ, as previously defined. The tube confines the entire length of the chain, excepting the ends of the chain. The chain may slide, snake-like, along the contour length, L_t, or reptate. Any deformation will deform the location of segments in an affine manner, storing the stress in the new chain conformation. The molecule relieves the stress by reptating to achieve a relaxed conformation. The diffusion is governed by the diffusion constant, derived from Einstein's diffusion discussion:

$$D_t \approx \frac{kT}{n\zeta} \qquad (5.88)$$

ζ is the friction factor, n is the chain length or degree of polymerization, and k is Boltzmann's constant. The chain contour length determines the length of the

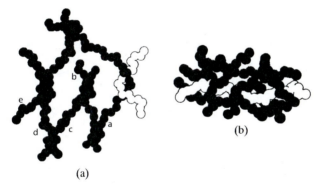

(a)

(b)

FIGURE 5.21 Models attempting to describe the molecular weight dependence of viscosity: (a) In the Bueche model, when a chain moves, the neighboring chains must also move. The first neighbor, a, experiences the strongest interaction. The interactions decrease as the neighbor distance increases from a–e. The network of chains thus exhibits cooperative flow. Note, as drawn, e. is both a 5th neighbor and a second neighbor interaction. (b) In the Reptation model, the chain is trapped in a tube created by neighboring chains. If the neighbors are short chains, the constraints are temporary, but for long polymer chains the tube is more or less permanent and flow can only occur along the chain length.

Box 5.2 (continued) Reptation and Beuche

tube, the distance necessary for travel to fully relax the memory of the applied stress:

$$L_t \approx n l_0 \tag{5.89}$$

l_0 is the monomer length. Diffusion is governed by the Einstein equation:

$$D = \frac{x^2}{2t} \tag{5.90}$$

The factor of 2 accounts for diffusion at either end of the chain. Substituting the appropriate definitions of L_t for x and rearranging, we come up with an expression for τ_n, the relaxation time for complete relaxation. The resulting diffusion time is:

$$\tau_n = \frac{L_t^2}{2D_t} \tag{5.91}$$

or

$$\tau_n = \frac{l^2 n^3 \zeta}{2kT} \tag{5.92}$$

Since $n = M/M_0$, this leads to the conclusion that the diffusion time is dependent upon the molecular weight to the third power. An expression for the viscosity can be drawn from the relaxation time using the expression previously defined at need in the continuum models, $\tau_n = G/\eta$. G is chosen over E in this application due to the use of a shear force, rather than extension:

$$\eta_0 = \frac{G l^2 \zeta}{2kT M_0^3} M^3 \tag{5.93}$$

This may also be arrived at via rubbery behavior with cross-links that break and re-form at equal rates, where $G = \nu kT$, where ν is the number of network chains. The zero-shear viscosity is obtained because the modulus as defined as superimposable for linear viscoelasticity. Shear thinning is not superimposable.

Self-diffusion describes the transport of the molecule in three-dimensions. During the time, τ_n, required to diffuse along the contour length, the entire chain will be observed to move a three-dimensional distance, r_g:

$$D = \frac{r_g^2}{2t} \tag{5.94}$$

so

$$D = D_t \frac{r_g^2}{L_t^2} \tag{5.95}$$

$r_g^2 = n l^2$, while $L_t^2 = n^2 l^2$, so the tube diffusion constant may be determined from the self-diffusion constant and the degree of polymerization.

---

ular weight at the change in behavior is remarkably consistent between polymer chemistries. The transition zone occurs over a small range of molecular weights. This may be due to polydispersity in the molecular weight distribution. The transition in behavior is generally ascribed to the onset of entanglements between chains. When a chain becomes long enough, the relaxation time competes with the flow rate.

Time Dependence

The previous corrections to the viscosity included power-law and Carreau-Yasuda. These were repairs to the viscosity to recover Newtonian flow. They do not include any time dependence, and so will be insufficient to describe a material with a high Deborah number. The introduction of reptation to explain the increased dependence on molecular weight above some entanglement limit re-emphasizes that polymers often will demonstrate some time-dependent viscous response. The deformation under flow will exhibit an initial response that will change as the liquid habituates to the stresses of flow. After the shear is removed, the original, equilibrium conformations will be recovered. The tendency to recovery introduces a potential restoring force, contributing an elastic-like component to flow. As with time-dependent mechanical deformation, a phenomenological model for the viscous response may provide useful empirical comparisons. Using a Maxwell model for the time-dependent relaxation yields an integrated form that includes memory of previous shear rates:

$$\tau = \int_{-\infty}^{t} \left[\frac{\eta_0}{\tau_0} \exp\left\{ \frac{-(t-t')}{\tau_0} \right\} \right] \dot{\gamma}(t') dt' \qquad (5.96)$$

Since t is the instantaneous time, t' is introduced as an integration time. η_0 is the zero shear viscosity, or the viscosity of the dashpot in the Maxwell element. τ_0 differs from τ_n in that the former is the relaxation time determined for the Maxwell element, and latter is the longest relaxation time. The variable forgetting function serves to weight the shear rate response. This function is a decreasing function of time. The longer ago the deformation occurred, the less memory is retained of it. The equivalent form for reptation is beyond the introductory purpose of this book.

Do It at Home 5.2

Carefully pour boiling hot oil into a new PET plastic juice bottle in a sink. The bottle will shrink, as the temperature increases the chain diffusion rate significantly. This allows the chains to flow back into the original conformation. Memory of the original configuration will fade with time, as the internal stress is relaxed away over time.

Dilatancy

Shear thickening behavior is characterized by an increase in viscosity as the shear strain rate is increased. Dilatant liquids will flow smoothly at low shears, but as the shear is increased, the shear strain rate will increase more slowly than is the case for a Newtonian liquid. An example of this type of behavior is particulates in suspension, such as quicksand. Polymer suspensions can also show this behavior. Dilatancy is often explained similarly to viscoelasticity. The molecules in a liquid have sufficient time to re-order under low shear. As the shear increases, the molecules are unable to relax sufficiently to easily respond to the force, and the force builds. Dilatant liquids can be modeled with a power-law or Carreau-Yasuda model, where the parameters are chosen to provide an increasing instead of decreasing viscosity as a function of strain rate.

Ideal Plastic Behavior

A final non-linear behavior is Bingham Plastic. In this type of flow behavior, the liquid will exhibit no flow under small shearing forces. After some threshold is passed, the liquid flows like a more traditional liquid. This corresponds with the ideal plastic behavior used in mechanics of materials texts. Examples of this type of behavior are toothpaste, gels, and gums. Such materials set up a temporary, force-sensitive network that can be overwhelmed under moderate shearing force. The network contributes no resistance to flow after it is disrupted, but may reform after the shear is removed. As a result, the behavior is solid at low shears and liquid at high shears or temperatures. This can be described by the piecewise function:

$$\eta(\dot{\gamma}) = \begin{cases} \infty & \tau \le \tau_y \\ \eta_0 + \dfrac{\tau_y}{\dot{\gamma}} & \tau > \tau_y \end{cases} \tag{5.97}$$

τ_y is the threshold shear. The η_0 can be replaced with a shear-thinning generalized fluid to give Bingham flow with pseudoplastic behavior.

Dynamic Shear

Many tests are conducted under small amplitude oscillatory shear. The applied $\dot{\gamma}$ has a periodic form:

$$\dot{\gamma}(t) = \dot{\gamma}_0 \cos \omega t \tag{5.98}$$

$\dot{\gamma}_0$ is the amplitude and ω is the frequency of the oscillation. The strain is 90° out of phase with the strain rate, or $\gamma_0 \sin\omega t$. The stress response to the strain will also be a sine wave with the same frequency as the input strain, but usually is not in phase with the strain:

$$\tau(t) = -\tau_0 \sin(\omega t + \delta) \qquad (5.99)$$

The new constant, δ, is the phase angle representing the lag in response between shear stress and applied strain. The shear can be rewritten using trigonometric identities:

$$-\tau(t) = (\tau_0 \cos\delta)\sin(\omega t) + (\tau_0 \sin\delta)\cos(\omega t) \qquad (5.100)$$

Written this way, there is a portion of the response that is in phase with the imposed strain and a portion that is out of phase. Again, the connection between the viscous behavior and viscoelastic response is observed. For a Newtonian fluid, the response is completely in phase with $\dot{\gamma}$. Thus $G' = 0$ and $G''/\omega = \eta$. For a purely elastic material, with no viscous response, the shear response is completely in phase with γ.

REFERENCES

Morrison, F., Understanding Rheology, Oxford University Press, New York, 2001.

Tadmor, Z. and C. Gogos, Principles of Polymer Processing, John Wiley & Sons, New York, 1979.

Hiemenz, P., Polymer Chemistry, Marcel Dekker, New York, 1984.

CHAPTER 6

Low-Viscosity Processing

Low-viscosity matrix materials have an advantage in manufacturing nanocomposites. The low-viscosity material can flow between the particles, creating a continuous matrix surrounding the reinforcements. In order to derive maximum benefit from the reinforcement, the matrix should make maximum contact with the particles. Further, there should be no residual porosity in the solid. There are several low-viscosity techniques for processing nanocomposites and composites, ranging from *in situ* polymerization of low-viscosity liquid monomers to dissolving the polymer in a solvent to lower the viscosity. Each of these approaches has governing physical principles. Understanding these principles is useful for: inspiration in nanocomposite invention; guiding product scaling from invention to a commercially viable product; troubleshooting structure and property considerations; and avoiding dead-end approaches to fabrication. This chapter focuses first on solvent processing, then on *in situ* polymerization.

Thermoplastic resins, as described earlier, can be melted, formed, and cooled to form devices with important performance characteristics and, often, complicated shapes. Thermoplastic resins such as nylon 6, acrylic, and polypropylene are commercially important resins that have been investigated as nanocomposite matrices in the laboratory. Traditional melt processes, such as extrusion, meet with processing limits on viscosity, beginning at about $5\%_{vol}$ nanoreinforcement. A simple technique for lowering the viscosity is adding a solvent. Appropriate solvents can range from supercritical CO_2 to exotic organic liquids. The requirements on the solvent are, generally, superior particle wetting and good polymer solubility.

Polymerizing directly in the presence of the reinforcing particles is another low-viscosity processing alternative. Some thermoplastic resins can be reasonably polymerized in a composite matrix. To qualify, the monomers must be liquid at the target processing conditions, and must also meet some of the requirements of a solvent for solvent processing, such as particle wetting. In particular, cross-linked and thermoset resins must be polymerized *in situ*, since they are intractable in melt or solvent processing. Fortunately, many commercially important polymers are available as liquid monomers, such as epoxies and urethanes, which flow readily and are compatible with numerous reinforcement materials. The two broad classes of polymerization, step-growth and chain-growth, are discussed here.

SOLVENT PROCESSING

There are many scales of particle behavior in solvent processing, ranging from the solvent molecule, through the polymer molecule, and up to the reinforcement. The polymer must be dissolved into solution, creating a suspension of particles in a solvent. To this is added the polymer, ceramic, or metallic-phase reinforcing particles of whatever size. The ability of the solvent to wet the particle and the motion of the particle in solution are both key phenomena to understand. The dissolution of the polymer, and subsequent precipitation and drying of the polymer to form a matrix, also are important concepts.

Particle Wetting

In order for a polymer to be reinforced by an inclusion, there must be sufficient interaction between the two materials to transfer the load between them. In solvent processing, the solution must also wet the reinforcements or the polymer will not adhere, forming a weak interface and poorer performance in the final part.

Despite the size, approaching atomic, of nanoparticles, standard wetting predictions have proven effective. Particle wetting is related to the contact angle between the liquid polymer and a substrate surface. The surface energies of both the substrate and polymer determine whether adhesion and wetting will occur. Indeed, these values make it possible to calculate the thermodynamic work of adhesion, W_A, theoretically. W_A is defined as the energy required to separate reversibly one unit area of interface to an infinite distance. With the work of adhesion and the surface tension of the liquid, the spreading coefficient, S, can also be calculated.

This coefficient predicts whether a liquid will totally wet a surface. Finally, with the polar and dispersion parts of the surface tension of the substrate and the equation of the work of adhesion, the wettability envelope of that particular substrate can be found. This envelope gives all the possible combinations of polar and dispersion components for the surface tension of a liquid that will, theoretically, wet that substrate.

Young Equation

Contact angle measurement is the most common technique of solid surface tension measurement. The equilibrium at the contact angle of a liquid on a surface can be considered the result of the mechanical equilibrium of a drop resting on a planar solid surface under the action of three surface tensions. The three surface tensions are: γ_{LV} at the interface of the liquid and vapor phases, γ_{SL} at the interface of the solid and the liquid, and γ_{SV} at the interface of the solid and vapor.

$$\gamma_{SV} = \gamma_{SL} \cos\theta + \pi_e \qquad (6.1)$$

The spreading pressure, π_e, represents the decrease of solid surface tension due to vapor adsorption. It is generally accepted that if the contact angle is larger than $10°$, the spreading pressure can be neglected.

Work of Adhesion, Spreading Coefficient, and Wetting Envelope

The thermodynamic work of adhesion (W_A) is the work required to separate a unit area of two phases in contact. W_A is related to surface free energies by the Dupré equation in dry air:

$$W_A = \gamma_1 + \gamma_2 - \gamma_{12} \qquad (6.2)$$

Where γ_1 and γ_2 are the surface tensions of the liquids and substrate respectively; and γ_{12} is the interfacial tension between the substrate and the liquid.

The surface energies of both phases are the sum of the dispersion (d) and polar (p) components. The former term includes dipole-dipole, dipole-induced dipole, and acid-base interactions.

The harmonic mean form of the work of adhesion expression best predicts the work of adhesion between low-energy materials such as two polymers. The harmonic mean form of Equation (6.2) is written as:

$$W_A = 4\left(\frac{\gamma_1^d \gamma_2^d}{\gamma_1^d + \gamma_2^d} + \frac{\gamma_1^p \gamma_2^p}{\gamma_1^p + \gamma_2^p}\right) \qquad (6.3)$$

Generally, the geometric mean form of the work of adhesion more accurately predicts the work of adhesion between a low-energy and a high-energy material such as a polymer and a metal:

$$W_A = 2\sqrt{\gamma_1^d \gamma_2^d} + 2\sqrt{\gamma_1^p \gamma_2^p} \qquad (6.4)$$

Equation (6.5) defines the spreading coefficient, S, for a liquid on a solid surface:

$$S = W_A - 2\gamma_1 \qquad (6.5)$$

γ_1 is the total surface tension of the liquid. If the spreading coefficient is greater than or equal to zero ($S \geq 0$), the liquid will spread spontaneously over the solid surface. Therefore, when the work of adhesion is twice the total surface tension of the liquid, the spreading coefficient will be equal to zero, and the liquid will wet the substrate. This coefficient predicts the way a liquid will act in contact with a material. To have good adhesion, a surface must be totally wetted by the liquid. The strength of adhesion between two materials will be greatest when the interface between the substrate and the liquid is as large as possible, creating more interactions between the two materials.

To see whether a given liquid will spread over the surface of a particular substrate, the wettability envelope can be used. The wettability envelope for a surface defines all of the possible combinations of liquid polar and dispersion components of surface tension that will cause the spreading coefficient to be equal to zero. To obtain the envelope, the values of the dispersion and polar parts of the surface tension of the substrate can be substituted into Equation (6.5), depending on the substrate. Consequently, all possible combinations can be calculated. It is then easy to determine whether a given liquid will spread spontaneously on the surface.

Particle Settling

The best composite behavior is obtained when there is a uniform distribution of reinforcement throughout the matrix. However, both the relatively heavy polymer chains and particles can settle out of solution. Particles larger than 1 µm will settle appreciably under the influence of gravity. On the other hand, for smaller particles and the polymer, Brownian motion becomes significant. Nanoparticles and polymer can be forced to sediment out of solution using centrifugation or by encouraging the flocculation of particles. Generally, the diffusion rate of the particles will affect the distribution of nanoparticles and polymer in solution.

When a large particle is suspended in solution, there are multiple forces acting on it (Figure 6.1). Buoyancy and viscous resistance to motion balance the force exerted by gravity. The terminal settling velocity of a particle is established very quickly, within seconds. The gravitational force is:

$$F_{gravity} = mg \tag{6.6}$$

For larger particles, the buoyancy force can be described by:

$$F_{bouy} = V\rho g \tag{6.7}$$

where V is the volume of the particle, which can be rewritten using the particle mass and density, which can be somewhat easier to measure:

$$F_{bouy} = \frac{m}{\rho_p}\rho g \tag{6.8}$$

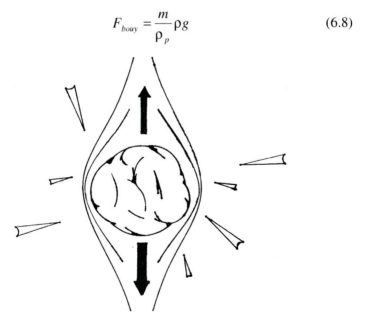

FIGURE 6.1 Forces acting on a particle include buoyancy, gravity, and viscous resistance from the surrounding liquid.

Viscous forces give rise to a drag resistance on the particle as it moves through solution. Stoke's law for spherical particles describes this resistance. Equation (6.9) gives Stoke's law at low Reynolds number, but there are modifications at higher Reynolds number, which should be applied as appropriate:

$$F_{viscous} = 6\pi\eta R v_s \qquad (6.9)$$

where η is the viscosity, R is the particle radius, and v_s is the solution velocity.

The stationary-state velocity of the particles in a still solution is a balance of these forces:

$$mg = \frac{m}{\rho_p}\rho g + 6\pi\eta R v_s \qquad (6.10)$$

The velocity of the particle through the solution can be rewritten as dx/dt, giving:

$$\frac{mg}{6\pi\eta R}\left(1 - \frac{\rho}{\rho_p}\right) = \frac{dx}{dt} \qquad (6.11)$$

or

$$\frac{dx}{dt} = \frac{g}{18\eta}(\rho_p - \rho)d^2 \qquad (6.12)$$

Equation (6.12) reflects that the settling velocity depends on the difference in density between the solvent and polymer, the square of particle diameter, and the inverse of the solvent viscosity. Sedimentation can also be affected by temperature. Since the solvent viscosity generally decreases as the temperature increases, the particles can settle faster in a warm liquid.

If the particles are not spherical, the unmodified Stoke's law equation is not applicable. This is usually overcome by replacing the real particle in calculations with an equivalent spherical particle. One approach is:

$$\frac{dx}{dt} = K_s \frac{mg}{6\pi\eta R}\left(1 - \frac{\rho}{\rho_p}\right) \qquad (6.13)$$

where

$$K_s = 0.843\log\left(\frac{\psi}{0.065}\right) \qquad (6.14)$$

ψ is the sphericity. The sphericity is the ratio of the surface area of a sphere having the same volume as the particle to the actual surface area of the particle. Disks show little deviation from spherical behavior, while cylinders can deviate more significantly. This result is particularly important for flake nanomaterials. These particles are expected to settle like spheres with a diameter equal to the diameter of the flake, not the thickness, and therefore much more rapidly than particulate reinforcements. Further deviation from this expression occurs at concentrations of particles exceeding about 10%. In this case, settling can be further hindered by interactions with other particles. In solutions consisting of identical spherical particles, an empirical approximation for the velocity is given by:

$$v = v_0 (1 - [P])^{4.65} \tag{6.15}$$

Where $[P]$ is the concentration of particles. This equation is less applicable for non-spherical particles, particles with a distribution of sizes, and if the concentration increases to the point of contact between particles. This form is also only relevant in the region at low Reynolds numbers, i.e., those below 0.3.

These equations assume that a continuum approach is valid, which is generally not true on very small size scales where diffusion will compete with settling. For smaller particles, the impact of molecules of a significant fraction of the size of the particles transfers energy, leading to Brownian motion. Einstein predicted that the particles would possess translation energy according to the temperature of the surrounding medium, $3/2kT$. Therefore, the average velocity of a particle can be evaluated from:

$$\frac{m\bar{v}^2}{2} = \frac{3}{2} kT \tag{6.16}$$

where $\bar{v}$ is the average particle velocity, k is Boltzmann's constant, T is the absolute temperature, and m is the mass of the particle. From the equipartition of energy, it is also possible to predict the number of particles with a given velocity using the Maxwell speed distribution:

$$N(v) = 4\pi N \left(\frac{m}{2\pi kT} \right)^{3/2} v^2 \exp\left(\frac{-mv}{2kT} \right) \tag{6.17}$$

N is the total number of particles in solution per unit volume, related to Avagadro's number, and v is the velocity of interest. Nanoparticles, below 100 nm, have a small mass, so diffusion due to the velocity of the particles may be significant. Einstein also posed this form of the diffusion constant. According to Einstein, the force exerted by gravity must be balanced by the osmotic pressure created by Brownian motion. The diffusion constant is related to the viscosity through the sedimentation velocity:

$$D = \frac{kT}{6\pi\eta R} \qquad (6.18)$$

where D is the diffusion constant. The reverse argument can also be made, allowing the calculation of the motion of particles under osmotic pressure, or a concentration gradient. This leads to Fick's law. If the rate of diffusion balances the sedimentation rate of the particles, an equilibrium gradient of particle density can arise:

$$n_{vol} = n_0 \exp\left[-\frac{N_A}{RT} V_{particle} (\rho_{particle} - \rho_{liquid}) gh \right] \qquad (6.19)$$

Note the absence of dynamic properties, including viscosity, from this equation. This equation describes a gradient in particle density that can be established in a manufactured part depending on particle size and density, solution density and part thickness. In nanocomposite fabrication, diffusion is seldom the dominant driving force in either polymer dissolution or particle distribution. The solution is stirred, resulting in a uniform distribution of mass. Sedimentation and diffusion begin at the cessation of stirring, and a sedimentation velocity, sedimentation equilibrium and diffusion can be used to determine the distribution of particles beyond this point.

Solubility

Actually, the first step to solution processing a polymeric nanocomposite is to create a solution of the polymer in an appropriate solvent. An appropriate solvent is one in which the polymer is soluble, with mixture properties that provide desirable processing characteristics. Typically, the solution should be concentrated in polymer, to minimize the amount of solvent that must be removed, and should be easily precipitated to fix the composite structure.

The stability of mixtures of the polymer is governed by the thermodynamics of solubility. The mixing of two materials, as any process, is governed by the free energy:

$$\Delta G = \Delta H - T\Delta S \qquad (6.20)$$

If the free energy change ΔG is negative, the process is spontaneous. Here, roughly speaking, there is an entropic factor from increasing the volume of space available to either component liquid during mixing. For ideal liquids, the volume increase is additive, but deviations in ideality often must be included. There is also an enthalpic contribution from the energy released by, or absorbed in, interactions between the molecular species during mixing. The entropy increase is always positive, however small, which makes solution spontaneous for ideal solutions, where enthalpy increases are zero and the volume of mixing is strictly additive. The enthalpy of mixing is usually positive, and so solubility is optimized, when the change in enthalpy is minimized. In those cases where the enthalpy of mixing is negative, that is exothermic, both terms are negative and mixing is always spontaneous. A negative enthalpy of mixing can occur in cases where there are specific interactions.

The entropy change, ΔS, can be determined through statistical thermodynamics. Assuming the liquid volumes are strictly additive, the enthalpy increase can be accounted for by the sum of the contributions from the respective volume increases for each component. The entropy can be calculated by counting all the different ways in which the two liquids can mix (Figure 6.2). However, for polymers, the volume increase is somewhat limited by the connectivity along the chain. By following the Flory-Huggins model of dissolution, for accounting for the connectivity during counting the different ways in which the solvents can mix, a simple modification using the volume fractions rather than the quantity, or mole fraction, of each component is revealed:

$$\Delta S_{mix} = R(x_1 \ln \phi_1 - x_2 \ln \phi_2) \qquad (6.21)$$

where x is the mole fraction and ϕ is the volume fraction of the two different components. The two components can be two solvents or a solvent and a polymer. In polymer nomenclature, the subscript '1' is always used for the solvent and '2' for the polymer. As stated above, the entropy is shown always to increase by combining two liquids with approximately similar molar volumes, and so in the absence of an enthalpy of mixing, two liquids will spontaneously mix. Polymer chains, however, experience less entropy increase during solution due to the chain connectivity, since individual elements have a more limited increase of available volume.

Solubility also has a strong enthalpic contribution. The Flory-Huggins theory continues by considering the average energies of interaction of any pair of species present, of the solvent with a piece of a polymer chain, of solvent with solvent, and of polymer segment with polymer segment.

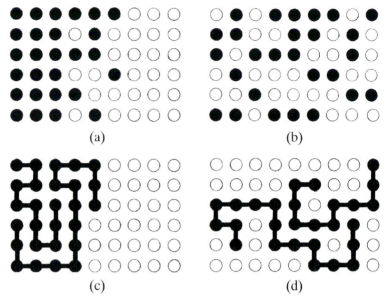

FIGURE 6.2 Examples of two liquids and a liquid plus a polymer mixing on a lattice. (a) Initial state of two unmixed liquids. (b) After mixing. (c) Initial state of a polymer in contact with a solvent. (d) After mixing.

The solvent-solvent interactions are determined by the energy of vaporization, the energy required to separate two molecules. The polymer-polymer and polymer-solvent interactions can be difficult to evaluate. The polymer-solvent interactions are estimated using the geometric mean of the interaction energies. Polymer-polymer interactions are sometimes estimated from small-molecule equivalents or by group approximations. In whatever manner they are obtained, by adding all of the interactions created in mixing the polymer with the solvent, the enthalpy of solution can be calculated. The theory introduces a new parameter, the Flory-Huggins interaction parameter χ:

$$\Delta H_{mix} = N\phi_1\phi_2\chi RT \qquad (6.22)$$

N is the number of molecules present, and all other parameters except χ should be familiar. The χ parameter is an estimate of the quality of mixing. χ is zero for athermal mixtures, in which there are neither positive nor negative interactions between polymer and solvent. The χ is negative for exothermic mixing and positive for endothermic mixing. The χ parameters for combinations of solvents and polymers can be determined by fitting, and some are tabulated in encyclopedias of polymer behavior. According to further developments of the theory, a χ of 1/2 will exactly balance the entropic contribution to mixing, leading to a Θ solvent or Θ conditions. In a Θ solvent, the polymer is at the edge of solubility and can precipitate out of solution under any influence, including increasing concentration or changing the temperature so as to raise the enthalpy of mixing. A χ of less than 1/2 is characteristic of a good solvent. The polymer will precipitate out of a poor solvent if χ is significantly larger than 1/2.

The enthalpy contribution to mixing is generally estimated, in the absence of specific energetic interactions between species, by equations from the work of Hildebrand and others:

$$\Delta H_{mix} = V_{mix}\left[\left(\frac{\Delta E_1^{vap}}{V_1}\right)^{1/2} - \left(\frac{\Delta E_2^{vap}}{V_2}\right)^{1/2}\right]\phi_1\phi_2 \qquad (6.23)$$

ΔE^{vap} is the internal energy required for vaporization, V is the molar volume, and V_{mix} is the volume of the mixed materials. The ratio of the heat of vaporization to the molar volume is called the cohesive energy density, CED. It is also convenient to define a new variable, the solubility parameter δ, which is $(CED)^{1/2}$. This is also called the Hildebrand solubility parameter. This definition is convenient since the energies are additive, and the solubilities are additive as squares. Note that the enthalpy of vaporization specifies the energy to sever the interactions between atoms and further energy to expand against the resistance of the surroundings. For the contribution to mixing, only the energy of vaporization is desired, making ΔE_{vap} more valid for the purpose than ΔH_{vap}.

The solubility parameter approach works best when the mixtures have similar sources for interactions, for example dipole-dipole or dispersive-dispersive. The equation also works poorly with exothermic mixtures. Attempts to generalize this approach, for example by Hansen, have led to the assignment of three components to δ, the first to dispersive forces, the second due to dipoles and the third to hydrogen bonding. The Hildebrand solubility parameter is related to the Hansen solubility parameters by:

$$\delta_2^2 = \delta_D^2 + \delta_P^2 + \delta_H^2 \qquad (6.24)$$

The components often can be determined for solvents by a variety of methods and again have been tabulated.

Solvent-polymer compatibility is maximized when δs match. The heat of vaporization is impossible to measure for polymers. The δ for high polymers must usually be inferred. If the polymer is lightly cross-linked,

TABLE 6.1 Solubility parameters for common solvents.

Solvent	δ (MPa$^{1/2}$)	Polymer	δ (MPa$^{1/2}$)
Hexane	14.9	Butadiene	14.6–17.6
Diethyl Ether	15.1	Polyethylene	15.8–16.9
Cyclohexane	16.8	Polystyrene	17.5–18.6
Carbon Tetrachloride	17.6	Polypropylene	18.8–19.2
Methyl Methacrylate	18.0	Polyethylene Oxide	20.2
Toluene	18.2	Polyvinyl chloride	20.3–21.4
Ethyl Acetate	18.6	Polylactic Acid	21.0
Tetrahydrofuran	18.6	Nylon 6	21.5
Benzene	18.8	Epoxy Resin	22.2
Chloroform	19.0	Polymethyl methacrylate	22.7
Styrene	19.0		
Methylene Chloride	19.8		
Acetone	20.3		
Acetic Acid	20.7		
Furfural	22.9		
Ethyl Alcohol	26.0		
Methanol	29.7		
Ethylene Glycol	29.9		
Ammonia	33.4		
Water	42.9		

Table derived from the Polymer Handbook.

the swelling in various solvents can be measured. The δ for the polymer can be taken as equal to the solvent causing the maximum swelling. Alternatively, the solution viscosity can be measured for a polymer in various solvents. The closest match in δ will expand the polymer the most, creating the maximum solution viscosity. Any size-related property, including swelling, solution viscosity, and permeability, can be used in similar fashion. Various measurement types have led to considerable uncertainty in reported values of polymer solubility parameters. Determining the individual contributions to solubility, ala the Hansen solubility, is particularly problematic.

Solubility parameters have been analyzed and broken down into group contributions. This approach uses the additive properties of the energies of vaporization. Each component of the molecule will contribute to the total energy of interaction. Dunkel proposed the first group contribution system based on the contributions to the enthalpies of vaporization:

$$\Delta H_{vap} = \sum \Delta h_i \qquad (6.25)$$

and

$$\delta = \left[\left(\sum \frac{\Delta h_i}{V} \right) - \frac{RT}{V} \right]^{1/2} \qquad (6.26)$$

The contribution from the volume expansion must be removed to calculate the solubility parameter.

The technique was refined by Small based on the additivity of the energies according to:

$$E = \frac{\left(\sum F \right)^2}{V} \qquad (6.27)$$

$$CED = \left(\frac{\sum F}{V} \right)^2 \qquad (6.28)$$

$$\delta = \frac{\sum F}{V} \qquad (6.29)$$

Therefore, the value of $(EV)^{1/2}$ can be calculated as the sum of individual fractional contributions. Group contributions are often used in molecular modeling packages.

TABLE 6.2 Dunkel group contributions to solubility parameters.

Group	Δh_i (cal/mol)
CH_3	1780
$=CH_2$	1780
$-CH_2-$	990
$=CH-$	990
CH	−380
O	1630
OH	7250
CO	4270
CHO	4700
$COOH$	8970
$COOCH_3$	5600
NH_2	3530
Cl	3400
F	2060

Table derived from Polymer Dissolution.

TABLE 6.3. Small attraction constants and polymer solubility.

*TABLE 6.3(a) Small's group interaction constants
for common functional groups.*

Group	F* (cal/cm³)	Group	F* (cal/cm³)	Group	F* (cal/cm³)
CH$_3$	214	C$_6$H$_5$–	735	Cl (mean)	260
CH$_2$	133	Conjugation	20–30	CF$_2$	150
CH	28	H	80–100	CF$_3$	274
C	–93	O (ethers)	70	S	225
=CH$_2$	190	CO (ketone)	275	SH	315
HC–C≡	285	COO (ester)	310		
–C≡C–	222	CN	410		

Table drawn from Polymer Dissolution.

*TABLE 6.3(b) Polymer solubility parameters calculated
using small contributions.*

Polymer	δ
Polytetrafluoroethylene	6.2
Polyisobutylene	7.7
Polybutadiene	8.4
Polystyrene	9.1
Polyvinyl Chloride	9.6
Polyacrylonitrile	12.8
Polymethyl Methacrylate	9.2

Table drawn from Polymer Dissolution.

The parameters for solubility should be related to the Flory χ parameter by the relation:

$$\chi = \frac{V_s}{RT(\delta_1 - \delta_2)} \qquad (6.30)$$

However, in experiments there is a 0.34 correction factor necessary to preserve the generality of the chi parameter. The relation has the general form of a line:

$$\chi = 0.34 + \frac{b}{T} \qquad (6.31)$$

The quantity 0.34 seems to be an entropic correction due to the free volume mismatch between a solvent molecule and a polymer segment, caused by the connectivity to neighboring segments in the latter. The *b* parameter represents the enthalpic contribution from the interactions between the solvent and segments of the polymer chain.

Not all polymers are soluble in all solvents at all concentrations at all temperatures. The solvent quality is related to the balance between the enthalpy, governed by the χ parameter, and the entropy. When the two contributions balance perfectly, the solution is at the so-called Θ conditions. A polymer will be dissolved readily in a good solvent, so that a large concentration of polymer can be dissolved quickly. On the other hand, a relatively poor Θ solvent can lead to easier processing. Small concentration increases through evaporation, or a temperature change, can cause the immediate precipitation of the polymer as a gel, stabilizing the device. This can prevent the mobility of reinforcing particles, for instance, leading to a uniform part.

Solvent quality does not depend exclusively on the solubility parameter. The solubility can also be changed by the presence of specific interactions, such as dipole interactions and hydrogen bonding. One three-dimensional graphing technique involves plotting the solubility parameter, dipole moment, and hydrogen bonding strength for various solvents. Once the available solvent space is well represented, the solubility of the polymer in these solvents is mapped as complete, partial, or insoluble, mapping out the entire solubility space of the polymer.

An alternative to the three-dimensional plots is a three-axis plot. Each corner of the plot represents a 100% contribution from dispersion, dipole, or hydrogen bonding forces. Solvent space is mapped according to mathematical relations derived using the assumption that all materials have the same total solubility value:

$$f_D = \frac{\delta_D}{(\delta_D + \delta_P + \delta_H)} \tag{6.32}$$

$$f_P = \frac{\delta_P}{(\delta_D + \delta_P + \delta_H)} \tag{6.33}$$

$$f_H = \frac{\delta_H}{(\delta_D + \delta_P + \delta_H)} \tag{6.34}$$

and

$$f_D + f_P + f_H = 1 \tag{6.35}$$

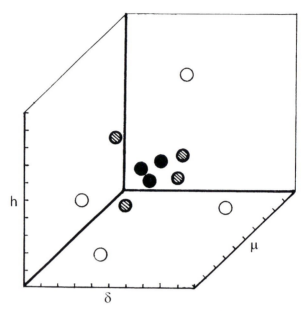

FIGURE 6.3 Schematic of a Crowley plot. Solvents are plotted based on three contributions: solubility, dipole moment, and hydrogen bonding strength. For a particular polymer, there will be a region of miscibility, a narrow surrounding region of marginal or partial solubility, and a large number of poor solvents.

The alkanes should be close to the dispersive apex. Water, which has a hydrogen bonding contribution as well as a dipole, will have a location with close to zero dispersion contribution. Again, the solvents that will dissolve the polymer are marked on the graph. One aspect of this approach is that mixtures of solvents can be made having specific positions on the graph based on additive, volumetric contributions to each component of the solubility parameters.

This plot can also be useful for predicting mixtures of non-solvents in which the polymer will be soluble. If a solvent will not dissolve the polymer because, for example, it does not have enough polar contribution, the addition of another non-solvent with a strong polar component can create a mixture that lies within the solubility window of the polymer. This works best if the non-solvents are close to the solubility envelope of the polymer.

Dissolution

Even when a polymer is soluble in a solvent, the dissolution process can be challenging. For example, creating an 8% solution of polyvinyl alcohol in water can require vigorous stirring at elevated temperatures for several hours. Polymers and small molecular solids tend to show different observable behavior during dissolution. When a molecular solid is soluble or partially soluble, the dissolved molecules will move directly, and rapidly, into solution. Molecules leave the solid to enter the solution, creating a diffusion layer with concentration decreasing with distance from the particle surface. Stirring the solution can accelerate the process, and may be necessary to complete dissolution. The dissolution rate of a stirred solution is controlled by mass transport, while diffusion controls the dissolution rate for a still solution.

The large size of polymer chains in solution modifies and slows the diffusion behavior at the dissolution boundary. Further, a chain must fully disentangle from neighboring material in order to enter the solution at all. There are generally believed to be three layers formed during polymer dissolution: solid polymer, solvated polymer, and dissolving polymer. The nature of this process probably is no different than for molecular solids. However, for molecular solids, mass transfer or diffusion is the sole rate determining feature of the process, while the kinetics of polymer chain motion can lead to slowing all phases of dissolution.

The solvated polymer, or gel phase, forms as the solvent infiltrates the polymer. The initial penetration occurs as the solvent occupies the free volume within the polymer, and is diffusion controlled. As the solvent

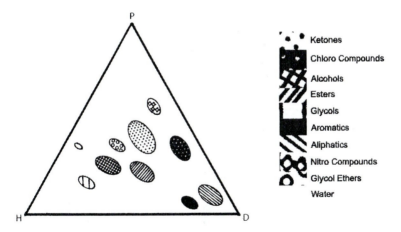

FIGURE 6.4 Schematic of the general locations of solvents on a Teas plot of solubility. The plot has axes consisting of dipole (P), dispersive (D), and hydrogen bonding (H) contributions to the solvent.

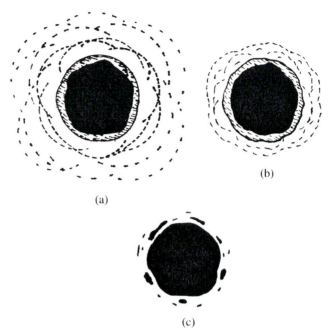

FIGURE 6.5 Schematic of a dissolving particle: (a) above T_g in a quiescent solution, (b) above T_g with stirring, and c) below T_g.

diffuses into the polymer, the free volume and mobility can increase, bringing an increase in polymer chain mobility. Proximity to the glass transition, previously described as corresponding to the ability of chains to move freely relative to each other, is a key feature of dissolution. When the material is at a temperature considerably below the glass transition temperature, the solvent will not diffuse readily into the polymer. In this case, a gel layer is usually not observed. The pressure on the solid exerted by the thermodynamic drive to mixing can interact with residual strains introduced during part fabrication. This combination of pressure and stress can lead the polymer to break up, sometimes causing eruptions of pieces of polymer into solution. This phenomenon is part of environmental stress cracking, and can contribute to some dissolution processes.

Above the glass transition temperature, sufficient free volume is intro-duced to allow solvent diffusion and the formation of a gel layer. The gel is a region where molecular motion is more liberal. In this region, under the influx of more diffusing solvent, the polymer can begin to release from the entangling forces of neighboring chains, exchanging these interactions for solvent interactions. The base time for this disengagement process is on the order of the reptation time, discussed previously. For self-diffusion of a pure polymer, the disentangling process is slow. In the limit where solvent diffusion is slow, dissolution will be controlled by the reptation time. As the solvent diffuses into the polymer, the localization of the polymer will be disrupted, and the friction factor will be modified by chain-solvent interactions. This allows enhanced dissolution of the polymers.

Once the chain is disengaged from neighbors, it can diffuse into the solution. In a still solution, this creates a diffusion gradient. Again, however, the overall diffusion velocity of the chain is controlled by mass and temperature, or more accurately by the energy of motion. Therefore, in a still solution, the concentration gradient can be very steep and diffusion is usually slow. The situation can be dramatically improved by agitation, such as stirring or flow through a pipe. The mass transfer rate of this process then dictates the dissolution rate. If the stirring rate is rapid enough, if the difference between the solvation and disengagement times is small, the gel layer can be stripped away as it is formed. This is the case with smaller molecules and generally for chains shorter than 2,000 carbons in length.

In quiescent solution, the polymer dissolution is diffusion controlled. Any feature of the system that increases mass transfer, such as decreasing molecular size or increasing solvent velocity with respect to the particle

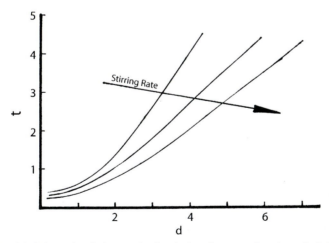

FIGURE 6.6 Schematic of changes in dissolution time as a function of stirring rate. Larger particles take longer to dissolve, but increased stirring decreases the dissolution time.

will increase the dissolution rate. However, there appears to be a particle size, approximately 20 μm, below which the dissolution time no longer decreases. This size increases with increasing agitation.

Solvent Removal

Polymer Behavior

Drying

The process of removing the solvent from a mixture of a dissolved polymer containing a reinforcing medium is complex. Several important issues include the process or processes governing the drying, the rate at which solvent can be removed under various conditions, and the path to achieving a film from which all of the solvent has been eliminated. In order to dry a composite film of any thickness, several common procedures may be employed. The liquid may be heated to raise the partial pressure and evaporate the solvent. An inert gas may be blown across the surface to remove the solvent. Finally, a vacuum may be drawn on the mixture to lower the partial pressure of the solvent above the liquid by lowering the pressure.

During the process typical of drying a solvated polymer/nano-composite, different stages of drying are likely to occur. In the initial stage, the polymer and particles are likely to be in dilute concentration. The particulates can settle by gravity, if large, or establish an equilibrium distribution if small. As the concentration increases, the polymer will begin to have a serious impact on the partial pressure of the solvent. Further, as the insoluble particles begin to occupy a significant portion of the available volume, the surface area can decrease. Eventually, the concentration will increase until either a solid bed of particles is created or the polymer solidifies. In the former case, evaporation of the solvent will show the features of a drying bed of solids. In the latter case, drying will become dominated by liquid diffusion through the polymer. The final stage of drying occurs when there is a quantity of solvent wherein the energy of specific interactions between the solvent and the polymer or ceramic particles exceed the energy available for drying. A solvent can be completely removed if and only if sufficient heat can be transferred to the drying film without damaging the polymer. The polymer or reinforcements also may have an equilibrium solvent content, below which the system cannot be dried further without some improved driving force.

The partial pressure of solvent above the liquid controls the drying. A purge gas is usually used to control the quantity of solvent in the atmo-

Box 6.1 A psychrometric chart usually uses the "dry" and "wet" bulb temperatures for water.

The dry bulb temperature is the traditional temperature, while the wet-bulb temperature measures the temperature in contact with a thin film of evaporating liquid. The traditional psychrometric chart is for water. Where a chart is developed for a solvent, such as toluene, lines of constant wet-bulb temperature are given to allow estimating quantities such as relative humidity of solvent, latent heat of evaporation, specific volume and other quantities useful to predicting and designing drying strategies or equipment.

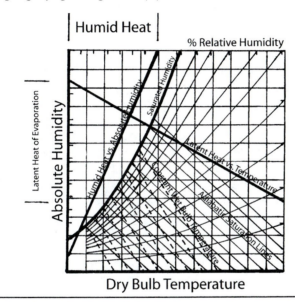

FIGURE 6.7 Sketch of a psychrometric chart for a solvent in air. The psychrometric chart provides a compact, systematic visualization of the relationships between: the temperature—dry bulb temperature; the amount of liquid vapor—absolute humidity; the saturated vapor temperature—wet bulb temperature; the relative humidity; the enthalpy of the air and solvent; and the latent heat of vaporization of the solvent. Psychrometric charts may also present data on volume changes due to solvent vapor.

sphere above the solution. The incoming gas will have little or no solvent, while the outgoing gas will have a measurably greater concentration. There are calculations or tables available for the latent heat of vaporization, vapor content, and specific volume of gasses at various temperatures and the other parameters necessary for calculating the mass flow rate of solvent removal above a liquid. These data are systematically plotted in psychrometric charts, which can be used to determine the drying rate graphically. Psychrometric charts are commonly available for water in air, but the equivalent diagrams can be made for other solvents using tables or calculations for latent heat of vaporization, vapor content, and specific volume for that solvent. The difference in solvent content between the inlet and outlet air can be read from the psychrometric chart to determine the mass removal rate. The purge gas flow rate will control how much solvent the exiting air will have. Psychrometric charts usually also contain enthalpy data useful for estimating the heat required to remove a given amount of solvent. Chemical engineering modeling packages for liquid-vapor equilibrium that will perform such calculations are also available. Care must be taken to account for the depression of the vapor pressure by the polymer in the solution. This is not a constant factor, rather the vapor pressure decreases as the mole fraction of polymer increases.

During the earliest part of the drying process, the warming up period, the mass removal is relatively slow. After this initial period, a constant rate period is entered, where the surface of the liquid is saturated and evaporation is controlled by diffusion through a stagnant vapor layer. If either the polymer solidifies or the level of the liquid falls below the surface of packed reinforcing particles, the solvent must travel to the surface in order to evaporate. At this point, the critical solvent content, the drying will enter the falling rate period. Here, the evaporation rate is controlled by other mechanisms, depending on the nature of the surface. If the polymer solidifies, the solvent must diffuse to the surface and then into the carrier gas. If the reinforcing particles pack together before the polymer solidifies, and the evaporating solvent front enters an effectively packed bed of solids, then drying can proceed by vapor diffusion, liquid diffusion, or capillary action. However, if this occurs there will almost certainly be significant porosity, leading to a part with poor performance.

In other words, the drying rate from the mixture should begin with a relatively fast, linear portion, where the liquid evaporates directly from the surface, preferably through a turbulent gas layer. Once enough liquid has evaporated so that liquid can only reach the surface by diffusion through the polymer, the rate should have a different, linear rate. Again, evaporation should be controlled at the surface side through turbulent flow. If the evaporation rate reaches a non-linear regime, this will usually indicate that the drying front will pass into a drying solids bed. This can lead to porosity.

Applying a vacuum lowers the total pressure, and therefore the partial pressure of the solvent, above the solution. The evaporation will follow the same steps as described previously, but the enthalpy of vaporization is lowered and drying can occur more rapidly and, sometimes, more thoroughly.

Precipitation and Phase Separation

During the solvent removal phase, the polymer concentration in solution will change dramatically. Increases in concentration can lead to important changes in the solution that may be important to the processing and performance. The polymer can precipitate out of solution, or can phase separate. As usual, the key to understanding when either of these events occurs is the free energy.

The complete equation for free energy actually includes a term for the chemical potential:

$$\Delta G = \Delta H - T\Delta S + \mu_1 \Delta n_1 + \mu_2 \Delta n_2 \qquad (6.36)$$

where μ is the chemical potential and n is the number of moles of each component. Therefore, the corresponding Maxwell relation is:

$$\mu_i = \frac{\partial G}{\partial n_i} \qquad (6.37)$$

The chemical potential is the rate of change in free energy with the increase in concentration of the polymer. Either precipitation or phase separation will change the nature of solvent processing. The chemical potential will feature prominently in evaluating when and in what form the polymer will leave solution.

The free energy of mixing can have different shapes depending on the nature of ΔH and ΔS. If ΔH is negative and ΔS is positive, the polymer is miscible at all compositions, as shown in Figure 6.8(b). A tangent can be drawn at any composition, and the intersection with the pure composition of each component provides information on the change in chemical potential for each species. If ΔH is very positive, however, the enthalpy can overwhelm the entropy at all compositions. Since the entropy term is temperature dependent, as temperature increases the miscibility can increase, giving rise to the second shape shown in Figure 6.8(c). At low and high concentrations, a unique tangent can be obtained, but at intermediate concentrations where the line fluctuates, there is a single, mutual tangent to the local minima in free energy. In the range of compositions between these points, the free energy can be improved by segregating into two solutions with the compositions described by the local minima. During the drying of a dilute solution, the concentration may increase into a similar thermodynamic condition, resulting in phase segregation.

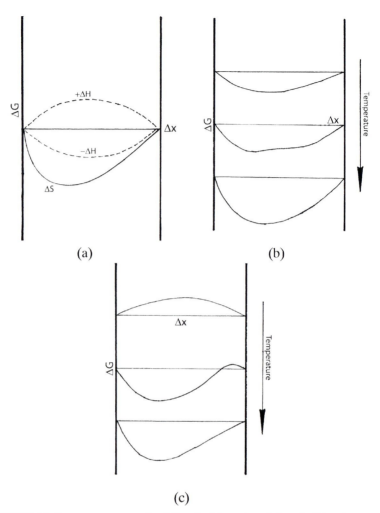

FIGURE 6.8 Free energy curves for fully miscible and partially miscible, phase-separating solutions. (a) The free energy at any composition is a combination of the entropy, which is always negative, and the enthalpy that can have any value, positive or negative, depending on the interaction between the two liquids. (b) If both the entropy and enthalpy contributions are negative, the combined, free energy curve will always be negative and the solution will always be miscible. At lower temperatures, the enthalpy will dominate miscibility. As the temperature increases, the entropic contribution to miscibility will increase, until at relatively high temperatures the entropy contribution to free energy will become dominant. (c) If the enthalpy is positive, the solution will not be miscible at all temperatures. At low temperatures, the solution will be immiscible at all compositions. As temperature increases, the entropic contribution will increase, and miscibility will become spontaneous at some compositions. At relatively high temperatures, entropy will drive miscibility at all temperatures, unless one or another of the liquids changes phase in the given temperature range.

Cooling the solution can also force phase separation. The nature of the two phases warrants further exploration.

The complete Flory Huggins theory provides an equation for free energy with substitutions for enthalpy and entropy of mixing according to:

$$\Delta G_m = RT[N\phi_1\phi_2\chi - (x_1 \ln\phi_1 + x_2 \ln\phi_2)] \qquad (6.38)$$

Recognizing that the number of solvent molecules $N_1 = V\phi_1$ and that:

$$\phi_2 = \frac{nN_2}{N_1 + nN_2} \qquad (6.39)$$

where n is the degree of polymerization of the polymer chain, Equation (6.38) for ΔG_m can be differentiated with respect to either n_1 or n_2 to give:

$$\mu_1 - \mu_1^0 = RT\left[\ln(1-\phi_2) + \phi_2\left(1 - \frac{1}{n}\right) + \chi\phi_2^2\right] \qquad (6.40)$$

or

$$\mu_2 - \mu_2^0 = RT[\ln\phi_2 + (1-\phi_2)(1-n) + \chi n(1-\phi_2)^2] \qquad (6.41)$$

These equations describe the tangents of the free energy curve. If the enthalpy is negative, these lines are identical. If the free energy curve fluctuates, then there may be circumstances at which the lines are not identical. As the temperature rises, the local minimum and maximum and the point of inflection between will all merge. This point is identified when the second and third differential with respect to composition equate to zero. The condition of this merged state is called the critical point, similar to the behavior of ideal gasses. A critical χ and solvent volume fraction can be determined in this manner:

$$\phi_{1,c} = \frac{1}{1 + n^{-1/2}} \qquad (6.42)$$

and

$$\chi_c = \frac{1}{2}(1 + n^{-1/2})^2 \qquad (6.43)$$

These are dependent on molecular weight. Therefore, the higher the molecular weight, the more solvent is required to completely dissolve the polymer. For an ideal, infinite chain, the χ parameter is 0.5. At n values of 10^4, 10^3 and 10^2, the χ_c is 0.510, 0.532 and 0.605, respectively. This can be restated that the higher the molecular weight, the less deviation from the ideal χ parameter, and the less soluble the polymer will be in a partially favorable solvent, slightly above or below the Θ conditions. In a system with a balance between ΔH and ΔS, the solubility may be improved by heating, since the entropy contributes to ΔG as $T\Delta S$. During solvent removal, the critical conditions eventually will be passed, resulting in polymer segregation and gel formation.

The implication of miscibility at low concentrations of solvents is different for a polymer and solvent pair than for two solvents. For two liquids, the solution is liquid at all compositions, even if two immiscible phases form. When one component is a solid, the tactile nature of the solution can change from a polymer in solution, to a gel, to a swollen polymer, to a plasticized polymer. Another implication is that the higher molecular weight polymer will separate out first, into the solution with the smallest volume. If a polymer solution phase separates, one component will have a higher molecular weight than the other, probably forming a precipitate. This precipitate retains some solvent, often resulting in gel-like behavior. Successive phase separation can result in stratified layers possessing a gradation in molecular weight, especially if solvent is removed very slowly or in stages.

PARTICLE BEHAVIOR

Packing

Reinforcing particles in solution often flocculate to minimize surface energy. Especially very small particles have a very large surface area, as well as a drive to reduce this area. The balancing forces are the kinetic energy of diffusion and the reduction in surface energy by the proximity of multiple particles. The surface energy of a spherical particle is the surface area times the surface energy of the particle:

$$E_s = \gamma_s \pi d^2 \tag{6.44}$$

The surface energy reduction caused by two particles interacting depends on the amount of the surfaces that can interact:

$$E_s = 2\left(\gamma_s \pi d^2 - \frac{1}{2}k\right) \tag{6.45}$$

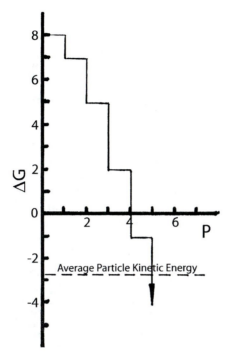

FIGURE 6.9 Schematic of the change in energy as the number of particles in a flock increases. Each particle interacting with the flock will result in a decrease in the free energy of free surface. When the flock becomes large enough, the flock will become stable, even to collisions with other particles.

where k is the percent of the particle surface that is occluded by a neighboring particle. If a third particle adds to the group, it will contact both of the previous particles, and the energy advantage per particle increases. The fourth particle contacts all three of the previous particles with a concurrent advantage. The total energy advantage depends on the number of new interactions within the flock. The size or distribution of sizes and shapes of the flocculating particles determines the energy gain. The interaction between any two particles is temporary, and can reverse at any time under the impact of solvent molecules.

Surfactants cover the surface of particles. In this way, the particle surface energy is replaced by the lower surface energy of the soap. The kinetic energy of diffusing particles can then be sufficient to break apart the agglomerates where the lower surface energy provides less drive toward interaction. A surfactant can disrupt group formation and prevent flocculation.

As the solution evaporates, several factors can influence the flocculation process further. First, solvent evaporation is usually endothermic, and the temperature of the solution will drop. This causes a drop in the diffusion velocity and can increase flocculation. Second, as the solution evaporates, the space available to the diffusing particles is decreased. This can have a localizing effect on the particles that can increase flocculation.

An additional influence on flocculation is the electronic state in solution. If the individual particles are all similarly charged, positively or negatively, the charge will stabilize the suspension and encourage solvation. There are no stabilizing influences on neutral particles, and so such particles will flocculate under normal conditions. Particles that typically pick up a surface charge, such as oxides, usually exhibit an isoelectric point in solution. The isoelectric point is achieved when the pH is adjusted to balance the native electronic charge formed. At the isoelectric point the neutralized particles will flocculate normally.

The surface area of non-spherical particles can lead to significant complications, and solutions of irregularly shaped particles are usually less stable. Fibers have a long surface, and can interact with the same or multiple fibers along this length. Any individual interaction is likely to be weak, and a surfactant can lower this energy still further. However, a second contact can maintain the fibers in proximity, even when one contact is broken. With increased proximity, the chances of a new interaction

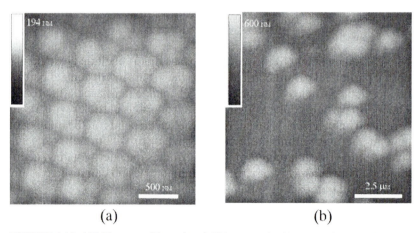

(a)   (b)

FIGURE 6.10 AFM images of flocculated, 300 nm nominal latex particles. (a) Tetramer structures represent a relatively stable structure. The regularity of these flocks still permits regular packing. (b) Flocks with less stability merge to form larger, irregular structures. The packing of these particles is far less regular.

forming are higher than normal. Therefore, the chances of a fiber long enough to form more than one contact breaking free from neighboring fibers is reduced. Once two fibers have come into multiple contacts, the clump formed has two important features. First, and most important, since the size of the combined fibers is larger than that of the single fiber, the clump is more likely to interact with new fibers, starting a tendency to form larger clumps. Second, the energy of two fibers interacting is maximized when they interact along the entire length. This tends to lead to fiber alignment.

The formation of flocculants has a lot in common with polymer chain crystallization. There will be a temperature above which, or concentration below which, there is too much energy and not enough time in proximity to allow flocculation. Under these conditions, too much motion occurs to allow the formation of stable groupings. At or below this transition temperature, stable interactions, however temporary, can form. These groups can cause catastrophic growth of flocs, especially with nanoparticles that can be diffusing with about the same velocity as the liquid molecules. Near the transition conditions, the particles within a stable floc will exhibit considerable relative motion, and will continue to move about within energy constraints until a minimal energy state is reached. Any packing arrangement that does not have the greatest number of interactions will introduce energy and create a packing flaw. As the floc gets larger, the impact, and pressure, applied by exterior particles to the interior of the floc will also encourage more elimination of packing defects. For particulates, this leads to a maximum number of neighbors; for fibers this will lead to a liquid-crystalline, parallel alignment of the axes of the fibers; and for plates, it will lead to a cholesteric, parallel alignment of the planes of the plates.

Disrupting flocs requires the opposite conditions. Plates, fibers, and particles may be split apart by the action of Brownian motion at the edges of the solid. A surfactant can accelerate this process by lowering the surface energy of individual particles. For fibers and plates, a surfactant may interact with, and penetrate, the solid without eliminating enough interactions to cause the individual particles to separate. The average distance between elements of the floc will get larger, and relative positions of individual components elements will become less certain. However, if enough interaction is preserved, the solvent will still be unable to penetrate and disrupt the floc structure and dissolve the solid. Dissolution is also better in a favorable solvent, when the solvent is present in excess, in a superheated solvent. That is, dissolution will be much more rapid if the liquid is well above the temperature required to disrupt the floc structure completely, and will be slower nearer to this temperature.

Flocculation can have a particular impact on drying suspensions. The impact depends on the nature of the drying. If the concentration of particulates is high enough to allow contact between the particles, regular or irregular particle shapes can affect the quality of the distribution. Flocks can be less regular in shape, resulting in irregular particle packing in the drying suspension. Soft particles, such as the polymer precipitating out of solution, with particle irregularity will tend to even out during the fusion process between the particles. Small particles fuse faster than

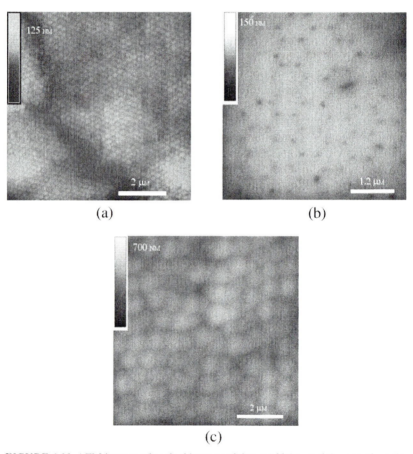

(a)   (b)

(c)

FIGURE 6.11 AFM images of packed latex particles. (a) If the particles are of regular size, packing is spontaneous on drying, forming an even, close-packed structure with limited defect structures. (b) For softer latex particles, fusion occurs as the film dries, flaws in the packing can result in large holes in the coating. (c) Irregular flocculated particles form a less organized packing structure, one that contains extensive defects.

large particles because of the drive to minimize surface energy, and size averaging will take place. However, as the particles become stiffer, via molecular chemistry or fusion temperature, fusion is hindered. The packing flaws can be maintained, leading to defects remaining in the fused solid. If the particles are rigid, they will fuse less readily. In this case, irregularity in shape can lead to regions where particles can be exceptionally close or exceptionally far apart. The average packing will be observed as constant, but on a local level there will be uncertain packing. This may have implications on properties. Elastic modulus is an average value, so bulk modulus will probably be unaffected except in exceptional cases. Permeability, which depends on the availability of diffusion paths from surface to surface, can be significantly increased by the presence of defects, even at the nanoscopic level.

In general, as particles become less spherical and less uniform the packing becomes irregular. Notice, however, that there are several exceptions to this generalization. One occurs if the particles are regularly irregular, such as cylinders (fibers) or plates. The presence of a long dimension can increase the particle packing density. Further, if two sizes of spherical particles are combined, where one particle size is approximately 4 times the diameter of the other, the smaller particles will be small enough to fit into the holes between a closest-packed structure of the larger particles. This can result in an increase in the maximum amount of particulate reinforcement that can be added to the composite.

IN SITU POLYMERIZATION

Examples of the *in situ* polymerization of nanocomposites are common. In *in situ* processing, liquid monomer can be mixed with reinforcing particles and then polymerized. Nylon 6, acrylic, and epoxy are nanocomposite matrices that have been made in the laboratory, and in some applications via the *in situ* polymerization technique. Nylon and epoxy are examples of a type of reaction called variously step or condensation polymerization. Acrylic is part of a family of reactions variously called radical, chain, or addition polymerization.

An *in situ* polymerization for a composite is usually conducted by mixing a liquid reaction mixture with the ceramic reinforcement. The reaction mixture is a formulation of a low-density solution of reactants with associated polymerization and processing aids. The reaction mixture, or reaction mass, contains liquid monomers, of course, as well as the required components for the reaction of the monomers. There may be required chemicals to control the polymer structure, such as, initiator for addition polymerization, catalysts, small quantities of special chemical species to promote shorter polymer chains or cross-linking, or inhibitors to keep the reaction from proceeding before desired. Other additives may include diluents, viscosity reducers, foam control chemicals, antioxidants, flame retardants, pigments, and so on. The following discussion of polymerization is restricted to components controlling polymer structure.

In situ polymerization has features in common with solution casting. However, this process has some particular advantages. The characteristics of the particle in the monomer solution can be determined in the same manner as solvent processing, including particle wetting, diffusion, and terminal or equilibrium settling. The low-viscosity of the monomer allows penetration of a dense bed of ceramic particles, providing that sufficient wetting is possible. In this way, a very dense composite can be created. There is no solvent, and so after the monomer is polymerized, there is usually little drying required. A second advantage is that there is less volume change overall during the process than in solution processing, since the monomer is polymerized after it is mixed with the ceramic particles.

There can be some volume change, due to two potential sources. The first is the reduction in the number of molecules common to all chemical bond-forming reactions. Since two molecules react with each other to form a single molecule, the distance is changed from secondary bond distances to primary bond distances, and the volume will decrease. If the ceramic reinforcement concentration is high, the particles will not allow the volume to change freely. This can result in the generation of stored energy along molecules stretching between and interacting with two different reinforcement particles. This stored energy creates internal stress. The magnitude of this stored energy is difficult to predict but can have an impact on the final properties, particularly dimensional stability and fracture toughness. Residual stress can usually be removed by heating the part above the T_g to allow molecular motion. This will result in a more stable part, albeit with the potential for observable dimensional deformation.

A second source of volume change is the formation of small chemical species, such as water, methanol, or other reaction products, during the chemical reaction. In this case, small species may evaporate during the reaction. Some quantity can also be trapped in the polymer as the chains become longer and the viscosity of the polymerizing mixture increases. Trapped quantities of small molecules can result in softening the part by acting as a diluent. Residual monomer can also remain after a reaction. Residual monomer is reactive, often volatile or flammable, and can pose

TABLE 6.4 Some monomers used for in situ polymerization of nanocomposites.

Monomer	Chemical Formula	Type	Functionality	Polymer Name(s)
Methyl Methacrylate	$H_2C=CCH_3COOCH_3$	Addition	2	PMMA
ε-caprolactam		Step	2	Nylon 6
Styrene	$H_2C=CH-C_6H_5$	Addition	2	Polystyrene, PS
Bisphenol A + Epichlorohydrin	$HOOC-C_6H_4-COOH$	Step / Step	2 / 2	DGEBA,Epon 828
Curing Agent Y, tetramine agent	with DGEBA	Step	2	Epoxy

significant health risks. Trapped solvent may have to be removed in a drying step. The trapped species are restricted in motion to diffusion, so this can be a slow process.

Any chemical with two or more reaction sites can form a polymer. Some reactive groups, with the associated number of bonds that can be formed by the group, are shown in Table 6.4. Single reaction sites, such as an acidic or basic group, can form one new bond. This does not introduce the possibility of forming a continuous chain. Two reaction sites, however, do introduce the possibility of a chain formation, as each molecule can react and bond with two neighboring chemicals, each of which can react to bond to a new neighbor, leading to chain formation and polymerization. Some types of functionality, notably unsaturated carbon bonds, are inherently difunctional. That is, the double bond must form two bonds in a reaction. If the reactant has three or more reaction sites, a network can form, leading to a cross-linked polymer if there are sufficient interconnections between the chains forming the network.

There are two primary classifications of polymerization. Addition polymerization works by initiating a chain reaction between multifunctional monomers, leading to long chains. Addition polymerization reactions are rapid and form high-molecular-weight, irregular chains. Condensation polymerization proceeds by acid-base chemical reactions. The kinetics of this type of reaction are generally slower, tend to occur simultaneously through the reaction mass, and increase the molecular weight slowly. Nevertheless, many important commercial resins, including nylon and polyester, are condensation polymers. The nature of the chemical reaction has a significant impact on the processing requirements, as well as on the structure and properties of the parts.

Addition Polymerization

The first type of reaction to be considered is addition polymerization, also called chain growth. In an addition reaction, a monomer with carbon-carbon unsaturated, double or triple bonds in the structure is polymerized to create a long chain. Typical types of double and triple bond-containing monomers are shown in Table 6.5. An unsaturated bond has the potential to open, allowing two new bonds to be formed. The first bond is opened by an initiation reaction of some type. The new, larger species can initiate a new neighbor, and so on, causing a chain reaction. This is a rapid process, and it proceeds until the chain hits something that stops the reaction, causing termination. The chain forms during this process, and chain length is statistical, representing a balance between initiation and termination, and propagation.

TABLE 6.5 Examples of monomers that polymerize by addition.

Monomer	Chemical Formula	Type	Functionality	Polymer Name(s)
Ethylene	$H_2C=CH_2$	Addition	2	Polyethylene, PE
Acetylene	$HC\equiv CH$	Addition	4	Polyacetylene
Vinyl chloride	$H_2C=CH-Cl$	Addition	2(1)	PVC
Acrylonitrile	$H_2C=CH-C\equiv N$	Addition	2	PAN
Isobutylene	$H_2C=CCH_3CH_3$	Addition	2	
Tetrafluoroethylene	$F_2C=CF_2$	Addition	2	Teflon
Ethylene glycol	◁O	Addition	2	Carbowax, PEO
Acrylimide	$H_2C=CHCONH_2$	Addition	2(1 or 2)	
Methacrylic Acid	$H_2C=CCH_3COOH$	Addition	2(1)	
Vinyl Alcohol	$H_2C=CHOH$	Addition	2(1)	PVA
1,3-Butadiene	$H_2C=CH-CH=CH_2$	Addition	4	
1-Butene	$H_2C=CHCH_2CH_3$	Addition	2	
Vinyl Acetate	$H_2C=CHOCOOCH_3$	Addition	2	
Allyl Ether	$H_2C=CHOCH_2CH=CH_2$	Addition	4	

When a monomer has functionality for both addition and condensation polymerization, the functionality in the addition reaction is reported first and for the condensation second, for example 2 (1).

Polymerization

The first step of the reaction is initiation. An initiator is a species that creates a highly reactive component under controlled circumstances. The reactive component can be a free radical or an ion that can add across the double bond. The mechanisms for ionic and radical reactions are similar, but radical reactions are unspecific and faster, while the ionic reactions tend to be much better behaved. This discussion will focus on radical reactions, since they illustrate chain growth mechanisms.

To begin a reaction, the initiator is broken open by some force, for example heat or light. The initiator will break into, usually, one or two active radicals, which in turn can participate in a further chemical reaction. Sometimes, the two radicals will run into each other or some other stable species, leading to immediate quenching of the radical. However, a portion, f, of the initiator will react with a monomer effectively, initiating polymerization. Most initiation reactions require only the initiator to be present, leading to a chemical reaction with a first-order rate law. The rate can be tracked by the disappearance of the initiator, or by the appearance of free radicals. For practical reasons, and mathematical reasons that will become apparent later, measuring the disappearance of the initiator is more convenient. The initiation reaction has a rate described by the rate law:

$$\frac{d[I\cdot]}{dt} = 2fk_i[I]$$
(6.46)

where f is the effectiveness of the initiator at starting reactions, $[I]$ is the concentration of the initiator, $[I\cdot]$ is the concentration of the initiator radical and k_i is the initiation rate constant. This form assumes that ineffective initiation reactions either do not form a radical, or that the radical is immediately quenched and is therefore not relevant. The initial concentration of initiator is the easiest quantity to determine, because this is usually a starting condition for the reaction mass.

The next portion of the reaction is chain growth. Once the reaction begins, there is little barrier to the addition of new monomers at the end of the chain. The rate of the propagation of the reaction is described by:

$$\frac{-d[M]}{dt} = k_p[M][M\cdot]$$
(6.47)

where $[M\cdot]$ is the concentration of the radical and k_p is the rate constant of propagation. Here, the rate is determined by tracking the disappearance of the monomer. Measuring the number of radicals in the reaction mass is

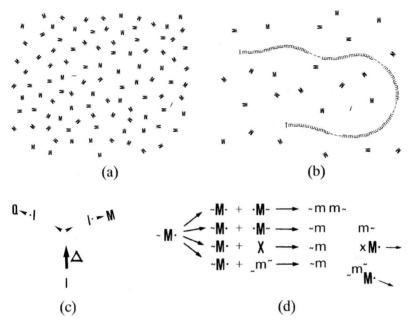

FIGURE 6.12 Schematic of addition polymerization: (a) A monomer solution with 2% initiator. (b) Approximately 10^{-8} seconds after an initiation event. (c) An initiator can start a chain reaction, or fail due to self-quenching or by reacting with an inhibitor or another radical. (d) Termination occurs by combination, disproportionation, chain transfer to chain, or chain transfer to chain transfer agent. A chain transfer agent is added specifically to shorten chains; sulfur-containing molecules work the best for this purpose.

challenging and usually inconvenient. The concentration of radicals will remain until some more easily determined quantity can be substituted.

Chain growth is observably finite. Propagation is in direct competition for radicals with termination events. At some point, the moving radical will interact with something that will absorb the radical without enabling further growth of the chain. The radical can run into another radical, for example on another chain, resulting in both radicals being stopped. Termination can be determined by the disappearance of free radicals or by the appearance of completed polymer. Both of these are difficult to measure. The rate of termination is described by:

$$\frac{-d[M\cdot]}{dt} = 2k_t[M\cdot]^2 \tag{6.48}$$

where k_t is the rate constant for termination. Termination will also be left, for the moment, in terms of the radical concentration. For two chain radicals running into each other, there are two possible results. If the radicals form a bond, called combination, the polymer will be twice as long as if the radicals exchange energies, called disproportionation, and bounce apart, forming two inert polymer chains. In either case, two radicals are consumed.

The radical can also transfer from the existing chain to some other component of the reaction mixture, such as a monomer, another polymer chain, solvent, or a chain transfer agent added specifically to facilitate this. The new radical can start a new chain, but the old chain will stop growing. Note that if the radical transfers from one chain to another, a branch will form on the pre-existing chain at the point where the radical transferred.

In order to find a rate of polymerization in terms of easily measured variables, an assumption is required. A simple concept is equilibrium between initiation and termination. Every radical formed will eventually quench, or terminate. The controlling factor is the length of time between creation and quenching. At low conversions, the mobility of radicals is high, and the balance between initiation and termination is reached quickly. This leads to the expression:

$$2fk_i[I] = 2k_t[M\cdot]^2 \quad at\ equilibrium \tag{6.49}$$

Rearranging this equation provides a way to estimate the concentration of free radicals:

$$[M\cdot] = \left(\frac{fk_i}{k_t}\right)^{1/2} [I]^{1/2} \tag{6.50}$$

The reaction scheme described above indicates that at any time, the reaction mass will include monomer, polymer, a small quantity of growing chains, and any processing aids.

The viscosity of the reaction mass is increased at higher conversion. This can cause the rate of quenching to change, slowing down. This occurs because the radicals are attached to large polymer chains with limited mobility. The initiation rate, however, continues to be controlled by the mobility of smaller initiator fragments and monomer. The result is autoacceleration of the reaction at higher conversions, known as the Trommsdorf effect. If the monomer is at a concentration in solution of below about 40%, the conversion with time is approximately linear, slowing as monomer is depleted (Figure 6.13). At higher concentrations, especially at 100% monomer, the conversion increases rapidly and non-linearly. In a 100% solids system, a reaction mass in which the only liquid is monomer, 100% conversion often is not possible. This happens because of the probabilities of radical quenching. The most efficient quenching mechanism may be interacting with another radical, but other mechanisms can occur as the opportunity for quenching by another radical is reduced. As new radicals initiate, they may be quenched without growing new polymer. Therefore, there will be either solvent and higher conversion or no solvent and significant residual monomer remaining in the polymerized part. This residual monomer must be removed.

Molecular Weight and Chain Structure

The simple equations above can be used to determine the molecular weight. Propagation and termination can be regarded as directly competitive reactions. The average length of the chain will be determined by combining the probability of chain growth and the inverse of the probability of chain termination. Since the probability of a reaction is well described by the rate, the size of a chain can be predicted by the ratio of the rates of propagation to termination. This can be expressed as follows:

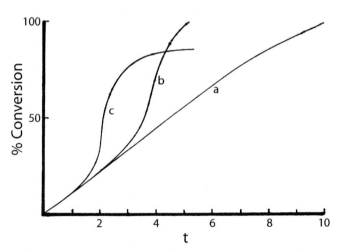

FIGURE 6.13 Schematic of the Trommsdorf effect. (a) Chains in solution remain mobile and full polymerization occurs after sufficient elapsed time. (b) As the monomer concentration increases, the long chains formed by polymerization increase viscosity of the solution. The heat generated by the polymerization reaction can no longer be fully dissipated, resulting in accelerated reaction rate and solidification. (c) In a pure monomer solution the reaction can accelerate, resulting in rapid solidification. However, the increase in viscosity can lower the mobility of the monomer sufficiently to suppress further reaction. This results in trapped monomer.

$$\overline{v} = \frac{k_p[M]}{2(fk_tk_i[I])^{1/2}} \tag{6.51}$$

where $\overline{v}$ is the kinetic chain length. This equation implies that increasing the concentration of monomer can increase the kinetic chain length. Increasing the rate of initiation can also reduce the chain length. Further understanding of the rates of initiation, termination and propagation, including changes in the kinetics caused by temperature, will provide a deeper understanding of the molecular weight resulting from any given polymerization. An important conclusion for *in situ* polymerization is that the chain length formed during the course of the reaction will change with time. This is an inevitable consequence of decreasing initiator and monomer concentration during the reaction as both are consumed during the polymerization. The kinetic chain length is an average, reflecting the statistical origin of the quantity. The kinetic chain length is a measurement of the polymer, and does not include the monomer or other species. The actual molecular weight of polymer being formed during a polymerization is a function of the kinetic chain length, which considers and combines all forms of termination. For example: if chains quench by combination, the molecular weight of that chain will be $2v$; if termination is by disproportionation, MW will be v. If termination is also possible by chain transfer to solvent, polymer, or chain transfer agent, additional competitive reactions are available for termination and the probability calculation is modified:

$$\frac{1}{\overline{v}_{tr}} = \frac{1}{\overline{v}} + \sum \frac{k_{tr}}{k_p}\frac{[TR]}{[M]} \tag{6.52}$$

where k_{tr} is the chain transfer rate constant and $[TR]$ is the concentration of the chain transfer agent, which can be polymer, solvent, or some species specifically added for the purpose, such as a thiol.

The kinetic chain is initiated, formed, and terminated in approximately 10^{-8} s. The rapidity of chain formation can cause isomers with distinctly different structures and different mechanical performance. Chemical isomerism includes: head-tail; *cis-trans*; atactic-isotactic-syndiotactic; branched, linear and linear low-density; and copolymerization. Some of these isomers can be controlled by catalysis, and some are basic to the type of polymer being formed.

First, if a monomer is asymmetric, there is a head and tail. The radical can attack via the head or the tail, providing one variation in chemical structure. The monomer generally has a bias for ion or radical formation, stabilized by the structure of the monomer. For example, the reactive site on a methyl methacrylate radical is much more stable if located on the multiply substituted carbon rather than on the unsubstituted end of the monomer. The amount of head-tail isomerism is controlled by the composition of the monomer, but is generally low.

A second type of isomerism is described by polymer tacticity. Tacticity describes the rotational isomerism of a polymer. Two types of rotational isomerism are possible for, e.g. polystyrene. Rotational isomers can form when one of the carbons along the backbone is asymmetric, that is, has four different groups attached to it. In a polymer, two of the groups are the front and back half of the chain, and if one group is hydrogen and the other a phenyl group, for example, as is the case for styrene, then the carbon is asymmetric. During polymerization, the phenyl group can add to the chain on either side, leading to the rotational isomers illustrated in Figure 2.8. There are three common types of tacticity. An atactic polymer is one in which the side groups are randomly ordered on the chain. An isotactic polymer has all of the side groups on one side of the chain, while a syndiotactic polymer has the side groups strictly alternating along the chain. The tacticity controls the regularity of the structure. An atactic polymer has little structural order and forms small, poor crystals, while a regular tacticity allows regular packing of polymer chains into crystals. Using a catalyst to control the stereochemistry usually forms isotactic or syndiotactic polymers. There are two general types of catalyst, as shown in Figure 6.16. Because of the impact of tacticity on the secondary and tertiary structures in a polymer, there has been considerable research into controlling the tacticity of polymerization and understanding the resulting structures.

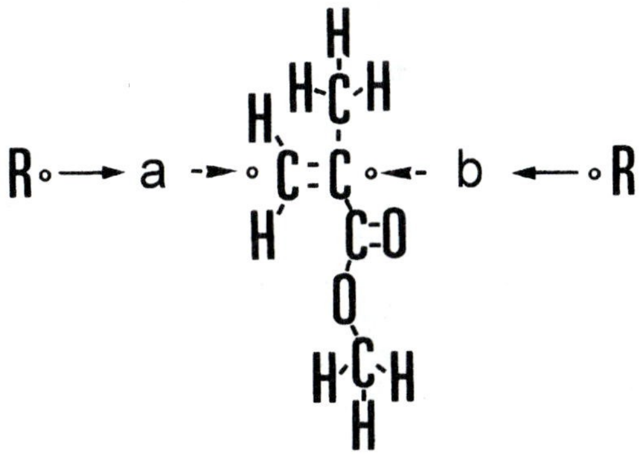

FIGURE 6.14 A schematic of (a) head-to-tail polymerization and (b) head-to-head polymerization in methyl methacrylate.

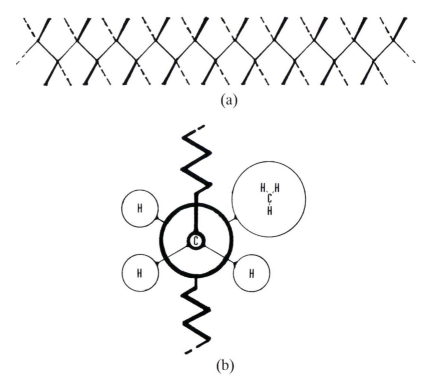

FIGURE 6.15   There are several ways to illustrate polymer chain structure, with specific advantages. (a) A three-dimensional projection of the planar zigzag allows one to show the relative location of all groups, but can complicate visualization. (b) The axial projection for polypropylene shows the potential for overlap and occlusion of the side groups, illustrating the energy cost of bond rotation. This view also illustrates the relative stability of specific chain conformations, such as the 3:1 helix of polypropylene in the crystalline phase, if the polymer chain has a regular isotactic or syndiotactic structure. The irregular orientation of side groups in an atactic polymer introduces rotational disorder and decreases the regularity of packing.

A third type of chemical isomerism is the formation of a *cis-trans* structure. If a carbon-carbon triple bond or a monomer with two double bonds is used for polymerization, a single double bond remains in the chain. The *cis*-isomer occurs when the polymer chain enters and exits the double bond on the same side of the bond, while the *trans*-isomer occurs when the chain enters on one side of the bond and exits on the other, so to speak. The *cis*-isomer traps more energy and is less thermodynamically stable, but also traps more space between chains. *Cis*-polymers tend to be more flexible. If a side group is present at the double bond, for example a CH_3 in *cis*-isoprene, or gutta percha, naturally occurring rubber, the space occupied and the flexibility of the chain is further enhanced. The *trans*- isomer is more stable, less easily degraded, and is more chemically resistant. *Cis-trans* isomer chemistry is very important in rubber manufacture and performance.

Another chemical isomerism of importance is branch and network formation. Branching occurs in free radical polymerization when a radical transfers from a growing chain to an already existing chain. The radical on the original chain is quenched, while the new radical on the chain backbone can initiate a new chain, forming a long branch. Chain transfer to a polymer backbone is a regular occurrence in a chain with an extra double bond. In this case, a growing chain can intersect any number of backbones and keep growing, resulting in a cross-linked, network polymer. In a catalyzed polymer, chain transfer is suppressed. Branches introduce free volume, however, which can cause useful changes in properties. Branches can be reintroduced in a catalyzed polymerization by adding controlled amounts of a large monomer with a double bond at one end, for example, dodecene.

Polymerizing with more than one monomer can be complicated, because multiple competitive reactions are introduced. The same concept used in single component polymerization used above, with ratios of probabilities, can be used to evaluate the most probable structure formed

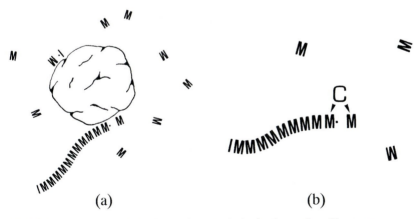

(a)   (b)

FIGURE 6.16 (a) A multiple-site catalyst. (b) A single-site catalyst. The monomer can only add to the chain in a specific direction, which can cause a chain with specific tacticity to form. By eliminating ineffective collisions, the catalyst particle accelerates the reaction.

(a)   (b)

FIGURE 6.17 (a) *Cis-* and (b) *trans-*isomers of butadiene. The zone occupied by rotation of the double-bond containing subunit of the chain shows that the *cis* isomer introduces more open volume than the *trans* isomer.

while different monomers are introduced during polymerization. If two monomers are involved in a polymerization, there are four competing reactions:

$$\frac{d[M_1]}{dt} = k_{11}[M_1 \cdot][M_1] \tag{6.53}$$

$$\frac{d[M_1]}{dt} = k_{21}[M_2 \cdot][M_1] \tag{6.54}$$

$$\frac{d[M_2]}{dt} = k_{12}[M_1 \cdot][M_2] \tag{6.55}$$

$$\frac{d[M_2]}{dt} = k_{22}[M_2 \cdot][M_2] \tag{6.56}$$

There can be an influence due to the next-to-last monomer on the reaction. The basic insight into understanding the structure formed during *in situ* polymerization can be derived from this simpler starting point. When polymerization is proceeding, the reaction of monomer 1 is a competition between further reaction with monomer 1 and or cross-reaction with monomer 2. The reaction of monomer 2 is a competition between further reaction with monomer 2 and crossing over to react with monomer 1. Therefore, the structure formed during polymerization is influenced in large part by the probability of the chain chemistry crossing over from one monomer to another. These are described by the reactivity ratios, based on the rate, or equivalently, the probability, of the polymeriza-

TABLE 6.6 r_1 and r_2 *for common monomers.*

Monomer 1	Monomer 2	r_1	r_2
Acrylonitrile	Butadiene	0.03	0.2
	Methyl Methacrylate	0.138	1.322
	Styrene	0.02	0.29
	Vinyl Acetate	4.05	0.04
	Vinyl Chloride	2.55	0.07
Butadiene	Isoprene	0.94	1.06
	Methyl Methacrylate	0.75	0.25
	Styrene	1.55	0.37
Methyl Methacrylate	Styrene	0.5	0.5
	Vinyl Chloride	8.99	2.97

Table from Hiemenz.

tion reaction continuing with the same chemistry—or of crossing over to the alternate monomer:

$$r_1 = \frac{k_{11}}{k_{12}} \tag{6.57}$$

and

$$r_2 = \frac{k_{22}}{k_{21}} \tag{6.58}$$

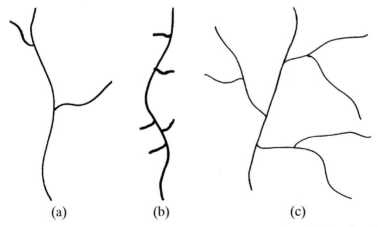

(a)   (b)   (c)

FIGURE 6.18 Sketches of chain structures: (a) standard branches, (b) linear low-density branches, and (c) highly branched.

where r_1 is the monomer cross-reactivity between monomer 1 and monomer 2, and vice-versa. There are some interesting effects on copolymer structure based on the sizes of the reactivity ratios. If r is much greater than unity, the monomer is much likelier to self-polymerize than to cross over. Stretches of blockiness would be expected with this polymerization. If r is much smaller than unity, the opposite is true, and the polymer will tend to incorporate isolated units of the monomer. Therefore, if both monomers have r-values less than 1, the copolymer will tend to have an alternating structure. If both monomers have r values much greater than unity, block copolymers will tend to form. In extreme cases, the two polymerizations will occur in parallel—with little crossover observed in the final structure. Predicting the structure becomes more problematic the closer the r-values are to unity.

The amount of either monomer incorporated in the chain can be determined by the ratio of the probabilities of the reactions for polymerizing monomer 1 to polymerizing monomer 2, and assuming the number of radicals remains constant:

$$\frac{d[M_1]}{d[M_2]} = \frac{[M_1]}{[M_2]} \frac{\left(\dfrac{k_{11}}{k_{12}}\right)[M_1] + [M_2]}{\left(\dfrac{k_{22}}{k_{21}}\right)[M_2] + [M_1]} \tag{6.59}$$

A more easily interpreted form of the equation is obtained by rewriting in terms of mole fraction of monomer in the polymer, X, and mole fraction of monomer in the reaction mass, x:

$$X_1 = 1 - X_2 = \frac{d[M_1]}{d[M_1] + d[M_2]} \tag{6.60}$$

$$x_1 = 1 - x_2 = \frac{[M_1]}{[M_1] + [M_2]} \tag{6.61}$$

and:

$$X_1 = \frac{r_1 x_1^2 + x_1 x_2}{r_1 x_1^2 + 2 x_1 x_2 + r_2 x_2^2} \tag{6.62}$$

The reaction ratios, together with the monomer feed composition, will govern the relative amounts of monomer incorporated into the polymer.

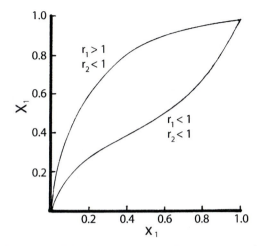

FIGURE 6.19 Schematic of X_1, the mole fraction of component one incorporated into the polymer chains, as a function of x_1, the mole fraction in the reaction mix, for two different cases of r_1 and r_2. If r_1 is > 1 and r_2 < 1, then the chain is biased to add blocks of component 1. If both r_1 and r_2 are <1, there is more balance between the monomer addition. Also, the monomers will tend to alternate.

However, the relative concentrations of the monomers will change with time as monomer is differentially incorporated into the polymer. Again, this is a potential concern for *in situ* polymerization, where the reaction mass is set by the initial conditions.

There are tabulated values of cross-polymerization constants. There are also some relations for determining these parameters from chemical structure. One useful method for predicting the relative cross-reactivity of the monomers is the Price-Alfrey Q-e scheme. This scheme recognizes different contributions to the reactivity of a monomer, recorded as Q and e values.

TABLE 6.7 Price-Alfrey constants for common monomers.

Monomer	Q	e
Acrylonitrile	0.48	1.23
Butadiene	1.70	−0.50
Butene	0.007	−0.06
Methyl Methacrylate	0.78	0.40
Styrene (definition)	1.0	−0.80
Vinyl Acetate	0.026	−0.88

Table from Hiemenz.

TABLE 6.8 Examples of monomers that polymerize by step.

Monomer	Chemical Formula	Type	Functionality	Polymer Name(s)
ε-caprolactone		Step	2	
Hydroxystearic Acid		Step	2	
Hexamethylene Diisocyanate	O=C=N–(CH₂)₆–N=C=O	Step	2	Urethane
Terphthalic Acid	HOOC–C₆H₄–COOH	Step	2	PET, polyester
Ethylene Glycol	HOCH₂CH₂OH	Step	2	PET, polyester
Sebacoyl Chloride	ClOC–(CH₂)₈–COOCl	Step	2/4	Nylon 6,10
Hexamethylene Diamine	H₂N–(CH₂)₆–NH₂	Step	2	Nylon 6,10
Bisphenol A + Epichlorohydrin	with DGEBA	Step	2	DGEBA, Epon 828
CO₂		Step	2	Polycarbonate
Amino Acids	H₂N–CHR–COOH	Step	2	Proteins
Alanine	R= –CH₃	Step	2/3	
Lysine	–(CH₂)₄–NH₂	Step	3/5	
Serine	–CH₂OH	Step	3/4	
Leucine	–CH₂CH(CH₃)₂	Step	2/3	
Glutamic Acid	–CH₂CH₂COOH	Step	3/4	
Succinic Acid	HOOCCH₂CH₂COOH	Step	2	
1,2-Dibutanol	HOCH₂CHOHCH₂CH₃	Step	2	
1,4-Dibutanol	HOCH₂CH₂CH₂CH₂OH	Step	2	

A primary amine is capable of forming two additional bonds, but usually only forms one, and so has a functionality of one or two (1/2).

260

Reaction Control

One necessary feature of an *in situ* polymerization is control. That is, the reaction should not begin until initiation is begun, and should proceed to create a polymer with the desired performance. The initiation of polymerization can be delayed using retarders or inhibitors. A retarder works by competing with the monomer for the radicals. An inhibitor works by absorbing the radicals, thus preventing productive polymerization reactions. Benzoquinone is a commonly used inhibitor added to monomers to prevent polymerization during transportation. Oxygen also is a very effective inhibitor, and chain polymerization usually requires an inert atmosphere. If the inhibitor is not removed from the monomer, there will be an induction period while the inhibitor is consumed, after which polymerization will proceed normally. A retarder also functions by reacting with radicals to create species ineffective in polymerization, but the retarder is less efficient and so will slow the reaction down without blocking it entirely.

As described above, as the concentration of polymer increases, the viscosity will increase and the mobility of the radicals will decrease. In a reaction mass with 100% monomer (barring the initiator and processing aids), the viscosity increase can lead to solidification and to the suppression of termination reactions. This can leave behind significant quantities of monomer. Solution polymerization allows more thorough control of the reaction, including rate, completion, temperature management, and chain length. This technique also introduces some features in common with solvent processing, including solvent removal and the potential for settling, and so forth. Another alternative is emulsion polymerization. In this case, the monomer is suspended as isolated droplets in water using a surfactant. The initiator is added in the water phase. As an initiator fragment nucleates a drop of monomer, the monomer in the drop is thoroughly polymerized. Latex polymer, or a polymer produced by emulsion polymerization, is characterized by low polydispersity. The particle sizes of the drops of monomer are typically around 200-300 nm, slightly larger than the particles in a nanocomposite. Paint formulations use mixtures of nanoscopic polymer and pigment particles, along with flow enhancers, biocides, ultraviolet light inhibitors, and other additives to create a stable liquid, which creates a composite film on drying.

Many commercially important monomers, especially propylene, are gaseous. These materials are polymerized using chemical vapor deposition for *in situ* polymerization. The monomer is directly added to growing chains to form the composite part. The industrial scalability of this technique is undetermined as yet.

Step-Growth Polymerization

In step-growth, or condensation, polymerization, a monomer with two or more acid-base groups is polymerized to create the polymer chain. Examples of monomers with functional groups appropriate for step polymerization are shown in Table 6.8. Polyesters, epoxies, and polyamides (nylons) are examples of commercially important polymers of this type. This reaction usually creates a small, inert molecule, a leaving group, which must be dealt with. There are several key features of step polymerization important for *in situ* polymerization in a composite. These include the process of the reaction, the chain length of the polymer as it forms, factors controlling the chain length, and the rate of the reaction.

The fundamental process of step-growth polymerization is the acid-base reaction. Figure 6.20 shows how a reaction mixture of diacids and dibases will react, step-by-step, to increase the length of the chain. In this type of reaction, most of the monomer reacts slowly in favor of early intermediate species, which in turn react again and again. The molecular weight increases slowly as the reaction proceeds. Monomer converts to dimers, dimers convert to trimers and quadrimers, and a constantly changing distribution of oligomers is formed. Eventually the molecular weight increases to above about 2,000 carbons in length, and the first molecules recognizable as polymers are obtained. As further time elapses, the average length of the chains will increase further, leading to commercially viable polymers. The contents of the reaction mixture at any time will be a statistical distribution of chain lengths determined by the number of groups that have reacted.

Molecular Weight

Determining the length of the chains is possible through a general extent of reaction. The extent of reaction, p, is defined as 1, when the reac-

TABLE 6.9 Molecular weight as a function of extent of reaction and stoichiometric imbalance.

	p			
r	0.95	0.97	0.99	1.00
0.95	13.5	18.2	28.3	39.0
0.97	15.5	22.3	39.9	65.7
0.99	18.3	28.7	66.8	199
1.00	20.0	33.3	100	∞

Table from Hiemenz.

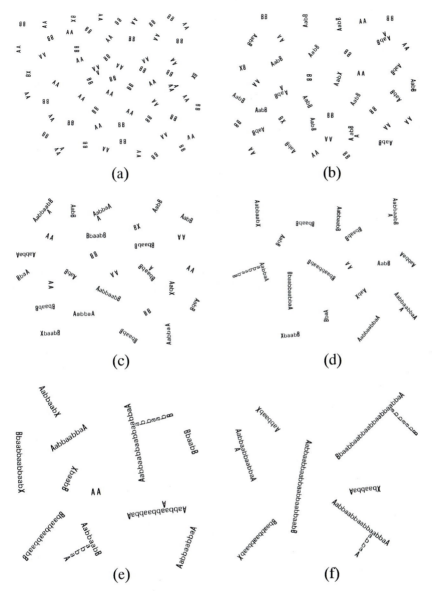

FIGURE 6.20 Six time steps in a condensation reaction between a diacid and dibase with a limited quantity of monomer. The solution shown contains 5% trifunctional A's and 5% monofunctional B's, 45% difunctional A's and 45% difunctional B's. At each stage shown, approximately 30% of the reactive groups react. The reaction must proceed to a significant fraction of completion before long molecules are common in the reacting mixture. The monomer, however, disappears rapidly. Further, the results of the stoichiometric imbalance, multifunctionality, and monofunctional groups become apparent at late stages of the reaction.

tion is complete. In reality, this state is never reached, but it can be approached infinitely closely with time. The extent of reaction for a given reaction can be determined by the amount of monomer consumed. Since the reaction requires one acid and one base, only one reactant must be followed, so:

$$p = 1 - \frac{A}{A_0} \qquad (6.63)$$

where A is the instantaneous concentration of one of the reactants, and A_0 is the original concentration of that reactant. A basic tenet of this view is that the reactivity of group A is the same, regardless of whether it is at the end of a small monomer or of a large oligomer chain. The smaller molecules are more mobile, but reactions between larger chains require only small rotations to allow the reactive groups to approach each other. The length of the chain also can be calculated based on the concentration of reactive species. The average length of the chain is the ratio of the original functional groups that have undergone reaction: that is, have been incorporated into a growing chain:

$$\bar{n}_n = \frac{A_0}{A} \qquad (6.64)$$

where $\bar{n}_n$ is the number average chain length. The number average is appropriate here because this is the equivalent of counting the number of end-groups relative to the total number of chains. Therefore, $\bar{n}_n$ is a function of the extent of reaction:

$$\bar{n}_n = \frac{1}{1-p} \qquad (6.65)$$

The chain length as a function of extent of reaction has been made reaction-independent by normalizing to complete reaction. The time required to reach this state for a given reaction will be discussed later. Table 6.9 shows the molecular weight as a function of extent of reaction. The ideal distribution of chain lengths also can be evaluated using the statistics of chain growth:

$$\frac{\bar{n}_w}{\bar{n}_n} = \frac{\left(\dfrac{1+p}{1-p} \right)}{\left(\dfrac{1}{1-p} \right)} \qquad (6.66)$$

where $\bar{n}_w$ is the weight average chain length. The polydispersity is this ratio of weight to number average chain length, yielding the relation:

$$PD = 1 + p \tag{6.67}$$

PD is the polydispersity, which is given many symbols in the literature. For an ideal condensation polymer, the dispersity tends toward 2.

Increasing molecular weight, therefore, requires a reaction going to a significant fraction of complete reaction. In order to go to complete reaction, or even close to it, the stoichiometric balance between the two monomers must be nearly exact. The stoichiometric imbalance, *r*, can be calculated by:

$$r = \frac{A_0}{B_0} \tag{6.68}$$

where *A* and *B* are chosen so that *r* is always less than 1. A stoichiometric imbalance will result in changing the statistics of the types of reaction that can occur, and in turn the length of the chain. The statistics with stoichiometric imbalance give:

$$\overline{n}_n = \frac{1 + r}{1 + r - 2pr} \tag{6.69}$$

If there is a stoichiometric imbalance, the growing chains will tend to terminate by preference for the species that is in excess. This effectively terminates the chains, leading to lower molecular weight. Table 6.9 shows the molecular weight, including stoichiometric imbalance. Even slight imbalances will have a dramatic effect on the maximum molecular weight. There are several ways to control stoichiometric imbalance, which will be discussed later.

Up to this point, the functionality of the monomer has been assumed as two. There are several important implications if the monomer has a different functionality. The first is including a small quantity of a monofunctional monomer. The effect of this is to end-cap the polymers, protecting and blocking further reaction. The principle of stoichiometric imbalance allows an estimate of the effect on chain length of the molecular weight for including a monofunctional group.

The addition of a polyfunctional monomer, where functionality is greater than two, is somewhat more complicated. The statistical probability of various types of reaction can be calculated using ratios of different types of monomers:

$$\rho = \frac{A_f}{A_{total}} \tag{6.70}$$

where ρ, Greek for "r" to keep the symbol separate from the previous *r* for stoichiometric imbalance, is the proportion of reactive groups that are multifunctional, A_f is the concentration of functional groups in multifunctional monomers, and A_{total} is the concentration of functional groups. If the number of multifunctional groups is small, the addition of a polyfunctional monomer will result in a branch forming. If the number of multifunctional groups is large enough, a significant fraction of growing chains will become capped at both ends in multifunctional groups. As a result, the growing chains can form a cross-linked network. Useful for this calculation is another variable, α, the chances that a chain has a multifunctional unit at either end, forcing further growth to have the form of a branch or network. Considering the probabilities of various reactions, an equation for creating such a network chain is:

$$\alpha = \frac{rp^2\rho}{1 - rp^2(1-\rho)} \tag{6.71}$$

When sufficient chains are formed, further growth will yield a cross-linked polymer. The conditions for this can be determined in two different ways. Following the statistical approach (above), the assumption is that every chain capped with two branch points will become part of an effective network. A critical concentration of branching occurs when:

$$\alpha_c = \frac{1}{f-1} \tag{6.72}$$

where f is the number of reactive groups in the polyfunctional monomer. This can be combined with the above equation to provide a relation for the critical extent of reaction.

$$p_c = \frac{1}{[r + r\rho(f-2)]^{1/2}} \tag{6.73}$$

As long as the extent of reaction is below this level, branched polymer will form, rather than a network. Above this, when the first molecule can form an infinite chain, a network is expected. Based on the assumptions, this relationship tends to predict gelation conservatively; that is, gelation usually happens in practice at an extent of reaction higher than this. Another approach to predicting the onset of an infinite network is through predicting when the average chain reaches infinite length. This can be calculated with:

$$p_c = \frac{2}{r\bar{f}} \tag{6.74}$$

where $\bar{f}$ is the average functionality of all monomers. This can predict gelation to occur at a point later than is typically observed.

Reaction Control

Acid-base reactions are equilibrium reactions, requiring sufficient energy for two appropriate monomers to collide, interact, and finally form new bonds. The reaction is considerably faster if the reacting species have been primed by the prior addition of an acid or base. The addition of an acid or base is catalytic, weakening the bond between the monomer and the small leaving group according to Figure 6.21. The rate law describing this simple, catalyzed acid-base reaction is:

$$-\frac{d[A]}{dt} = k_c [A][B] \qquad (6.75)$$

where $[A]$ is the concentration of one functional species, $[B]$ is the concentration of the other, and k_c is the catalyzed rate constant. If $[A] = [B]$, integrating this equation to develop an equation for determining the length of time required to reach any stage of the reaction gives:

$$\frac{1}{[A]} - \frac{1}{A_0} = k_c t \qquad (6.76)$$

The equation is somewhat changed where stoichiometric imbalance is present, but the conclusions of the rate discussion are not significantly different. The equation can be converted to determine the extent of reaction:

$$\frac{1}{1-p} = 1 + k_c [A]_0 t \qquad (6.77)$$

If the reaction is uncatalyzed, the acid and base in the reaction mass can supply the necessary components. However, this makes the reaction third-order in $[A]$ instead of second-order:

$$-\frac{d[A]}{dt} = k_u [A]^2 [B] \qquad (6.78)$$

Leading to the relation:

$$\frac{1}{(1-p)^2} = 1 + 2k_u [A]_0^2 t \qquad (6.79)$$

This equation shows that the polymer will grow more slowly without an acid or base catalyst added.

Experiments show that with the exception of very early in the reaction, when the predominant species is monomer, the rate is a constant. This is

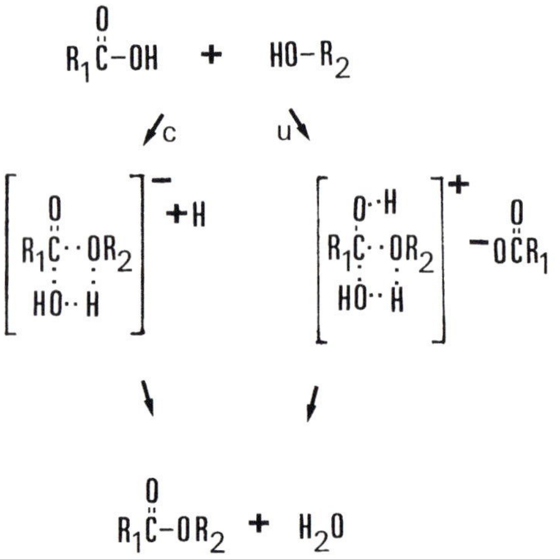

$$R_1\overset{\overset{O}{\|}}{C}-OH \quad + \quad HO-R_2$$

$$\overset{c}{\diagup} \qquad \overset{u}{\diagdown}$$

$$\left[\begin{array}{c} O \\ R_1\overset{\cdot\cdot}{\underset{\cdot\cdot}{C}}\cdot\cdot OR_2 \\ H\overset{\cdot\cdot}{O}\cdot\cdot\overset{\cdot}{H} \end{array}\right]^{-} +H \qquad \left[\begin{array}{c} O\cdot\cdot H \\ R_1\overset{\cdot\cdot}{\underset{\cdot\cdot}{C}}\cdot\cdot OR_2 \\ H\overset{\cdot\cdot}{O}\cdot\cdot\overset{\cdot}{H} \end{array}\right]^{+} \; {}^{-}O\overset{\overset{O}{\|}}{C}R_1$$

$$\diagdown \qquad \diagup$$

$$R_1\overset{\overset{O}{\|}}{C}-OR_2 \quad + \quad H_2O$$

FIGURE 6.21 Schematic of catalyzed (c) and uncatalyzed (u) step-growth reaction.

not to say that the reaction will not slow down as the reaction proceeds. In fact, as the viscosity increases, the rate of reaction will drop as a result. Another assumption used in the above derivations is that the polymer is soluble in the monomer. The higher the molecular weight, the less likely this assumption becomes.

From the discussion, maximizing the polymer molecular weight requires catalysis and stoichiometric balance. There are several ways to control stoichiometric imbalance. First, when a polymerization occurs in a large reaction, small differences in measurement are minimized. This is a significant concern for *in situ* polymerization, because the part size determines the maximum initial concentration of monomer. In the absence of this advantage, several tricks can be employed. If the polymer can be formed using a monomer that has one acid and one basic group on the same molecule, such as caprolactam, the stoichiometry is automatically insured. In the nylon rope trick, a common chemistry demonstration, the reactants are kept in separate, immiscible solvents, with polymerization occurring in a diffusion-controlled form at the interface. In polymerizing polyethylene teraphthalate, the first industrial step is usually to react the diacid, teraphthalic acid, with an excess of ethylene glycol, end-capping all of the monomer. The polymerization uses this modified monomer. The end-capped monomer is heated under vacuum, which causes transesterification with elimination of ethanol as a leaving group. Removing the small ethanol molecule with vacuum reduces the chances of the reverse, depolymerization reaction.

Not related to stoichiometric imbalance, but also important to insuring high molecular weight is assuring that the chains stay in solution, so that the reaction can continue. The reaction can be conducted in solution to keep the polymer dissolved, accessible to monomer and growing.

There is seldom any monomer remaining in the polymer after step-polymerization. The small molecular weight species created during the reaction are another matter. The presence of a small molecule can lead to plasticization of the polymer, affecting the mechanical and permeability properties. If the small molecule is water, water may unzip the polymer, reversing the acid-base reactions that formed the backbone. If the small group is a monomer or a solvent, such as methanol, standard concerns include toxicity and flammability as the polymer outgases.

POST-FORMING

Once the composite device has been formed, further processing is somewhat limited by the high viscosity that was avoided by choosing a low-viscosity processing technique. Several processes do avoid significant energy requirements. For example, the composite film can be heated to above the melting temperature and compression molded to form simple shapes. Thin films created by solvent processes can be stacked together and welded in the same manner to create thicker parts. To avoid adding significant processing cost, the parts created by low-viscosity processes should be at near net shape, that is, similar in volume and shape, to the desired final part.

HAZARDS OF SOLVENT PROCESSING

Many common solvents and monomers are flammable, that is, have a combustion temperature below 200°C. If the partial pressure of the vapor reaches a minimum level, ignition can occur. Flammable chemicals also have a maximum pressure, above which burning can no longer occur. The flammable concentration is usually high, for example, between 6.2% and 15.9% for ethylene dichloride. These materials should be handled in an environment such that the local concentration is outside of these limits. The local concentration is the important concern. If the solvent is stored in a closed environment, the partial pressure can rise to dangerous levels.

The autoignition temperature, the temperature at which the energy of the system without outside influence to cause combustion, is 413°C for ethylene dichloride, while the open cup flash point shows an ignition temperature of 18°C. Below this temperature, igniting the liquid is predicted to be impossible. Above 413°C, no external ignition source is necessary. Note that combustion has often been noted to occur at lower temperatures than indicated by literature values.

Combustible materials are capable of burning, but do not generate the vapor pressure necessary for classification as a flammable material. The distinction between a flammable and a combustible material is sometimes drawn based on the temperature of the flash point. Finely divided combustible powders can form a dust cloud that can be ignited, making them particularly dangerous.

The chemicals used for polymerization are inherently reactive. There are potential hazards in a polymerization reaction due to various features, including the monomer and any species added to the reaction mass. The reaction, potential side reactions, and heat of reaction should be understood. Each reaction has a potential for pressure build up, flammability, exotherms, and shock sensitivity or other form of explosiveness. Radical-forming initiators are explosive, solvents are often flammable, polymers are often combustible, acid or base catalysts are corrosive hazards. The specific hazards of a polymerization reaction should be considered in designing the process and instituting safety protocols.

There are several potential health risks with exposure to the monomers and solvents necessary for low-viscosity processing. Toxic effects can be local or systemic; acute, or chronic; immediate or latent; and transient or persistent. The mode of entrance for these chemicals can be through skin absorption, inhalation, ingestion or injection, with absorption and inhalation being the most common. Some particular classes of health effects include poisonous, carcinogenic, mutagenic, and teratogenic. Most organic materials target the kidney and liver, where the body concentrates and eliminates toxins. Other target organs include the skin, the nervous system, and the heart. The measure of toxicity is the lethal dose that kills 50% of the animals to which it was given (LD50). LD50 is usually reported for several routes of contamination, including oral ingestion, subcutaneous, and inhalation. Key safety parameters for inhalation toxicity are the threshold limit value (TLV), which is the level below which routine exposure should be limited, and the short-term exposure limit (STEL), which is the largest single exposure allowed by safety guidelines. These levels are usually reported in parts per million. The dangers of any solvent or monomer are set out in the material safety data sheet (MSDS). Most states have right-to-know guidelines, which require that

MSDS for all chemicals in use in a laboratory be stored in a single location near the entrance to the laboratory.

It is important to read the MSDS and prepare for any hazard by using a fume hood, adequate ventilation and protective clothing. Notice that most rubber gloves are meant for casual contact, and should be changed immediately if contamination occurs. Many solvents pass through rubber gloves very quickly. Any material containing a leachable, toxic substance cannot be used for food contact.

REFERENCES

Perry, R., D. Green, J. Maloney, eds., Perry's Handbook of Chemical Engineering 7th International ed., McGraw Hill, New York (1997).

Brandrup, J., E. H. Immergut, E. A. Grulke, eds., Polymer Handbook, 4th ed., John Wiley and Sons, New York (1999).

Hiemenz, P., Polymer Chemistry, Marcel Dekker, New York (1984) pp. 273–658.

Miller-Chou, B., and J. L. Koenig, A Review of Polymer Dissolution, Progress in Polymer Science, vol. 28 (2003) pp. 1223–1270.

CHAPTER 7

Melt, High-Shear, and Direct Processing

This chapter describes various approaches to melt processing nanocomposites where the matrix is a high-viscosity material, such as a thermoplastic. Generally, the solid matrix material must be converted to a melt, the reinforcement must be added and mixed with the polymer, and then the target device must be formed. This chapter will explain some of the phenomena involved in this sequence.

The melting behavior of crystalline polymers is a kinetic and statistical phenomenon due to the slow response times of the long chains and the distribution of a polydisperse system. As with viscosity and dissolution, underlying phenomena determine the melting process. In addition to considering crystalline matrixes, amorphous plastics and ceramics may be chosen for particular applications. Amorphous polymers, and ceramic glasses, are processed relative to the glass transition temperature, rather than to melting.

Direct processing is another way to describe the techniques discussed in this chapter. The family of processes described does not require a solvent, melting to a low-viscosity liquid, or polymerization of a lower-viscosity monomer, in order to convert the polymer and reinforcement to a finished material.

The stages of direct processing are melting or softening, flow, and cooling. During melting, the native structure is disrupted, and a melt is formed. Melting occurs over a range of temperatures in a polymer. Softening occurs approximately 50°C above the glass transition temperature, over a range of temperatures. The flow through process equipment and molds involves the time- and shear-dependent viscosity discussed previously. The properties of the flow channel, including wall temperature

273

and shape, play a role requiring further consideration. Solidification, including crystallization, is a kinetic phenomenon requiring under-cooling to ensure timely and thorough transformation from a soft melt to a solid that can be physically handled without damage. However, the solidification process influences the quality of the crystal structure. The detailed steps leading to the finished part is called the process history.

This chapter considers each of the steps in a process history. The melting phase is considered first. The features of important processes will then be discussed, including low-shear techniques, such as compression molding, and high-shear techniques, such as extrusion. Thermokinetic processing, using thermal spray, is also given separate attention. Mixing the nanoreinforcement into the matrix requires consideration of flow fields during mixing processes. The elements of crystallization processes are also explored.

MELTING AND SOFTENING

In order to add reinforcing particles and nanoparticles to a bulk thermoplastic matrix, the polymer must be melted sufficiently to allow process flow rates at reasonable applied pressures. Melting requires a temperature sufficient to disrupt the order within a crystal. When the average melting temperature is exceeded in a metal or ceramic crystal, the atoms will be driven to separation, and the crystals will melt. The melting and crystallization temperature of metals and ceramics is a unique material property and a very weak function of the heating or cooling rate. On the other hand, melting polymeric crystals, whether organic or ceramic, is a kinetic phenomenon. When the average melting temperature is exceeded, only half of the atoms in a chain have enough energy to leave the crystal. A higher temperature is required to drive melting, regardless of the time allowed for the processing. The kinetics of diffusion also affect the rate of chains leaving the crystal.

Other factors also affect the melting behavior of a plastic resin. The kinetics of crystallization indicate that practical crystallization rates are achieved only within a reasonable range of super-cooling, ΔT. The greater ΔT is, the less perfect will be the crystals formed during cooling, and the lower the temperature at which they will melt. The greater ΔT is, the more defects will be present, and the closer to the crystal will come to being disrupted. In this situation, less additional energy is required to sufficiently disrupt and melt the crystals. There is also a distribution of chain lengths, wherein shorter chains may disengage from crystals more easily than longer ones. On the other hand, during the crystallization pro-

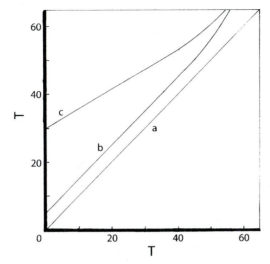

FIGURE 7.1 Melting range of a polymer, with observed temperature vs. crystallization temperature. (a) The freezing line, $T = T_c$. (b) The temperature at which melting is first observed for a given crystallization temperature. Notice that there is always a ΔT between the freezing and melting points. (c) The temperature at which melting is complete for a given crystallization temperature. The least perfect crystals melt first; the more-perfect crystals require a higher melting temperature. The lower the crystallization temperature, the more imperfect crystals form. (After Hiemenz.)

cess, a range of crystal perfection may be achieved within a single piece. As a result, the melting behavior will have a range, depending on the polymer structure and the ΔT used during part forming, as seen in Figure 7.1. During the process of melting, the last of all elements to disappear from the melt will be self-formed, homogeneous, nuclei. Over time, the homogeneous nuclei will most probably disappear with standard zero or first-order rate kinetics in the melt. Heterogeneous nuclei, externally added nucleation sites, will not melt. They will remain in solution, ready to nucleate new crystals on cooling.

Polymers and ceramic glasses have no melting temperature, yet cannot be processed below the T_g. Near the glass transition, the melt viscosity is still excessive. To achieve practical processing viscosities, a higher temperature must be attained. The temperature dependence is usually described by a standard thermal activation equation, such as the one below, given for Boltzmann statistics:

$$\eta(t) = \eta_0 e^{\left[\frac{\Delta E}{R}\left(\frac{T_0 - T}{T_0 T}\right)\right]} \tag{7.1}$$

Here, η_0 is the viscosity at a reference temperature T_0, and ΔE is the activation energy for flow. The WLF equation provides a relationship for viscosity changes relative to a reference temperature:

$$\log\left[\frac{\eta(T)}{\eta(T_0)}\right] = \frac{-17.44(T - T_0)}{51.6(T - T_0)} \qquad (7.2)$$

These values apply to most polymers, regardless of composition. The viscosity increases very rapidly approaching T_g. The simple shear flow is extremely limited below glass transition, rubbery between about T_g and $T_g + 100°C$, and freely flowing above this temperature. T_m is usually at or above this threshold, and there is little barrier to flow once the crystals melt. There may still be temperature-dependence for the viscosity, however, and an increase in temperature will usually lower the viscosity, ease processing, and increase the oxidation and degradation of the plastic.

The energy for melting may be imparted by heat conduction, compression, mechanical deformation, chemical reaction, electrical dissipation, and ultrasound. The most common is heat conduction, but the others do

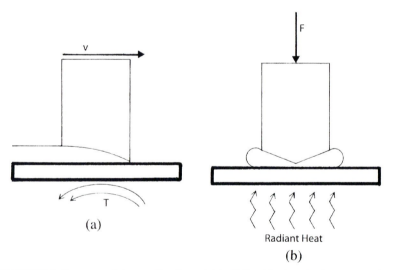

(a)

Radiant Heat

(b)

FIGURE 7.2 Contact melting applications. (a) A sliding block melts in contact with a hot surface, with melted material removed by the surface as it moves past. Here the surface is heated by forced convection with hot gas. (b) By applying pressure to the heating block, the melt formed in contact with the hot surface is forced out of the melting zone by the applied backpressure. Here the surface is heated by a radiant heat source, such as an infrared lamp or inductive heater.

have commercial application. For example, ultrasound is used to fuse layers of polyester nonwoven fabrics together. Composites formed with discrete or continuous, conductive reinforcements may be heated by resistive dissipation of electrical energy. Radio frequency energy can be used for radiative heat transfer.

Heat conduction can be applied in one of three basic modes. A solid can be added to an already molten material, with heating and agitation, called mix melting. Hot gasses may be circulated around the solid in order to cause convective heat transfer. Finally, the solid can be placed in direct contact with a hot surface, called contact melting. Within the context of contact melting, there are additional geometries with specific importance for process dynamics. Some are shown in Figure 7.2.

During the heating and melting process, a temperature gradient builds up through the material, as shown in Figure 7.3. Eventually the solid will melt, with extra heat consumed by the phase transition. A temperature profile, governed by conductivity, develops with time. Because of the high viscosity of the polymer, the molten polymer must be physically removed from the interface. Thus, generating melt for part production is driven by a balance of heat flow to generate molten material and the rate of mechanical removal of melt, along with the heat stored therein.

For Newtonian liquids with constant properties, a simplified heat transport equation in three dimensions governs temperature change:

$$\rho C_p \left(\frac{\partial T}{\partial t} + v_x \frac{\partial T}{\partial x} + v_y \frac{\partial T}{\partial y} + v_z \frac{\partial T}{\partial z} \right) = k \left(\frac{\partial^2 T}{\partial x^2} + \frac{\partial^2 T}{\partial y^2} + \frac{\partial^2 T}{\partial z^2} \right) + \Delta \dot{H}$$

$$+ \left\{ 2\eta \left[\left(\frac{\partial v_x}{\partial x} \right)^2 + \left(\frac{\partial v_y}{\partial y} \right)^2 + \left(\frac{\partial v_z}{\partial z} \right)^2 \right] \right.$$

$$\left. + \eta \left[\left(\frac{\partial v_x}{\partial y} + \frac{\partial v_y}{\partial x} \right)^2 + \left(\frac{\partial v_y}{\partial z} + \frac{\partial v_z}{\partial y} \right)^2 + \left(\frac{\partial v_z}{\partial x} + \frac{\partial v_x}{\partial z} \right)^2 \right] \right\} \quad (7.3)$$

The left side of the equation represents the heat stored at a point in the form of a temperature change, along with heat carried away from the point by flow. ρ is the density, C_p is the heat capacity at constant pressure, T is the temperature, t is the time, v_i is the velocity in the ordinal x, y or z directions. The right side has three components. The first is for heat conduction, where k is the thermal conductivity constant. The second term describes the heat generated by viscous dissipation, where η is the viscosity. The final term, $\Delta \dot{H}$ accounts for point sources generating heat.

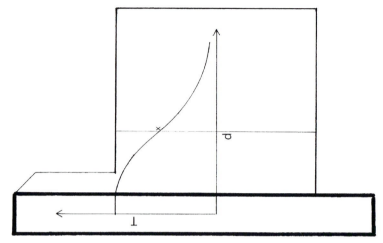

FIGURE 7.3 Temperature in a sliding block. The temperature, T, is highest in contact with the hot surface, and drops as a function of distance, d. There is an inflection in the temperature profile at x.

There are applications in which some of the terms in the complete expression are unnecessary. For example, in the absence of flow, as would be the case when heating a solid, the flow terms drop out of the expression. In the absence of chemical reactions, melting, or other heat generation terms, ΔH can be eliminated. Often, for complex geometries, in processing systems with point heat sources, such as a phase change, in systems with variable transport properties, or under many other circumstances, the heat equation becomes too complex to be solved analytically and must be solved numerically.

Phase changes and second-order transitions also may change the thermal properties of the material. The property changes may be incorporated in the heat transfer equation using an appropriate empirical or analytical equation form. The properties may be step changes, such as the density change at melting, continuous functions of temperature, such as the viscosity above melting, or discontinuous, as the heat capacity around T_g.

Examining some of the simple, explicitly solvable applications of the equation will help illustrate important features of heat transport. The first case to examine is steady-state heat transport across a flat plate, as shown in Figure 7.4. Heat transport is in one dimension; there is no flow, no heat generation term, and no time-dependent temperature change. The heat flow can be solved by the simple Fourier heat conduction equation, as follows:

$$Q = -kA \frac{dT}{dx} \qquad (7.4)$$

where Q is heat flow and A is the cross-sectional area across which heat is flowing. Integrating gives a linear temperature profile:

$$T - T_s = -\frac{q}{kA} x \qquad (7.5)$$

where x is the position through the thickness. The temperature varies linearly within the solid between the temperatures at either surface. The amount of heat passed by the wall depends on the temperature differential between the two surfaces and the thickness of the wall. This gives us a picture of the basic nature of heat conduction and the developing temperature profile in the part from which to build.

An assumption of one-dimensional steady-state conductivity requires perfect contact. This may be difficult to achieve, particularly for two rigid materials in contact. Protruding features on the surface may be the only regions of thermal contact. In the gaps between the surfaces, conductivity is governed by the properties of the gas or liquid trapped there. If these gaps are large enough, a circulating cell can form. If this occurs, convection can play a role in heat transfer at the boundary.

Heterogeneous materials such as nanocomposites do not have a single value for thermal conductivity. In principle, the conductivity can be calculated as shown in Figure 7.4, taking proper account of particle shape and distribution. For randomly distributed, uniform, spherical particles with good thermal contact with the matrix where the volume fraction of the reinforcement, x_V, is below about $10\%_{vol}$ an effective conductivity can be determined:

$$k_T = k_m \left[\frac{2k_m + k_r - 2x_V(k_m - k_r)}{2k_m + k_r + x_V(k_m - k_r)} \right] \qquad (7.6)$$

where k_m is the conductivity of the matrix, and k_r is the conductivity of the reinforcement. The conductivity will have a different form at higher concentrations, different shapes of reinforcement, very different values of conductivity, such as metal particulates in a polymer matrix, or a non-uniform distribution. The conductivity of a powder is also a function of the powder's composition, the packing efficiency, the effective contact between particles, and the conductivity of the gas or liquid in the voids. These expressions are beyond the scope of this introduction.

When the solid is effectively infinite, the approach to the solution changes in character (Figure 7.5). This is the problem of heat conduction

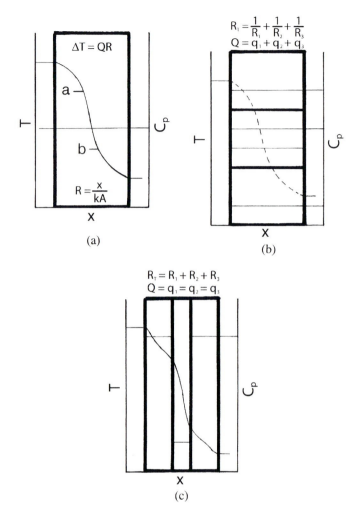

FIGURE 7.4 Schematics of thermal profiles in finite solids. (a) The temperature depends on heat flux, Q, and thermal resistivity, R. For a solid block, there is a single heat capacity. The temperature profile as a function of thickness varies between the temperature at surface 1 and the temperature at surface 2 with inflection points at a and b. (b) For a composite block with the blocks in series, the heat flux and resistivity are composites of the individual components. The top and bottom blocks in the illustration have a lower heat capacity; the middle block has a higher heat capacity. The majority of the heat conduction occurs through the center block, but the average thermal profile as a function of thickness profile looks similar to the solid block. (c) For a composite block with the blocks in parallel, the resistivities and heat fluxes add differently. The middle block as shown has a lower heat capacity than the outer blocks. The average temperature as a function of thickness goes through the most rapid change in the center block. The temperature profile in a composite solid can be difficult to evaluate explicitly. However, R and Q may add as if the particles are in series or parallel.

280

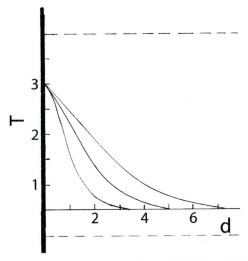

FIGURE 7.5 Temperature profile for heating in a semi-infinite solid. Initially, the entire block is at a lower temperature than the applied temperature at the outer surface. The temperature at the surface of the block quickly reaches the applied surface temperature, but the temperature inside the block increases only slowly.

in a semi-infinite solid with a step change in surface temperature. This type of problem describes the temperature variation occurring when heating a typical part. The solution to this problem provides insight into what happens to the temperature of a solid on a hot plate, when immersed in a fluid heating bath, or when the surface is heated using other techniques. The heat transfer equation can be solved analytically. Assuming constant material properties, the governing equation is:

$$\rho C_p \frac{\partial T}{\partial t} = k \frac{\partial^2 T}{\partial x^2} \tag{7.7}$$

where ρ is the density, C_p is the heat capacity at constant pressure, and k is the thermal conductivity. A cluster of thermal constants, the thermal diffusivity $\alpha = k/\rho\, C_p$, may be defined and tabulated. The explicit solution to this equation provides a description of the temperature profile:

$$\frac{T - T_s}{T_0 - T_s} = erf\left(\frac{x}{\sqrt{4\alpha t}}\right) \tag{7.8}$$

where x is the depth into the solid, T_0 is the initial, homogeneous temperature in the solid, and T_s is the temperature at the surface, which was changed at $t = 0$. The temperature profile has a constant shape and changes with time according to the error function. The heat flux, Q, as a function of time may also be defined as follows:

$$Q = \frac{k}{\sqrt{t\pi\alpha}}(T_s - T_0) \tag{7.9}$$

Q is a transient measure. The heat flow rate, $\dot{Q}$, would be the time differential of this expression. The heat flow is initially infinite, but is inversely dependent on $\sqrt{t}$. The heat flux drops rapidly with time, and the efficiency of heating is reduced. Therefore, the molten material must be removed in order to retain the efficiency of conductive heating. The heat flux is also directly dependent on $\sqrt{k}$, $\sqrt{\rho}$ and $\sqrt{C_p}$, so increases in any of these properties will increase heat flow. A numerical solution is necessary for the case where the properties are not constant, however.

So far, the solutions have not included the melt transition. Introducing the desired phase change creates a still more complex situation. After the phase change, the density changes, resulting in an increased thickness in the melt boundary. This corresponds to a change in the distance required for heat conduction, as shown in Figure 7.6. At the boundary between the solid and melt, as it advances through the solid, the temperature must be T_m for both the liquid and the solid in contact with it. The location of the melt boundary is displaced by the ratio of the densities:

$$\frac{x_l}{x_s} = \frac{\rho_s}{\rho_l} \tag{7.10}$$

where x_l is the thickness of the liquid layer, and x_s is the thickness of an equivalent amount of unmelted solid. Each phase is a semi-infinite material, with a matched boundary condition at the phase front. The temperature profile in the melt follows a specific functional form, the error function:

$$\frac{T - T_m}{T_s - T_m} = 1 - \frac{erf\left(\dfrac{x}{2\sqrt{\alpha_l t}}\right)}{erf\left(K\dfrac{\rho_s}{\rho_l}{2\sqrt{\alpha_l}}\right)} \tag{7.11}$$

The temperature profile in the solid is:

$$\frac{T - T_m}{T_0 - T_m} = 1 - \frac{1 - erf\left(\dfrac{x}{2\sqrt{\alpha_s t}}\right)}{1 - erf\left(\dfrac{K}{2\sqrt{\alpha_s}}\right)} \tag{7.12}$$

where K is an unknown constant related to the location of the melt front:

$$x_s = K\sqrt{t} \tag{7.13}$$

The amount of material melted is seen to be directly proportional to the square root of the elapsed time. The rate of melting per unit area is related by:

$$R_m = \frac{\rho_s K}{2\sqrt{t}} \tag{7.14}$$

The rate of melting, much like the heat flow, is inversely proportional to the root of elapsed time. Again, the value of mechanically removing the melt is clear. The analyses of systems including melt removal do not provide simple relations for melt thickness, melt rate and so forth. This level of detail again is beyond the scope of this text.

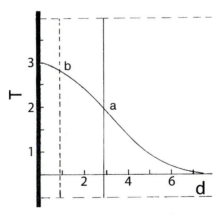

FIGURE 7.6 Phase change in a solid during conductive heating. The temperature drops off to T_0 as a function of depth. The temperature at a. is sufficient to melt the solid, representing the melt front. However, the melt expands, causing the surface to expand past the original, solid face at b. Thus, as the melt front advances, the heat must penetrate farther, due not only to the moving melt front, but also to the increased distance caused by the melt. The only way to overcome this is by removing the melt.

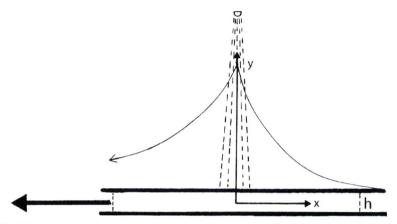

FIGURE 7.7 A moving heat source creates an asymmetric heating profile. Conduction in the solid causes some heating in the solid ahead of the source. The greatest heating occurs in the heating zone. However, the heat loss after the heat source is removed is slower, and the temperature decreases slowly.

Moving Heat Source

Another real situation is a sheet of material passing under a heat source, as shown in the diagram in Figure 7.7. Behind the heat source, the material loses heat to the surroundings as h_c, where c is the perimeter area and h is the convective heat coefficient of the gas. The material is preheated via conduction through the sheet as kA, where k is the thermal conductivity of the material and A is the cross-sectional area. The energy equation is:

$$\rho C_p V_0 \frac{dT}{dx} = k \frac{d^2 T}{dx^2} - \frac{hc}{A}[T(x) - T_0] \qquad (7.15)$$

V_0 is the velocity of the moving sheet or the moving heat source. The most interesting value is the maximum temperature increase that can be obtained. The expression after solving the differential equation is:

$$[T(x) - T_0]_{\max} = \frac{I}{2k\sqrt{\left(\frac{hc}{kA}\right) + \left(\frac{V_0}{2\alpha}\right)^2}} \qquad (7.16)$$

I is the intensity of the source, and the maximum temperature rise decreases with increasing velocity, increasing thermal conductivity, and heat transfer coefficient.

MELT PROCESSES WITH SMALL SHEARS OR LOW-SHEAR SHEAR-RATES

The first family of processes we will discuss is lower in applied shear stress. The viscosity of the melt is typically high, and the addition of the reinforcement particles, particularly nanoparticles, will make the viscosity still higher. However, instead of trying to displace melt, requiring high applied shear stress, these techniques accomplish part production with smaller displacements and therefore more modest shear forces.

Powder Coating and Rotational Molding: Sintering

Powder coating and rotational molding are very low-shear melt processes that fuse the solid particles directly, in a fashion similar to that previously discussed for latex film fusion. However, the powder is not carried in a solution, but is directly deposited on a hot surface. The powder accumulates as a semi-molten or high-viscosity liquid against the surface. The particles reach melt temperature, and may in turn attach new particles. The particulate agglomerates and then fuses by sintering. The nanoparticles can be co-mixed with the matrix powders, and will be incorporated at the boundaries of the fusing polymer particles. Solid, unmelted reinforcements will almost certainly have an impact on the agglomeration process.

Powder coating can be accomplished by dipping a hot substrate into a polymer or composite powder. A constant stream of air may be used to fluidize the powder bed. The air may be heated in order to pre-heat the powder, improving the dynamics of particle attachment. An alternative powder delivery system is electrostatic coating: the powder is attracted to the surface of the substrate due to an electrical charge placed on the part. Heating the substrate and the coating together then fuses the coating. The particle accumulation occurs under effectively zero shear.

A method for creating hollow parts with a solid surface is rotational molding. In this technique, a hollow mold with the target shape is heated to above the melting point of the polymer. Powdered polymer is introduced to the mold cavity, which is spun on three axes. The powder melts and sticks on contact with the hot mold surface. After the surface is coated, new particles will contact and stick to the new plastic surface, accumulating the part thickness. Again, the part builds with effectively zero shear.

The coatings must sinter after the powder is deposited on the surface. In sintering, the viscous flow of the melt is driven by the energetic advantage of decreasing the surface area. Distinct particles will merge and fuse as shown in Figure 7.8. First, a bridge will form between particles in contact. The bridges will thicken after forming during the coalescence phase. An expression for the rate of fusion during this period follows the form:

$$\frac{x^2}{R} = \frac{2}{3}\frac{\gamma_s}{\eta}t \qquad (7.17)$$

where x is the neck radius, R is the particle radius, γ_s is the surface tension, and t is the time. The specifics of the form for polymers are different than they are for glasses, ceramics and metals, which is due somewhat to the non-Newtonian melt viscosity. The equation does not have an explicit dependence on temperature. Until the ratio of the neck to the parti-

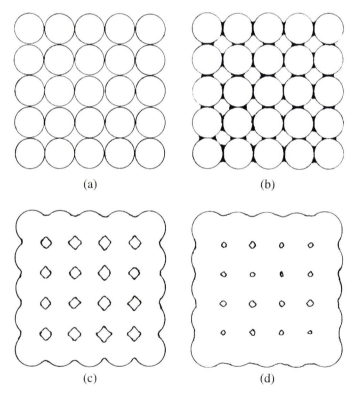

(a)   (b)

(c)   (d)

FIGURE 7.8 Sintering schematic: (a) Beads initially packed in a regular array. (b) The beads begin to merge at the initial points of contact. (c) The initial contact points expand to form bridges. (d) At some point, the sintering part is more akin to a solid with imperfections in the form of holes than to a bed of merging beads. The solid can only continue to sinter by diffusion of the holes out of the solid.

cle diameters reaches about 0.5, the particles are considered to be coalescing. When the particles have fused to this point, the particle bed may be better described as a solid with large holes. During the rest of the sintering process, densification occurs. Densification follows a different rate expression:

$$\frac{r}{r_0} = 1 - \frac{\gamma_s}{2\eta r_0} t \qquad (7.18)$$

where r_0 is the initial radius of the holes left in the coalesced film, and r is the radius at time t. Again, the time and temperature dependence of viscosity, and the time dependence of rate, are not explicitly considered. Nanoparticulate reinforcements, particularly at the boundaries, will also hinder powder fusion. The source of the energy drive for flow during sintering is also not addressed. Nevertheless, these expressions provide some insight into the relative impact on the expected rates of fusion for matrix powders.

Compression Molding

To compression mold, a charge roughly the desired shape of the finished part is placed in a mold. The charge can be either a solid piece or powder. The part is heated, and pressure is applied to fill out the mold. Shown schematically in Figure 7.9, the charge is placed in the center of the mold, and the force causes even spreading of the plastic to fill all parts of the mold. Upon completing the part, the piece is removed, allowing a new cycle to begin. There are several advantages to compression. First, the finished part is usually near net shape; there is little excess material requiring removal in a second finishing step. The small flows required also make compression a good process for very high-viscosity materials such as nanocomposites. Processing challenges arise from the same features. The thickness of the part is controlled by how perfectly the mold closes. This depends in turn on exactly how much material was in the initial material charge, making tolerance control particularly challenging. The process also is generally not good for intricate parts, since flow is not normally sufficient to allow large undercuts or complex geometries to fill.

Thermosets and short-fiber or heavily filled sheet molding compounds (SMCs) are often compression molded. A partially cured thermoset maintained in the B-stage can undergo the relatively small flow required for mold filling soon after the initial heating. Following mold filling, the heat transfer can complete the cure. The curing step can be accounted for using a heat source term for the reaction. Thermoplastics are occasion-

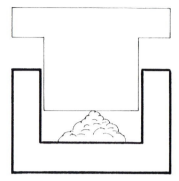

FIGURE 7.9 Compression molding. A powder is placed into a hot mold. After the polymer is sufficiently softened through melting, high pressure can be used to close the mold. The polymer will flow to fill the entire mold. Nanoparticles will increase the viscosity, and consequently the pressure required to close the mold.

ally compression molded, as is the case, for example, with margarine tubs, but this is less common because of the limited part geometry and because removing excess polymer to improve tolerances is challenging.

The central force, F, required to close the mold during compression molding of a power-law fluid at a constant mold closing rate is given by:

$$F = m\pi \frac{\left(\dfrac{2n+1}{n}\right)^{n}}{2^{n}(3+n)} \frac{(-\dot{h})^{n} R^{3+n}}{h^{1+2n}} \tag{7.19}$$

where m and n are the power-law constants, h is the mold separation or charge height, $\dot{h}$ is the mold closing rate, and R is the final radius of the part after molding. R may be replaced by another characteristic distance within the mold for noncircular parts. Off-center placement of the charge would not have the same value for the closing force. The resulting gap thickness, or how much closure has occurred, for a constant force as a function of time is predicted to be:

$$\frac{h(t)}{h_{0}} = \left[1 + \frac{2\left(\dfrac{n+1}{n}\right)\left(\dfrac{3n+1}{n}\right)^{\frac{1}{n}}}{\left(\dfrac{2n+1}{n}\right)} \left(\frac{F}{m\pi R^{2}}\right)^{\frac{1}{n}} \left(\frac{h_{0}}{R}\right)^{\frac{n+1}{n}} t \right]^{\frac{-n}{1+n}} \tag{7.20}$$

where h_{0} is the initial thickness of the charge.

Thermoforming

Sheets of plastic can be thermoformed into simple part geometries as well, which is shown in Figure 7.10. In thermoforming, a sheet of thermoplastic or B-stage thermoset is heated until softened. The sheet is then applied to a thermoforming tool, using a combination of gravity, vacuum draw, mechanical plunger and/or air pressure. The film cools on contact with the mold, and the part can be removed. During thermoforming, the part experiences elongational flow. The relatively lower temperatures and higher viscosities of thermoforming can result in significant molecular orientation, with commensurate impact on structure formation during cooling.

Particular thermoforming challenges are also present, most of which are exacerbated by reinforcement particles. There are practical minimum and maximum temperatures set by sufficient softening to comply with the tool and insufficient to permit excess sagging, respectively. In addition, since the extruded sheet charge may be large and relatively thick,

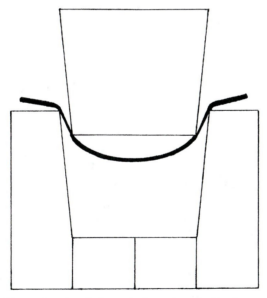

FIGURE 7.10 Thermoforming. A sheet of polymer is melted by, for example, radiant or convective heating. A vacuum is drawn in the mold cavity, drawing the molten polymer into the mold. A plug can be used to assist the vacuum forming. Composite particles can increase the viscosity, making the vacuum insufficient to draw the film into the cavity effectively.

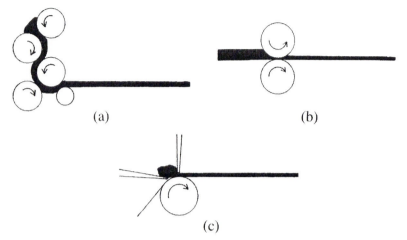

FIGURE 7.11 (a) A sheet of molten polymer is transferred from roll to roll in a roll mill. The rolls may be used to supply heat to melt the sheet or apply shearing pressure to thin the sheet, which is called calendaring. (b) Counter-rotating rolls, with or without heat, may be used to thin a sheet of polymer or to apply a coating to a surface in a calendar. (c) A reservoir of polymer may be passed through a narrow gap, created by a doctoring blade, to create a thin coating.

achieving a uniform temperature can be difficult. The part usually exhibits thickness variations that arise through differential heating across the part, differential drawing due to part geometry, and the sequence of mold contact. Material on the side of the part contacts the mold early, freezing further deformation. Meanwhile, other parts of the mold continue to be filled. The mold filling and shape issues are sometimes addressed by creating an uneven heating profile for the charge.

Rolling, Calendering, and Doctoring

Rotating cylinders can provide the heat for transforming a polymer or composite prepreg into a melt or a sheet. Pressure to change the dimensions of the material charge is then applied, using a second roll or a doctoring blade. Rolling, calendering, and doctoring are therefore different but related processes. A schematic of these processes is shown in Figure 7.11. In a roll-mill, two rolls run at different speeds, and the material is passed between the two rolls. A sheet of material is usually wound around one of the rollers. The purpose of the roll mill is generally to melt the material. In a calender, only one pass occurs between any pair of rolls. The calender is usually used to shape the product into a sheet form. If a doctoring blade is used to do the shaping, the process is blade over roll instead of a roll over roll process.

Cylinders for roll mills or calenders are generally polished to be very smooth. If melting is desired, the rolls can be heated using circulating fluid, resistive elements, or radiative heat transfer. The material is passed between two rolls, or a knife and a roll, to transform the thickness of a molten, semi-molten, or cold-workable solid charge of material. The key to controlling thickness is to maintain sufficient pressure and gap distance to ensure the desired thickness change. Figure 7.12 shows a schematic of the flow through a gap between two rolls.

The process flow rate between the cylinders per unit width is constant and independent of position:

$$Q = 2h_r \, 2\pi\Omega R \qquad (7.21)$$

where Ω is the rotational velocity of the cylinder, R is the radius of the cylinder, and h_r is the film thickness at release from the roll. The processing rate is therefore directly proportional to drum speed, drum diameter, and film height. The pressure under the pinch point is not constant, but

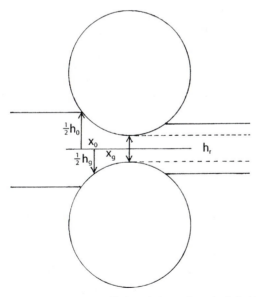

FIGURE 7.12 The gap between two cylinders, h, is not the only definitive parameter for calendaring rolls. The thickness where the sheet enters the rolls, h_0, and the gap where the rolls grip the sheet, h_g, are also important gap sizes. The distances, x, relating these points can be defined relative to the center of the gap.

builds from the initial bite at x_g according to an equation made somewhat complex by the boundary conditions:

$$P = \frac{6\pi\Omega R\eta}{4h_0}\sqrt{\frac{R}{2h_0}}\left\{\left[\frac{d^2 - 3d^2 d_r^2 - 5d_r^2 - 1}{(1+d^2)^2}\right]d\right.$$

$$\left. +(1 - 3d_r^2)\tan^{-1}(d_r) + \frac{(1+3d_r^2)}{(1+d_r^2)}d_r - (1 - 3d_r^2)\tan^{-1}(d_r)\right\} \quad (7.22)$$

h_0 is twice the gap height at the closest approach between the cylinders, entered in this form to allow for symmetry across the centerline of the gap. Further, the equation can be made somewhat simpler to write by defining normalized, unitless distances along the total gap length:

$$d^2 = \frac{x^2}{2Rh_0}, \quad d_r^2 = \frac{x_r^2}{2Rh_0}, \quad d_g^2 = \frac{x_g^2}{2Rh_0} \quad (7.23)$$

where x is the distance away from the origin, defined for convenience as the center of the gap; d is the normalized distance from the gap; d_r is the normalized distance to the point of film release from one of the forming surfaces, upstream of the gap; and d_g is the normalized distance, negative in the x direction and downstream of the gap, to the point of "bite" where the charge contacts the cylinder. The cylinders are taken to be large relative to the gap height, so curvature is neglected. From this expression, the pressure is observed to be atmospheric at both x_r and x_g. The pressure differential with respect to x is zero at x_r and $-x_r$. The maximum pressure occurs when $d = -d_r$. Extremes in shear rate and shear stress occur at the gap where $x = 0$:

$$\dot{\gamma}_m = \frac{3\pi\Omega Rd_g^2}{h_0} \quad \text{and} \quad \tau_m = \frac{3\eta\pi\Omega Rd_g^2}{h_0} \quad (7.24)$$

The total force required to maintain the gap in the absence of normal forces is the integral of Equation (7.22) from the point of bite to the point of release:

$$P = \frac{3w\eta\pi\Omega R^2}{4h_0}\left\{\frac{d_r - d_g}{1+d_g^2}[-d_g - d_r - 5d_r^3(1+d_g^2)]\right.$$

$$\left. + (1 - 3d_r^2)[d_r\tan^{-1}(d_r) - d_g\tan^{-1}(d_g)]\right\} \quad (7.25)$$

where w is the width of the charge as it passes through the pinch point. The dramatic increase in viscosity caused by the addition of nanoparticles to a melt will notably increase the pressure required to maintain the gap height. Positive normal stress coefficients also indicate the formation of normal forces. The normal force will result in both an increase in the pressure required to maintain the gap distance and swell in the exiting film, as the pressure drops back to atmospheric after release from the gap. Therefore, normal force will move the release point farther from the pinch point.

There is little shear force applied to the matrix if the calendering rolls are large. However, there will be significant elongational flow, with corresponding material flow and the potential for particle and molecular alignment.

FLOW

In Chapter 5, viscosity and flow were considered, since viscosity and viscous loss are part of all material transport, regardless of the liquid velocity. In producing and transporting the feed for direct processing methods, additional features of flow may be important. Particle conveying, flow fronts and flow mixing, and mixing high-viscosity materials with each other or incorporating small particles will be discussed.

Moving Particles

Most reinforcements and raw matrix materials are particulate. The primary forms are pellets, powders, beads, or granulates, which can have diverse shapes: smooth or irregular spheres, cylinders, disks, lenses, ellipsoids, or regular or irregular multi-lobed blobs. The shape will affect the flow properties of the solid material. These particulates are usually previously formulated with all necessary additives for processing. Powders are formed by, for example: pulverizing a solid; melting the solid, forming a wire or rod and cutting it up; or dissolving and precipitating the solid.

The particulate solid flows, after a fashion similarly to liquids, for convenient storage and transportation. Particles can be piped and poured. The flow of particles often exhibits dilatancy when there is a free surface. The shear stress in particulate solids, unlike liquids, is proportional to the normal load. Particles do behave differently from liquids in other ways. Applied pressure can compress the solid bed rather than initiating flow. Particulates tend to stack in metastable structures, forming cones, which

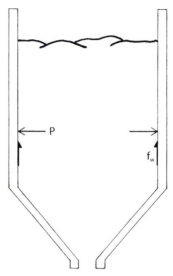

FIGURE 7.13 A bin, or hopper, filled with powder. The powder above a given point applies pressure to the side of the bin. The walls of the bin apply a friction force to the powder, preventing flow along the walls. The balance of forces determines the emptying behavior of the bin.

then give way in an avalanche to form a new metastable cone. The particles can agglomerate into aggregates, if there are attractive forces or interactions between them. Agglomeration is undesirable in flowing systems, and is desired for tableting or caking processes.

Particles are often stored in bins or hoppers, Figure 7.13. The pressure exerted in the hopper is given by the same $\rho g h$ as with a liquid, although the value of density may be more difficult to ascertain, and may be a function of depth. Further, there is friction along the walls that can divert some of the force to the walls of the storage container. Finally, the powder may arch or dome. For a cylindrical bin with no adhesion and no applied pressure at the top, an approximate equation for the pressure may be obtained:

$$P = \frac{\rho_b g D}{4 f_w K}\left\{1 - \exp\left[\frac{4 f_w K(h - H)}{D}\right]\right\} \tag{7.26}$$

where ρ_b is the bulk density, D is the diameter of the hopper, H is the particle column height, f_w is the coefficient of friction with the wall, and K is the ratio of compressive stress in the horizontal direction to that in the vertical direction. The bulk density and the ratio of vertical to horizontal stresses are constant and independent of depth, the vertical compressive stress is constant along any horizontal plane, and wall friction is in a condition of incipient flow. This expression has a maximum pressure, as H approaches infinity:

$$P_{max} = \frac{\rho_b g D}{4 f_w K} \qquad (7.27)$$

Most of the weight is borne by the walls of the bin. The maximum pressure is inversely proportional to the friction constant at the wall: the ability of the force to be passed into the wall. This friction effect is also responsible for bridging and arching. The discharge of solids from a bin or hopper has a flow rate, flow kinematics, and disturbance or disruption. Flow disruption may occur in cohesive powders, or when the hopper is improperly designed.

Flow in Channels

Once the particulate has entered the process stream, flow must continue through a channel until reaching a melting zone. After melting, liquid flow behavior takes over. The solid may be pushed through by a ram or dragged forward by a moving surface, by pressure, or by shear. Posi-

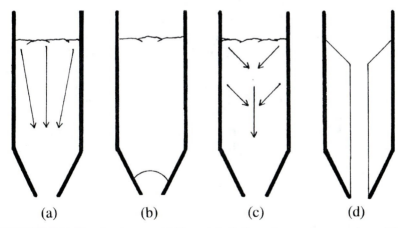

    (a)   (b)   (c)   (d)

FIGURE 7.14 Flow in a hopper: (a) Normal, bulk flow where mass moves toward the center drain, with material near the walls moving last. (b) Bridging can occur when an arch of powder blocks regular powder movement. (c) Funneling occurs when the material directly above the drain can flow, and everything feeds towards this center region rather than moving evenly toward the drain. (d) Piping occurs where the material in the center of the hopper drains rapidly, leaving the mass of particles around the edges behind.

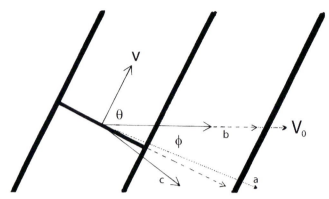

FIGURE 7.15 Schematic of pressure and drag flow in a channel, such as in an extruder. V_0 is the velocity of the screw, v is the velocity of the polymer in the channel, and θ is the angle between these two, the screw pitch. a, b and c represent three different conditions for resolution between the screw and the velocity of the screw. ϕ is the angle between the moving screw and the force applied to the moving polymer. The faster the solids move, from c to a, the greater the force applied to the moving polymer.

tive displacement drives cause both conveying and compaction. Friction is a cause of reduced flow rate in positive displacement flow. Drag causes conveying, and compaction occurs at any walls crossing the direction of flow. Friction is the source of flow in drag flow.

The pressure required to drive a ram in a cylindrical channel depends on the friction coefficient and the output pressure required. If an outlet pressure is required, for example, to fill a mold or to produce sufficient driving force to carry the solid into the process, then the ram must do the job of building pressure, overcoming compaction, and generating forward flow. Certain assumptions must be made, including: flow is steady and streaming, with no radial variation in velocity; the ratio of radial to axial flow, K, is constant; the friction coefficient is constant; and temperature effects can be ignored. A form for the pressure drop is observed:

$$\frac{P}{P_0} = e^{\left(-\frac{4L_0}{D} f_w\right)} \tag{7.28}$$

where L_0 is the initial length of the powder charge, D is the diameter of the channel, P is the discharge pressure, and P_0 is the initial pressure. The pressure drop is observed to be exponential with length, rather than linear, as observed with liquids. The flow rate is determined by the positive displacement of the ram, dx/dt.

The most common example of drag flow, caused by sliding a moving plate across an angled channel is found in single-screw extruders. For drag flow, the flow rate is governed by the geometry of the system:

$$\tan\phi = \frac{v\sin\theta}{V_0 - v\cos\theta} \qquad (7.29)$$

where ϕ is the angle between the velocity of the plate, V_0, and the force applied on the flowing solid, while θ is the angle between the flow of the liquid, at velocity v, and the velocity of the plate. This expression relates the velocity of the solid material, dx/dt, and the flow rate to the angle between the channel and the dragging surface. The downstream component of the velocity, only, can result in flow. The cross-channel force is that exerted by the walls to counter the component of the force directed into the wall. The pressure drop in this system, including the cross-channel force exerted by the walls, can be found using the same assumptions as for a ram extruder:

$$\frac{P}{P_0} = e^{\left\{ \frac{\left| f_{w_1 Kx}\left[\cos(\theta+\phi) - f_{w_2}\sin(\theta+\phi) - \frac{f_{w_1}}{f_{w_2}}\left(1 + \frac{2H}{W}\right) \right]\right|}{H} \right\}} \qquad (7.30)$$

where H is the channel height, W is the width of the channel, f_{w_1} is the friction coefficient with the moving surface and f_{w_2} is the coefficient of friction with the stationary wall. The cross-channel force reduces the amount of solids conveyed at a given pressure drop.

Mixing

Mixing can occur through a variety of processes. Molecular diffusion is driven by concentration gradients. This occurs in low-viscosity gases and liquids. If, additionally, there is turbulence in the liquid body, then molecular diffusion is superimposed on eddy diffusion. If, further, there is flow, then both of these occur within the convective flow mixing process. The large molecules in a polymer hinder molecular diffusion. In addition, since flow is largely laminar in these systems, there is little eddy diffusion, and mixing is dominated by convective flow. Particles further impact the flow profile. Figure 7.16 shows two different types of mixing, random and ordered, that might occur during a processing step.

Mixing is a challenging aspect of processing. Good mixing requires a combination of an adequate starting condition, both orientation and ar-

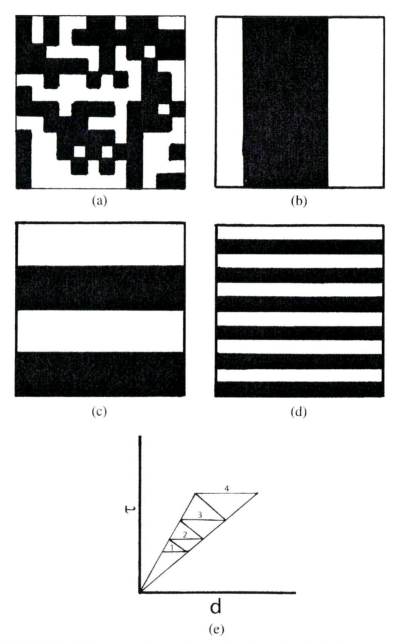

FIGURE 7.16 (a) Random mixing requires complete disruption of both phases, such as occurs in a high-shear blender. (b)–(d) Ordered mixing occurs, for example, in stationary mixers. With each rotation, the two components, originally distinct, are sheared and folded to cause mixing. (e) The shear to displacement of a particular point during each rotation of the solid follows a step-wise pattern.

298

rangement, and a sharp distribution of shear force. A poor initial distribution can hamper the reorganization of the smaller, discontinuous component in the larger component, as shown in Figure 7.17. A broad shear distribution also will not disrupt the flow front sufficiently for satisfactory mixing. In addition to including localized shearing zones, mixing can be assisted by changing the direction of shear.

If the unmixed material begins as layers with a given thickness, then a rule of thumb for complete mixing is a 4- order-of-magnitude reduction in layer thickness. The real goal is to reduce the size scale to the point where diffusion mechanisms can do the remainder of the randomization. Mathematically, one may begin with randomly arranged cubes of a minor component in a major component with equal sized interfacial areas. The average thickness of these cubes, t, under a shear, γ, is described by:

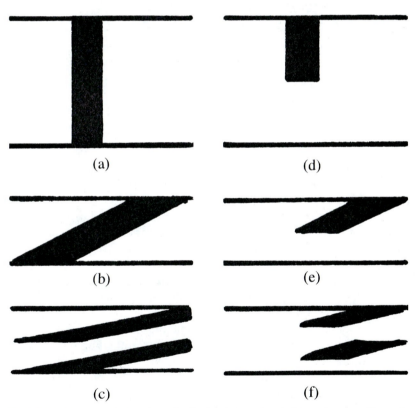

(a)   (d)

(b)   (e)

(c)   (f)

FIGURE 7.17 (a)–(c) If the initial distribution in an ordered, shear mixer is uniform, the final distribution will be uniform. (d)–(f) However, a poor initial distribution will result in poor mixing.

FIGURE 7.18 Liquid elements near the upper, moving surface have the shortest dwell time and experience the least shear. Liquid elements near the lower, stationary surface have a very long dwell time and experience a great deal of shear.

$$T = \frac{2L}{3\gamma x_V} \qquad (7.31)$$

where x_V is the volume fraction of the minor component and L is the length of the cube side. The thickness of the layers depends inversely on the shear and the volume fraction of the minor component, and directly on the size of the layers. This leads right to the understanding that mixing small amounts of a minor component is more difficult than mixing larger amounts. Larger zones of the minor component are also more difficult to mix thoroughly.

Mixing is not only a matter of concentrated shear stress. There is also an effect due to the residence time in the high-shear zone in the mixer. In Figure 7.18, a single pass of our putative cubes through a high-shear pinch-point results in large distortions near the stationary wall and small deformations near the moving surface. However, the number of cubes passing through the pinch point is the sum of all cubes passing through or over the cross-sectional area. The flow rate near the moving surface is high, that is, the residence time is low. Close to the wall, the residence time is higher, the deformation is higher and the flow rate is lower. The longer the residence time is in the shear zone, the greater is the deformation, and the slower is the processing. Thus, there is a balance between residence time in the mixing region and the material flow to the next stage of processing.

If we turn to a drop, or an approximately spherical agglomerate of nanoparticles, instead of a cube, the process of homogenization can be reformulated as deformation of the drop until it breaks into smaller pieces. The deformation of the droplet, D, can be described by the change in shape induced by the shear flow:

$$D = \frac{m_1 \dot{\gamma}}{\zeta_s} \frac{19 \frac{\eta_2}{\eta_1} + 16}{16 \left(\frac{\eta_2}{\eta_1} + 1 \right)} \tag{7.32}$$

where η_1 is the viscosity of the continuous phase, η_2 is the viscosity of the secondary phase, $\dot{\gamma}$ is the shear rate, and ζ_s is the interfacial energy between the two materials. The drop will become longer and thinner, until the energy difference between breaking the particle and further elongating the particle is insignificant. At this point, the droplet will break. Rupture seems to be limited by the viscosity ratio, $0.005 < \eta_1/\eta_2 < 4$, and the critical rupture value seems to be about $D = 1/2$. If the viscosity of the

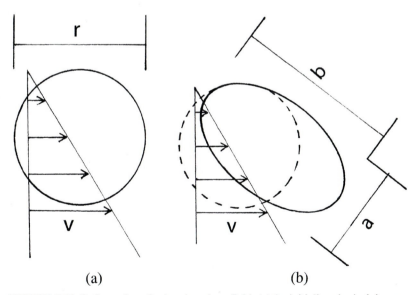

(a)   (b)

FIGURE 7.19 Deformation of a drop in a shear field: (a) An initially spherical drop experiences more displacement at one end. (b) The result is an elongated drop.

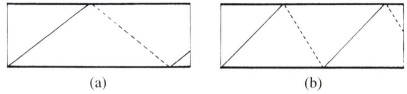

(a)   (b)

FIGURE 7.20 Particle path in a screw: (a) A high shear flow and a low pressure at the exit allow rapid progress along the barrel. The particle executes a helical path along the screw. (b) If the exit pressure is increased, the progress of the particle is slower, the helical path has a shorter period, and more mixing occurs.

second component is too high relative to the surroundings, the particles behave as solids, rotating if small and aligning with flow if elongated. There is also a minimum particle size to allow particle rupture.

Particles tend to clump during storage. Various surface forces contribute to agglomeration, including water adsorption, chemical bonding or electrostatic interactions. To properly reinforce the matrix, any agglomerates must be broken up during processing, particularly in the mixing stage. In order to break up agglomerates of particles, sufficient shear force between two particles in contact must exceed the adhesion strength between the particles. The maximum force acting on a pair of particles in contact is given by:

$$F = 3\pi\eta r_1 r_2 \qquad (7.33)$$

The maximum force depends on the viscosity of the dispersing medium, the shear rate, and the product $r_1 r_2$. So, de-agglomeration is best accomplished in high-viscosity media, at high-shears, and with large agglomerates.

The single screw extruder is one example of a successful approach to mixing. There is a constant change of direction in flow, as shown in Figure 7.20. The amount of churning depends on the relative importance of the contributions due to pressure and drag flow rate.

Flow Fronts

Laminar flow has advantages and disadvantages. Laminar flow does not allow backflow of material: flow channels and molds are filled evenly and smoothly. Figure 7.21 shows many features of a center-filled channel, such as an injection-molded disk. At the inlet, as the liquid pours into an open area and begins to build sufficient pressure to initiate flow through the channel, the liquid will mix. A short distance from the mold entrance, the flow will become fully developed, demonstrating the familiar parabolic distribution of velocity profile. At the advancing flow front, the liquid at the center of the channel is traveling more rapidly, the material near the walls is traveling slowly, and mass transport will be flow from the center of the channel, then toward the edges of the channel. At the channel wall, the material will begin to move more slowly. The distribution in flow yields a distribution in shear forces, highest at the surface of the channel and minimal at the center. This can cause orientation or alignment of reinforcements. Further, the liquid will freeze first at the surface of the channel, creating a skin, and freezing in any orientation. Many molded parts have different structure in the center of the part than at the surface.

Interruptions in the flow pattern are another challenge in molding heterogeneous, high molecular weight, high-viscosity liquids. Figure 7.22 shows some changes to flow caused by disruptions in channel height, by obstructions, and by direction changes. Where a channel changes dimensions, for example, becomes thinner, the flow is slowed until sufficient local backpressure builds to drive the flow front forward. This results in a change to how the liquid flows forward during the filling process. The liquid will flow around an obstruction, but this will result in a weld line where the liquid must then flow back together. Due to the high-viscosity of the liquid, the liquid will not immediately mix. Unless sufficient backpressure can build to drive intermingling of the liquid, a weak point with little interpenetration of the liquid across the weld line boundary may form. Weld lines do not require a complete obstruction of the channel. Sharp changes in channel dimension can dramatically influence the

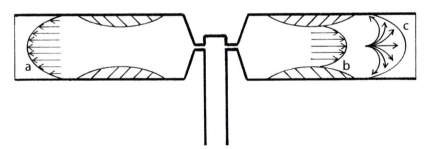

FIGURE 7.21 Center filling in injection molded parts: (a) the advancing flow front for a polymer is typically laminar. (b) As the polymer at the cold mold surface freezes, the melt flows through a constricted channel. (c) Since melt freezes in contact with the mold, new melt fountains from the rapidly moving central line.

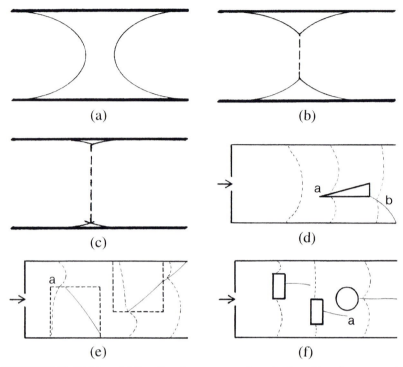

FIGURE 7.22 Welding. (a) Two flow fronts approach each other, (b) make contact, (c) and fill in to create a complete weld. However, the high viscosity of the polymer melt prevents complete mixing at the flow front, and the boundary is often weaker. (d) An interruption in the flow, a, results in a weld, b, downstream of the interruption. (e) Even a narrowing point in a flow channel, a, results in interrupted flow and a weld line. (f) Flow around an object can result in a compressive force, which can cause the weld to close, a.

rate of flow, and backpressure along an edge can change the flow direction in surprising patterns.

MELT PROCESSES WITH LARGE DEFORMATIONS OR HIGH-SHEAR RATES

The next family of processes induces large deformations or requires high-shear rates. Extrusion, already introduced in part when describing the process of mixing, denotes the use of a ram or screw in order to drive the liquid through a die former. The extruder is also a convenient source of molten material for filling injection molds. Finally, injection-molded parisons or extruded tubes are potential feed components for blow-molded shapes. The high-shear processes are common for bulk polymer processing, but in composites applications are more often used for particle/matrix mixing, to create finished parts with relatively smaller concentrations of reinforcements, or to infiltrate lower-viscosity melts into composite powder beds.

Extrusion

There are two basic forms for the extruder, ram and screw. The ram extruder uses positive pressure to drive molten liquid through a channel, usually a capillary. The viscosity describes the drag-related energy loss during flow. In a screw extruder, the liquid is dragged through a long, continuous channel, usually a rectangular channel. The viscosity affects the amount of motion generated in the liquid by the moving screw. The direct pressure ram extruder requires very large pressures in order to process materials with significant viscosity. The workhorse of the processing line is the screw extruder.

The screw extruder has three zones with separate effects or purposes (Figure 7.23). The conveying zone is that part of the channel where solid particles are being compacted for melting. In the transition zone, the melting and pressurization occurs. In the metering zone, the melt is pumped up to full pressure to pass through the die. Heat is applied through the barrel of the extruder, sometimes in several zones, in order to provide individual control of the three processing zones along the screw. After passing through the shaping die, the liquid can be cooled and used in whatever form has been achieved, or enter any new process. The liquid

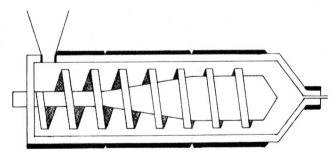

FIGURE 7.23 An extruder with a central, rotating barrel. Pellets are fed through a hopper. In the solids melting zone, the channel is typically deeper. In the transition zone, the channel depth decreases. In the pressure-building zone, the channels are shallow. The melt is extruded through an exit die.

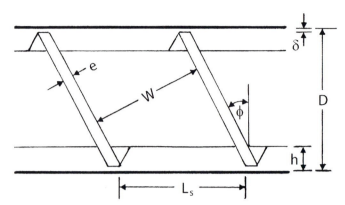

FIGURE 7.24 The extruder screw, with important parameters: the screw width, *W*; the channel depth, *h*; the barrel diameter, *D*; the screw pitch, ϕ; the barrel clearance, δ; the screw period, L_s; and the flight width, *e*.

also can feed a film blower, wire coater, blow mold, fiber drawing, or injection molding process.

The extruder may be used to pump low-viscosity polymer matrix into a reinforcement bed, or to mix a matrix with reinforcement. The pumping properties of the screw depend on the geometry, as illustrated in Figure 7.24. The clearance between the screw and barrel is defined by *h*. This may change along the length of the screw in order to maintain desired pressure profiles or flow rates. The clearance between the screw flights and the barrel are described by δ. This is usually kept as tight as can be reasonably attained in order to prevent backflow. Particulates will often cause wear in the barrel that will affect the clearance. The width of the channel is *W*, and may change in order to control melting or flow rate. ϕ is the average screw pitch or helix angle. *e* is the width of the flights. The screw length is L_s and the diameter of the screw is *D*. The length/diameter ratio is one characteristic commonly used to define an extruder. Most screws and barrels have standard measures for simple manufacturing, interchange of screws or barrels with special properties, for controlling the behaviors in the specific zones, and repair.

The maximum flow rate through the solids conveying zone can be calculated by assuming a smooth barrel and screw, with negligible friction at the screw surface compared to the barrel, and no pressure building in the conveying zone:

$$q_c = \rho \pi D v_s \left[\frac{\pi}{4}(D^2 - D_s^2) - \frac{eh}{\sin\phi} \right] \tan\phi \qquad (7.34)$$

where ρ is the density of the bulk solid, D_s is the diameter of the screw, and v_s is the screw velocity in revolutions per unit time. This is a geometrical equation, wherein the flow rate is controlled primarily by the size of the barrel and the rotation speed, D and v_s, but secondarily by the channel depth, h, with a component removed for the thickness of the flights, $e \times h$.

In the melting zone, the number and size of the solid particles gradually diminishes and the melt fraction increases. The melt forms against the heated barrel, and is swept to the backside of the channel. A melt pool accumulates there. The solid bed maintains an open connection all the way back to the hopper. In this way, bubbles can be removed from the melt as it forms, providing a high-quality melt without complex venting.

The pressure to drive flow is created in the metering zone. The pressure will force the melt through a capillary or channel of a desired shape to create a finished part, for example, extruded rod, sheet, tube or other continuous profiles.

For a Newtonian fluid, neglecting the flow across the flights, an analyt-

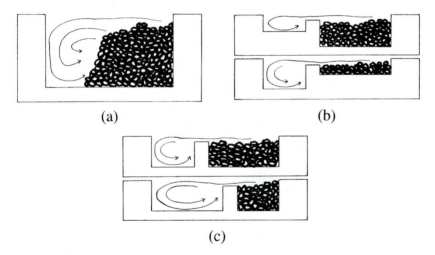

(a)   (b)

(c)

FIGURE 7.25 (a) Flow pattern in the transition zone: a decreasing volume of solid pellets is conveyed along the channel, melting through contact with the hot barrel. (b) An alternative design for the transition zone, with a channel for the melt removed from the pellets. Early in the transition zone, the melt channel is shallow, getting deeper farther along the transition zone. (c) Another design has the melt channel starting narrower than the solids channel at the beginning of the transition zone. The width of the melt channel widens while that of the solids channel narrows as the transition zone progresses. These designs remove the excess melt, increasing heat flow into the solids and enhancing melting.

ical equation for the flow rate in terms of the pressure difference and drag flow can be obtained. It is:

$$Q = \frac{V_b W h}{2} F_d + \frac{W h^3}{12\eta} F_p \left(-\frac{\partial P}{\partial z} \right) \tag{7.35}$$

where V_b is the velocity of the barrel, which is related to the velocity of the screw.

$$V_b = v_s \pi D \cos\phi \tag{7.36}$$

F_d and F_p are shape factors for the drag and pressure components of the flow due to the shape of the channel, converting from flow between infinite parallel plates. The shape factors are calculated using an infinite series. Approximate values have been found for shape factors, where the ratio of channel depth to width is sufficiently small: that is, where the channels are shallow:

$$F_d \approx 1 - 0.571\frac{h}{W} \quad \text{and} \quad F_p \approx 1 - 0.625\frac{h}{W} \tag{7.37}$$

The flow rate is controlled by the geometry of the flights, the diameter and velocity of the screw, and the helix angle. The pressure developed will never result in flow of material backward along the screw, unless the outlet is at a higher pressure than the screw is generating. However, melt will often flow between flights. There are many refinements to the flow equations, which go beyond an introductory discussion.

Extruders produce melt from a die that can be used alone or as the feed for various other processes. The output melt can be shaped using the outlet die. The most common observations at the extruder outlet are die swell, and melt instability or melt fracture, as shown in Figure 7.26. Melt instability and melt fracture occur when the normal forces exceed the forward flow. This generally occurs under high extrusion rates or with short die lengths. Longer dies can create more orientation in the melt and reduce stored energy contributing to normal forces. Particulates will also accelerate wear in the die.

Extrusion dies are available in various shapes (Figure 7.27). A circular die creates a rod of a constant diameter. This may be cooled using a water bath, cold gas cooling or long drawing distances. The resulting rod may instead be drawn off at a higher rate than the extrusion, which is called fiber drawing, resulting in drawing and alignment. An annular die can feed melt into a tube, or to coat a wire. A slit die can extrude a continuous

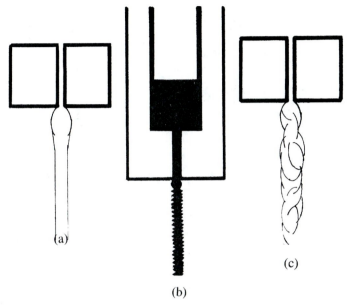

FIGURE 7.26 Extrusion output. (a) normal melt flow, exhibiting die swell at the die exit. (b) melt instability, resulting in an extrudate with an uneven, rough surface. (c) melt fracture, resulting in an irregular, or even discontinuous extrudate.

plate, or the melt can be drawn to create a film. Other, continuous profiles such as I- and T- beams, rectangular rods, and lobed structures can be continuously extruded.

Injection Molding

Injection molding is the processing action for creating parts with complex geometry, in net or near net form, using a positive mold. Injection molding is really a multi-step process, as shown in Figure 7.28. A melt is created with slightly greater than sufficient pressure to completely fill a positive mold. The melt is injected into the mold and held under pressure in such a way as to insure complete filling. The part is allowed to cool while the next shot of molding material is melted and pressurized. The part is removed when sufficiently cool to withstand the mechanical forces required for discharging from the mold. Finally, the process begins again.

Due to the high viscosity of nanocomposite melts, injection molding can be problematic. However, a modification called resin transfer injection molding is available, where a low-viscosity pre-polymer is injected

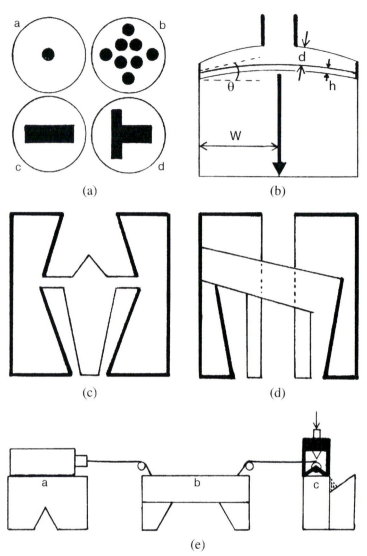

FIGURE 7.27 (a) Extruder dies. a. a cylinder or fiber die. b. a multi-channel spinneret, used for example for multi-strand fibers. c. a slot die and d. a T-shaped die for extruding rods with simple, continuous profiles. (b) a coat-hanger die for extruding sheet. The die is characterized by width, w, die angle, θ, the channel diameter, d, and the opening width, h. (c) a spider die for extruding tube. Narrow buttresses hold the central mandrel. Narrowing downstream of the buttresses causes pressurized flow for closing the weld lines formed. (d) A tangentially fed die, allowing wire or air for blow molding to be fed centrally. (e) An example of an extrusion line. a. The extruder, producing for example a composite extrudate, or a shaped extrudate. b. The extrudate is drawn through a cooling bath. c. The cooled extrudate is chopped to length and deposited in a collection hopper.

into a mold loaded with a reinforcing material bed. The injection fills the void space in the bed, hopefully completely. The composite part is consequently heat treated to achieve full polymerization. The fundamentals of resin transfer molding are the same as those for injection molding.

A schematic for a typical system is shown in Figure 7.29. The mold is clamped with a large pressure required to remain closed during filling. The clamping pressure of the machine indirectly determines the maximum size of the parts that can be molded on the machine. Thus, clamping pressure is a primary defining parameter of injection molding machines. A screw extruder with a movable barrel usually provides the melt. The extruder creates a melt pool in front of the screw, with the aid of valves designed to prevent backward flow of the liquid. After the mold is clamped, the screw drives forward, providing more force for melt injection. The polymer flows through and fills the mold. Meanwhile, the screw continues to provide liquid with the continuously increasing backing pressure required to drive the melt throughout the mold, and to supply any excess material needed if the liquid shrinks during cooling. After the hold/packing period, the screw is withdrawn to allow the liquid pool reserve to recover for the next shot. Meanwhile the material in the mold

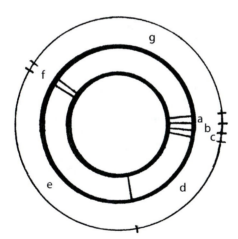

FIGURE 7.28 The injection molding cycle presented as a repeating cycle: (a) The mold is closed; (b) The extruder is moved forward to apply filling pressure. (c) The melt reservoir is injected into the mold. d) Backing pressure is applied to pack the mold, insuring complete filling. (e) The part is cooled. (f) The injector is moved back, and the part is ejected. (g) The extruder pressurizes the melt reservoir. Prior to the advent of the microprocessor, the timing of injection molding was controlled by a rotating mechanical system. Choosing appropriate gearing controlled the timing of each step. Therefore, this diagram also illustrates the logic flow of the process. (After Osswald.)

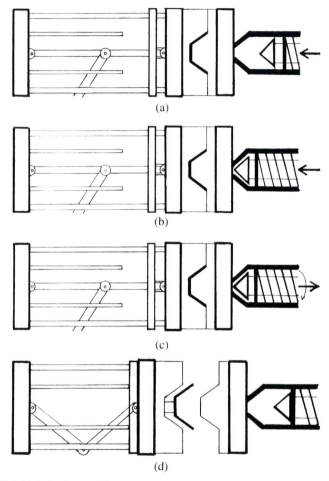

FIGURE 7.29 Injection molding steps: (a) The screw moves forward to apply filling pressure. (b) The extruder continues to supply backing pressure to pack out the mold cavity. (c) The screw is retracted. (d) The mold is opened, and the finished part is removed. (After Osswald.)

cools and freezes. When the part has solidified sufficiently to reach mechanical stability, the mold is opened, usually with automatic pin ejection to remove the part. This completes the cycle.

The mold closing and injection periods of the cycle are usually quite short, as is mold opening and part ejection. The pressurization time depends on how long filling requires. The cooling time is generally the longest part of the cycle. The time for cooling can be estimated using:

$$t_{plate} = \frac{h^2}{\pi\alpha} \ln\left(\frac{8}{\pi^2} \frac{T_M - T_W}{T_E - T_W} \right) \tag{7.38}$$

for thin plate parts with thickness h

or

$$t_{cylinder} = \frac{d^2}{23.14\alpha} \ln\left(0.692 \frac{T_M - T_W}{T_E - T_W} \right) \tag{7.39}$$

for cylindrical parts with diameter d

T_M is the temperature of the melt at injection, T_E is the temperature at ejection, and T_W is the temperature of the mold wall. Ejection time is improved using a large temperature differential at the mold wall, but this also increases the chance that the liquid will freeze prematurely, causing incomplete mold filling. The target temperatures for melt, wall and ejection are a matter of experience and art. However, some exploration of the crystallization process, later in the chapter, can provide insight.

Mold design is arguably the most critical part of injection molding. The mold can be used to create complex parts. The goal is generally to have all sections of the mold fill at approximately the same time. If one part is filled very quickly and remains under backpressure for a long time, a flashing failure may occur. On the other hand, if a small part, requiring high pressure to fill properly, is at the end of the filling scheme, this segment of the mold may not fill reliably. The general features of a mold include the sprue, runners, gates, parts, and vents. The sprue is where the melt is first introduced to the mold (Figure 7.30). The sprue may open directly into the part cavity. The air from the mold is displaced from the part cavity during filling through very fine holes, in effect, vents, which the melt generally cannot penetrate under mold-filling pressures. If the sprue leads directly into the mold cavity, a separate machining operation to remove the sprue is required, and this does create a simple system for mold making. Multiple cavity mold and near-net molds usually have a runner system leading from the sprue to the mold cavity or cavities. The liquid will enter the part cavity through a gate, such as a pin gate. To remove the part from the runner system, the pinch point at the gate is the only connection that must be broken. Specialized gates, such as the diaphragm or film gate, create a barrier that will not be penetrated until the backing pressure along the entire entrance passes some threshold. The value of this is that the part will fill more evenly. In

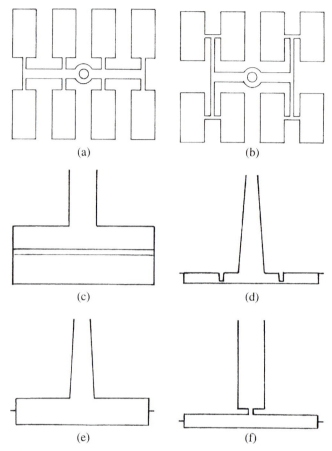

(a)   (b)

(c)   (d)

(e)   (f)

FIGURE 7.30 Two runner layouts: (a) The mold cavities fill sequentially. (b) All of the cavities fill simultaneously, but the geometry is complex. (c) In a film gate, the narrow point assures that the sheet cavity is not filled until sufficient pressure is built along the entire width of the mold sufficient to force flow through the gate. (d) In a diaphragm gate, the narrowing point assures even filling of a circular part. (e) In a sprue gate, the melt enters directly into the mold cavity. The cavity fills quickly and thoroughly, but part separation requires finishing to remove the sprue. (f) In a pin gate, the part is easily removed at the narrowing point, but cavity filling may be blocked if the melt freezes in the pin.

any case, sufficient material to fill the mold cavities, runners and sprue must be injected during the mold filling injection.

There are several failures typical of injection molding. In a short shot, the mold is incompletely filled. A short shot occurs under conditions providing for insufficient pressure, insufficient melt feed, premature melt freezing, or improper mold design. Flashing occurs when material flows out of the mold at the mold lines where the mold cavity parts meet. Flashing is caused by overpressure that exceeds the mold clamping force, too low a viscosity in the melt, or improper mold design. In fact, with improper mold design, a part can both flash in cavities near the sprue and short-fill cavities far from the sprue. Mold filling problems will be exacerbated by reinforcement particles, which in addition to increasing the viscosity can cause bridges and block flow in the mold. Other molding issues include sink points where the part is thick and weld lines. If the melt is heated too much, more volume reduction can occur during cooling, leading to sink lines, which follow the thickest lines in the parts. During crystallization, a semicrystalline polymer will shrink considerably, which may lead to sink lines even if the mold is well designed. Reinforcement particles, which usually have a lower coefficient of thermal expansion, will generally suppress sink marks at molded ribs. Weld lines occur anywhere flow is split, for example, in multi-entry cavities: cavities with more than one gate-filling point. The final problem with molds is service life. After many uses, a mold will wear, especially if the liquid is especially corrosive or abrasive. Composites and nanocomposites can be very abrasive and often result in short mold life. As the mold wears, the chances of flashing increase significantly. Degradation may also oc-

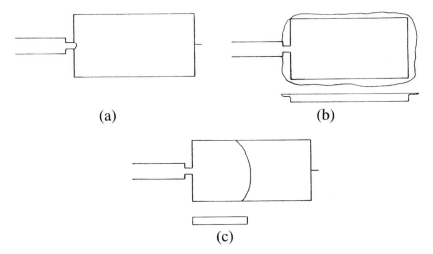

(a)   (b)

(c)

FIGURE 7.31 Injection mold failures. (a) Gate freezing blocks filling of the mold entirely. (b) In flashing, insufficient clamping force allows melt to bleed out of the mold cavity at the join lines. (c) In a short shot, insufficient filling pressure results in incomplete filling of the mold cavity.

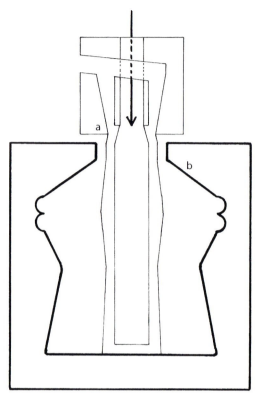

FIGURE 7.32 Diagram of combined injection-blow molding. (a) Melt is injected through a circumferential die, into a mold cavity to form a preform, called a parison. (b) The molten parison is inflated by pressurized gas through the center of the die. The expanding melt freezes on contact with the cold mold surface.

cur, as usual, when the melt is sustained at too high a temperature for too long, given the processing limits of the liquid. Alternatively, poor formulation with antioxidants may provide an unstable melt.

Blow Molding

The general steps of blow molding are creation of a parison, placement in the mold, blowing the parison with or without stretching to fill the mold, freezing on contact with the mold, and finally mold opening with part removal. The interruptions to smooth flow and the viscosity increase caused by particles can add challenges to designing a blow molding system for composites.

The key to blow molding is design of the parison. The parison must be designed so that it expands evenly and smoothly on blowing to evenly contact all of the mold surfaces as close to simultaneously as possible. A drink container parison looks a lot like a test tube with a screw lid. The parison is fit to or molded around a blowing tube. The parison is then melted in a large mold of the desired shape and inflated to fill the mold. Cooling rates are very rapid, and can be predicted with the same film-cooling equation used to predict injection molding cooling cycles.

Mini-systems

There are miniature extruder and injection molding systems that have slight changes to configuration to accommodate their reduced dimensions. In order to reduce the size of an extruder, the screw can be placed in the barrel instead of on the rotating screw. In order to shrink the injection molder, a ram extruder often is used. The solid is loaded into the ram, melted and then injected by direct pressure. These systems are useful for small or single part prototyping, but the results generally will not scale to pilot-scale or full-scale machines.

Cooling and Structure Evolution

After the melt is in the desired conformation, the product must cool to room temperature. All materials will experience a volume change during cooling. Semicrystalline materials may also have a few special features that affect the properties and performance of the finished piece. After the part has been formed, crystallization occurs slowly, as the molten material freezes into crystals.

Glassy State

Glassy materials cool from the melt in a straightforward manner. The volume decreases constantly according to the thermal expansion coefficient of the liquid. There may be a liquid-liquid temperature, a temperature above which there is a change in the thermal expansion coefficient. The occurrence of a liquid-liquid transition is not universal and may represent a subtle effect. At the glass transition temperature, there is a sudden change, a drop, in the rate of shrinking. In cooling a glassy material, the backing pressure may be removed with the least impact on the part once the part temperature has reached the glass transition.

The glass transition, as discussed previously, is a kinetic phenomenon. The cooling rate of the liquid, controlled by both the mold temperature

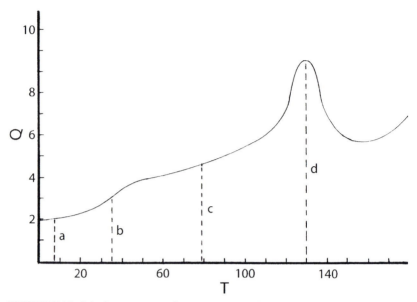

FIGURE 7.33 Calorimetry curves show some types of transition that may have implications on processing, but does not show others. (a) A low temperature transition resulting from crankshaft motions. (b) The glass transition. (c) A liquid-liquid transition, which might manifest as a softening point. (d) A melting transition.

and cooling rate, is a balance between rapid cooling and long-term dimensional stability. If the part is cooled very rapidly, orientation and residual stresses can be trapped in the structure. These will relax very slowly and change the dimensions of the part over time, a process called aging. Aging may also result in increased part brittleness over time. Again, the importance of long-term stability must be balanced with the desired production rate and the desired properties and part lifetime. Particulate reinforcements generally suppress dimension changes caused by residual stress.

Crystal Formation

The crystallization of simple atomic and molecular solids is a little more complex than the changes occurring for amorphous materials. Crystallization occurs at a well-defined temperature, and structure evolves at specific temperatures at predictable rates. There are tables, pressure-volume curves, and time-temperature-transition graphs for most common metals and ceramics (Figure 7.34). The volume change occurs predictably over a small temperature range, and the final structure is usually mapped clearly as a function of time and temperature.

The formation of crystal structure in semicrystalline, long-chain materials is often considerably more complex. The crystallization occurs over a range of temperatures, and if cooling is rapid enough crystallization can be completely suppressed. The quality of the crystals formed is a function of the super-cooling temperature, ΔT. The greater the ΔT, the

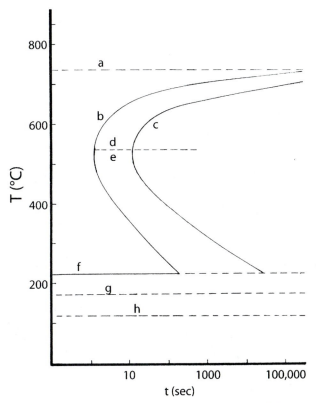

FIGURE 7.34 A schematic of time-temperature-transformation (TTT) curve for steel. (a) The homogenization temperature. Above this temperature, a uniform austenite structure is observed. (b) The onset of structure formation. (c) The completion of structure formation. (d) Above the maximum crystallization rate, pearlite is formed. (e) Below the maximum in crystallization rate, the bainite structure forms. If the steel is rapidly cooled to below this temperature, an instantaneous, non-kinetic transformation to Martensite occurs. (f) Initial appearance of Martensite. (g) 50% transformation to Martensite. (h) 100% of the structure will convert to Martensite. By manipulating the time the steel is held at a given temperature, various internal structures can be obtained.

more crystals will nucleate and the faster and less perfectly the crystals will be. A larger ΔT also results in less volume change. This also pertains to the cooling rate. Higher cooling rates will generally result in poorer crystals and less volume change. The time-temperature transition curves for polymers are much less commonly available, and molding remains firmly grounded in industrial art and experience.

Crystallization is governed, except during initiation and during the very last stages, by the Avrami equation:

$$X_{cr} = 1 - \exp(-Kt^m) \tag{7.40}$$

where X_{cr} is the percent crystallinity, and K and m are constants that depend on the material and conditions of the crystallization process. Factors affecting the Avrami equation include the mobility of the liquid in the melt, the presence of crystal nucleation sites, and the geometry of the crystals. Crystal nucleation sites can be forced using nucleating agents, and sometimes reinforcements will act as nucleation sites for crystal growth. As discussed earlier, a nucleating agent, such as a particle with one or more surface features corresponding with a crystallographic growth face, will result in maximum effectiveness. However, almost any surface can serve as a nucleation site in some systems. In addition to different textures, many polymers are also observed to have different crystal structures that can form under specific conditions (Table 7.1). Typical conditions include high-pressure crystallization, crystallization under deformation, crystallization in an electrical field, and crystallization of oriented melts.

Oriented melts, or melts in contact with surfaces, often crystallize differently than disordered melts (Figure 7.35). Ordinarily, a random melt would be expected to form typical 3-D spherulites. If the melt is highly constrained, such as in very thin-walled parts, or if the melt has been highly oriented, the geometry of the crystallization process will change, resulting in very different crystal structures. Oriented melts often form a drawn, highly aligned stalk, or shish, with lamellar plates spaced nearly equidistantly along it, forming kebob-like structures. In constrained melts, or melts near nucleating surfaces such as composite fillers, the proximity of the surface results in row-nucleated structures. The Avrami constants would not be expected to be the same for a quiescent melt as for an oriented or constrained melt.

The final properties of a solid may be further changed through annealing. Annealing requires heating a part to near but below the melting temperature. The exact temperature required to produce crystal transformation depends on the thermal characteristics of the material.

TABLE 7.1 Crystal structures for metals, ceramics and polymers.

Material			a	b	c	α	β	γ
Al	cubic	FCC	4.05					
Au	cubic	FCC	4.08					
Cu	cubic	FCC	3.62					
W	cubic	BCC	3.16					
Zn	hexagonal	HCP	2.66		4.95			
Fe_2O_3 (hematite)	rhombohedral	corundum	5.03				13.7	
Fe_3O_4 (magnetite)	cubic	spinel	7.96					
MgO (periclase)	cubic	rock salt	4.21					
TiO_2 (rutile)	tetragonal		4.59		2.96			
SiO_2 (quartz α)	hexagonal		4.91		5.40			
SiO_2 (coesite)	monoclinic		7.15	12.4	7.15		120	
SiO_2 (cristobalite α)	tetragonal		4.97		6.92			
SiO_2 (melanophlogite)	cubic	clathrate	13.4					
$CaCO_3$ (calcite)	rhombohedral		7.14		8.52			
Al_2O_3 (corundum)	rhombohedral	corundum	4.76		13.0			
ZnO (zincite)	hexagonal	zincite	3.25		5.21			
Muscovite	monoclinic	2M2 mica	5.20	9.00	20.0		94.5	
Diamond	cubic	diamond	3.57					
Polyethylene	orthorhombic		7.40	4.93	2.53			
Polytetrafluoroethylene	pseudo-hexagonal	<20°C	5.54	5.54	16.8			
	hexagonal	>20°C	5.61	5.61	16.8			120
Polyethylene teraphthalate	Triclinic		4.56	5.94	10.8	98	118	112
Nylon 11	Triclinic		9.6	4.2	15.0	72	90	64
	Triclinic		4.9	5.4	14.0	49	77	63

Table from CRC Handbook and Geil.

321

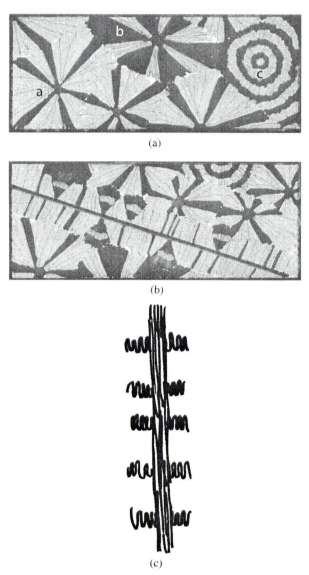

(a)

(b)

(c)

FIGURE 7.35 (a) Spherulites: a. crystal type I. b. crystal type II, similar to crystal type I but with inverted phase contrast. c. crystal type III, which appears to be a completely different texture from crystal type I. (b) Reinforcement may cause crystals to nucleate. Reinforcements with one or more long dimensions may cause row nucleated structure, where the nucleation sites are so close together that the crystals can only grow directly away from the surface. This transcrystalline layer persists some finite distance away from the nucleating surface. (c) Under shear, a shish kebob crystal structure forms. The shish kebob consists of a highly ordered central stalk and regularly spaced lamellae propagating away from the stalk.

322

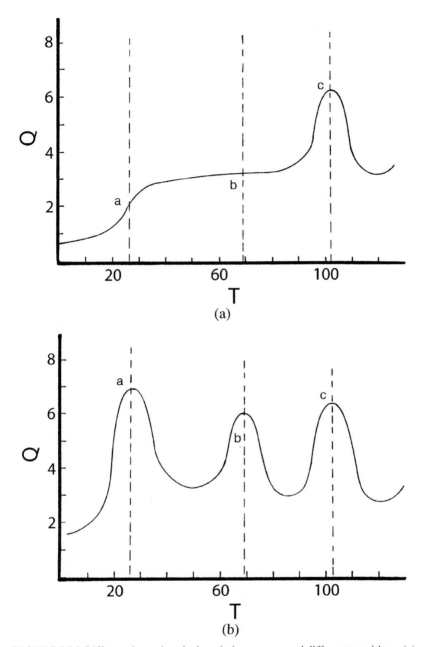

FIGURE 7.36 Different thermal analysis techniques can reveal different transitions. (a) Calorimetry, is sensitive to transitions absorbing significant heat, including a. the glass transition and c. the melting transitions. (b) In thermally stimulated current, which senses the motion of electrical dipoles whether significant heat flow occurs or not, b. a liquid-liquid softening transition is observed.

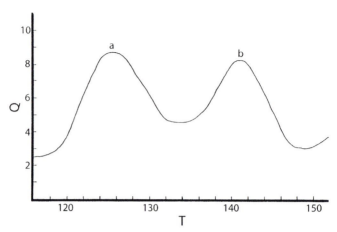

FIGURE 7.37 Calorimetry may reveal two crystal melting transitions, characteristic of two types of crystal structure with different stability and thus different melting temperatures.

Some materials have a transition temperature above which residual strain can be annealed out, as shown in Figure 7.36. Some materials exhibit metastable crystal structures (Figure 7.37), which can be removed by heating to the melting temperature of the lower-temperature, less stable phase until the undesired phase is removed. This temperature is below the melting and crystallization temperature of the stable phase, which spontaneously forms with additional time at the annealing temperature.

THERMO-KINETIC PROCESSES

There are processes that do not employ melting exclusively. These techniques work only on materials with sufficient molecular mobility, combined with the ability to re-form interactions after deformation. Two examples of solid or semisolid processes, here referred to as thermokinetic processes, are cold drawing and thermal spray. These techniques use a combination of mild heating to enable some molecular mobility and physical deformation to displace the material. The shear force in these techniques is thus expected to be high.

Cold Forming

Materials with two tangents to the true stress-true strain curve exhibit two deformation mechanisms. These materials can be transformed using cold drawing. In metals, cold working results in harder materials. Cold working introduces many defects and breaks apart the grain structure. Future deformation is restricted by these defects. Cold-drawn polymers also become very rigid. After the intrinsic yield stress, the polymer begins to neck, drawing more material into the neck. The neck has highly oriented backbones densely packed into aligned crystal structures. Cold drawn fibers can show mechanical properties approaching the ideal strength of a carbon-carbon bond. Stamping, forging, and cold calendering of metals and polymers are common processes that take advantage of cold forming. Cold forming can be used on composite parts.

Thermal Spray

A completely different approach to creating coatings is thermal spray, a family of techniques that propels particles of material at a target using a high velocity gas jet. The particles may be melted, partially molten, or unmelted. The particle velocities range from 50 – 1500 m/s. The building block of a thermal spray coating is the "splat," a deformed piece of material. Figure 7.38 shows the basic building process of thermal spray coatings. A particle is heated by the gas stream and accelerated toward the surface. The goal is to form a very thin splat that conforms with and sticks to the surface, building a coating by accretion. There can be defects introduced by the process. Oxidation is possible by exposure to oxygen at elevated temperatures. Porosity is an occasional occurrence when a poorly melted particle creates a shadow. Thermal spray is a line-of-sight process, and is not appropriate for complex surfaces. However, the family of techniques creates coatings with very rich microstructures. In addition, coatings are deposited through a mixture of melt flow and cold working. Materials with very high viscosities, such as polymer composites with a high concentration of ceramic reinforcement, can still be formed into well-densified coatings using these processes.

Figure 7.40 shows schematics of several important thermal spray processes. Flame spray was the original thermal spray process, invented in about 1910. The process is simple, with powder introduced to a gas jet produced by combustion. The particles tend to be slower, the deformation is less complete and only easily melted materials can be sprayed using this process. High-velocity oxy-fuel (HVOF) uses a constrained flame to produce a faster, hotter jet that can process many more materials. The very fast gas jet also results in very short residence times for particles. Plasma processes create a hot plasma, if a little slower moving than seen in HVOF. Plasma processing can melt very high temperature melting materials such as carbides and cermets. If the spray is conducted in a

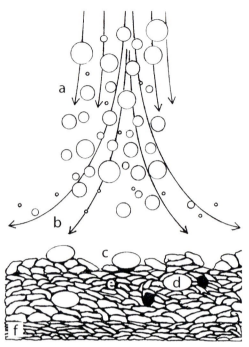

FIGURE 7.38 Thermal spray coating process: (a) particles are carried along in a high temperature, rapidly moving gas stream. The particles are heated and accelerated toward the surface. (b) At the surface the gas is diverted, while the particles continue on to be deposited on the target surface to form a coating. (c) The surface coating may build unevenly. Features of a sprayed coating may include: (d) unmelted particles and (e) trapped porosity. (f) A bond coat of the bulk material applied under different conditions or of a different material may be used to enhance adhesion.

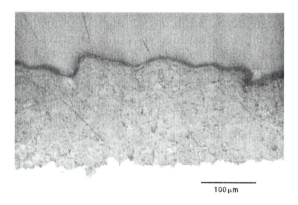

100 μm

FIGURE 7.39 A thermally sprayed coating: a nylon-11 matrix reinforced with 7 nm fumed silica particles.

326

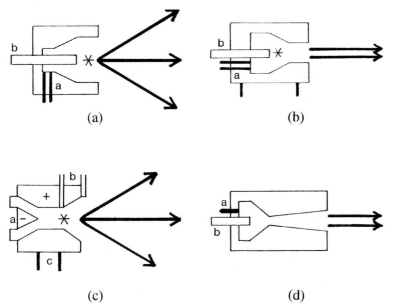

FIGURE 7.40 Sketches of several thermal spray processes: (a) Flame spray. Combustion gases are fed a. into a combustion chamber (*). The particles are fed b. into the combustion zone. Flame spray is characterized by a relatively cold, relatively slow flame. (b) HVOF spray. a. The combustion gases are fed into a high-pressure combustion chamber. b. The particles can be fed into the combustion and expansion chamber, or at the exit of the flame. c. Cooling water is required to prevent the spray torch from melting. HVOF spray is characterized by a very hot, very fast flame. (c) Plasma spray. The flame is caused by electrical formation of plasma in the chamber. Particles are fed forward of the plasma chamber. Plasma spray is characterized by a very hot but relatively slow flame. (d) Cold spray. Gas and particles are fed through a convergent-divergent nozzle to accelerate the gas. While hot feed gas is usually used, there is no combustion process. Cold spray is characterized by a low temperature, very fast gas stream.

vacuum or an inert atmosphere, oxidation can also be prevented. The final process, cold spray, is the most recent addition to the family. Cold spray focuses on maximizing the cold working of impact.

REFERENCES

Tadmor, Z. and C. Gogos, Principles of Polymer Processing, John Wiley & Sons, New York, 1979.
Osswald, T., Polymer Processing Fundamentals, Hanser/Gardner Publications, Cincinnati, 1998.
Hiemenz, P., Polymer Chemistry, Marcel Dekker, New York, 1984.
Knight, R., Personal Communication, 2005.

Nanocomposites Characterization Science and Technology

This final section introduces materials characterization. The goal of characterization is to provide a quantitative, accurate representation of the structure and properties of a nanocomposite. There are three chapters. Chapter 8 discusses general features of characterization, including sample preparation, controls, experiment design, and visualization. Chapters 9 and 10 focus on structure and property characterization, respectively. Chapter 9 discusses the various approaches to characterizing dispersion and crystal structure. The limits of resolution are important in nanocomposite structure analysis. Chapter 10 introduces approaches to characterizing different physical properties. These include general characteristics, and thermal, mechanical, surface mechanical, and barrier properties.

CHAPTER 8

Introduction to Characterization

This final section of the text is dedicated to the characterization techniques necessary to understanding composites and nanocomposites. The characteristic bulk properties of the matrix provide only partial information about the solid. The interaction of the discontinuous phase with the matrix is particularly difficult to assess. The very small size scales in nanocomposites make evaluating the distribution of the reinforcement even more difficult. Further, the performance of nanocomposite materials is complicated with the same features of typical composites, with the additional difficulty of evaluating changes on a nanoscopic scale.

Chapter 9 will introduce many of the techniques appropriate for characterizing the structure at all levels. Chapter 10 will turn to analysis of physical performance, measuring mechanical properties, diffusion and permeability, adhesion and other parameters. In this chapter, however, the focus is on topics that are more universal for experimentation. This includes experiment design, sample preparation, image formation, and recording results.

EXPERIMENT DESIGN

At the end of a characterization effort, the investigator should know something new about the material tested. How are the reinforcements distributed? Are the particles agglomerated? Has the matrix completely cured? How scratch-resistant is the new formulation? To insure that the desired questions are answered, the questions should be posed logically and thoroughly, a plan for obtaining the answers should be developed and executed, and the results should be evaluated against the original question. This is the process of experiment design. Designing experiments correctly is important at all levels of investigation, from student learning to high-level scientific research. When the features of interest are nanoscopic and the performance changes very small, experiment design takes on even more importance. Experiment design is not difficult. A list of common steps can help assure that each stage is at least considered appropriately.

The first step is identifying the question being asked. Asking the question includes: determining the knowledge needed and why it is needed; quantifying the question, if possible, including delineating the input and output variables; and understanding how accurate and precise an answer is desired. The question may be as simple as "what is the modulus of this composite". However, as polymer composites have rate-dependent moduli, there are input and output variables needed. The purpose of the question may help guide the quantitative delineation of the problem. If the modulus is needed for a material that is expected to be in service for 50-year sewage transport pipes, the thoroughness required during characterization is easier to understand. The final part of the question is the desired precision. How certainly must the answer be known? An experiment cannot really provide a value of, for example, the modulus. A series of experiments can provide an estimate of the modulus in the form of an average and error. Common choices of precision are 5%, or a five percent chance of being wrong about the answer, and can also be 1% or lower.

The second step of the process is to pick a test or tests to evaluate the property. The method must be sensitive enough to determine the property to within the desired precision. A visualization technique that has a maximum resolution of 0.1 μm cannot provide quantitative information regarding the distribution of 50 nm particles with a precision of ± 1 nm. The method chosen should be evaluated against potential competitive techniques. A method should be chosen to provide the best information, and this cannot be assured without considering alternatives. However, a large number of methods should not be used in the hope that one will turn up something of interest. Often, different methods have different sensitivities for the variable of interest, and considerable time can be spent in resolving the differences between test procedures that provide conflicting results. Scatter testing also has a greater chance of turning up testing artifacts.

Choosing to conduct tests according to a standard protocol has strengths and weaknesses. In testing laboratories, standard operating procedures (SOPs) are used to ensure that consistent results are obtained from tests. SOPs are an excellent tool for conducting routine quality control on well-characterized products. Here, well-characterized means that the performance is expected to be uniform, and what is really sought is to detect variation from a normal, expected behavior. SOPs are written based on methods created in laboratories to be: accurate in measuring the target property, with a specific precision; and tested to assure that the test can be run in any laboratory with any set of appropriate equipment (repeatability, reproducibility, and robustness) by any technician. Some organizations, such as the American Society for Testing Materials (ASTM), the American National Standards Institute (ANSI) and other organizations, produce standards that delineate procedures and guidelines for conducting routine testing. These procedures often serve as a basis for creating new, in-house standard test methods. Standard tests may not be appropriate for testing products that are experimental or otherwise do not conform to the product for which the original test was designed. In other words, international or national standards are great, if they work, because they have larger recognition, but they should never be applied blindly.

The third step is to perform the test or tests and evaluate the results. Decisions will rest on the results of the tests.

Controls

Included with the test sample should be both positive and negative controls. A positive control is a test specimen with a known quantity of whatever property is of interest in testing. A negative control is a material that should give a quantifiable, known, negative response in a test. Examples of controls depend on test type. For an assessment of the amount of reinforcement in a matrix, a positive control may be a physical mixture of the two components prior to processing, and a negative control might be a sample of the pure resin with no reinforcement. In mechanical tests, a negative control is obtained by reading transducer outputs without any load, while attaching a known load to the transducer provides a positive control.

There are organizations that create exhaustively tested standards. NIST, the National Institute for Standards and Testing, is one such. There are several levels of standards. The primary standard is often securely stored and seldom directly used. Secondary standards are then prepared and evaluated against the primary standard. The secondary standards then comprise the certified material for comparison. The secondary standards are acceptable as reference materials as long as they provide a traceable record to a primary standard. There is a market for "NIST-Traceable" materials for use as control standards in quality control and in regulated industries. Examples of regulated industries are aerospace, biomedical and biomaterial, government suppliers, safety industry manufacturers, and quality testing organizations. The aerospace and biomedical industries are target markets for new nanocomposite technologies.

A good experiment is usually conducted blind. That is, the test is conducted without any knowledge of the initial status of a test specimen. This helps to assure that the results are not measured with any desire to achieve a specific result. By putting the positive and negative controls into the evaluation schedule in a random fashion, the test method can be fairly evaluated.

Accuracy and Precision

Test accuracy describes the ability to measure the actual value of a certain property. Precision describes within what limits the actual value of the property is known. Testing relies on the central-limit theorem, that given enough tests the average value measured will approach the actual value. The central-limit theorem assumes that the results of the test will be described by a Gaussian, or normal, distribution:

$$ y = \frac{e^{\left[-\frac{(x-\mu)^2}{2\sigma^2} \right]}}{\sigma\sqrt{2\pi}} \tag{8.1} $$

where μ is the average, most likely value, i is the standard deviation, x is a measured value, and y is the deviation from the average. The actual form is not critical. What must be noticed is that variations from the real value are described by the standard deviation. Good estimates of the value of the real value and the standard deviation are needed. Narrow standard deviations provide a narrower range of knowledge about the measurement. The standard deviation can be estimated by the following:

$$ s = \sqrt{\frac{\sum_i (x_i - \bar{x})^2}{n-1}} \tag{8.2} $$

where $\bar{x}$ is the estimated average value, i is an individual measurement, and n is the number of independent values of x used to estimate the standard deviation. A minimum of six replicates is generally recommended for determining a standard deviation ($n - 1 = 5$). The standard deviation gives an estimate of the interval containing the real value with a specified degree of certainty (Figure 8.1). This is also a way to set error bars on a data graph.

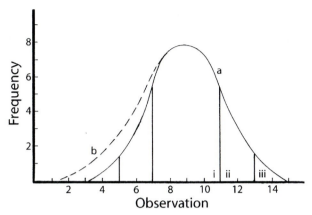

FIGURE 8.1 (a) Normal distribution of, for example, particle sizes. i. One standard deviation, containing 68% of the distribution. ii. Two standard deviations, containing 95% of the distribution. iii. Three standard deviations comprise 99.7% of the distribution. (b) A larger number of small particles will distort only one tail of the distribution.

Error can creep into any test. The most common and simplest to identify is gross error. Gross error is a mistake, such as contaminating a sample, a piece of equipment malfunctioning, or operator error. There are also systematic errors, for example, an offset in a piece of equipment. Systematic errors are constant or proportional. A zeroing or baseline error will usually result in a constant offset in the results. A proportional error is one that increases with an increasing value of the experimental measurement, such as would be expected from an error with the slope of a calibration line. The final type of error is random. Random error is present in all measuring equipment and in human measurement. The primary method for checking for error is to compare the measurement with known standards. Usually, a test will include evaluating a blank and a known calibration material to test the zero condition and the accuracy, respectively. Gross errors should be rejected as unsuitable for evaluation. Systematic errors may be correctable. Random error cannot be removed, and is part of the uncertainty in any result reported.

An indirect concern with error is identifying the consequence of mistakes. If a part does not meet criteria, and is identified as such, then the consequence is a failed part. More significant are the consequences of incorrectly identifying failure. Unidentified errors come in two types: a point which is misidentified as being incorrect (false negative), and a point which is misidentified as belonging to the correct set (false positive). When designing a part, the consequences of false results should be considered as the basis for a correct decision about precision.

TABLE 8.1 The t and Q distributions.

Degree of freedom (Sample size)	t Test			Q Test
P value	0.10	0.05	0.01	0.05
1	6.31	12.71	63.66	
2	2.92	4.30	9.92	
3	2.35	3.18	5.84	
4	2.13	2.78	4.60	0.831
5	2.02	2.57	4.03	0.717
8	1.86	2.31	3.36	0.524
10	1.81	2.23	3.17	0.464
20	1.72	2.09	2.85	
50	1.68	2.01	2.68	
∞	1.64	1.96	2.58	

Table from Haswell

Another concern with error is the presence of outlier data. Some measurements are subject to gross error, giving results that contaminate the data. Outlying data must be excluded from the data set in order to obtain an accurate measurement. Two tests for excluding outliers are the t-test and the Q test:

$$t = \frac{(x - \bar{x})\sqrt{n}}{s} \quad \text{t-test} \tag{8.3}$$

$$Q = \frac{\text{suspect value} - \text{nearest value}}{\text{largest value} - \text{smallest value}} \quad \text{Q-test} \tag{8.4}$$

These tests compare measurements to data that would be expected if t statistics or Q statistics apply to the measurement set. The t and Q values for a given value of n and significance are tabulated (Table 8.1). Only if the calculated statistic is larger than the tabulated value is a point in error. The t statistic can also be used to compare the average determined by the test to the known value of a standard test sample, or to set confidence limits on a data set.

If the distribution is not normal, especially if it is multi-modal, then the median is a less sensitive estimate of the correct value of the measurement. The median is the middle value of a ranked set of an odd number of measurements, or the average of the two middle values of a ranked set of even numbers of measurements. A data set will be distributed around the correct value according to some distribution. The median should therefore be the correct answer. For a normal distribution, the mean and median are the same.

Calibration

Many techniques use single-point calibration. In this procedure, a well-characterized material with a known value of the target parameter is tested with the method in question. A test value is obtained and compared to the known value. If there is significant deviation from the known value that is greater than expected from the random error imparted by the technique, there are two courses open. First, the test can be "fixed": the machine may be adjusted, the zero may be adjusted, or the solvents may be replaced. The known must then again be tested. Alternatively, the deviation may be recorded and used to adjust later results reported for the test.

Multipoint calibrations follow the same general procedure just given, but use multiple known data points to create a linear, or higher-order polynomial, calibration line. A new measurement on an unknown generates a result based on this calibration curve. This is a powerful approach.

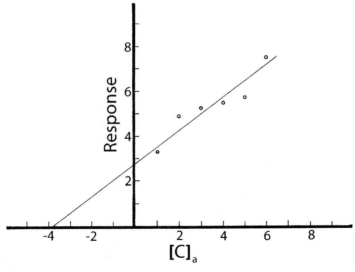

FIGURE 8.2 Standard addition procedure: Measured amounts of the analyte, that is, the variable one wishes to measure, are added to the unknown. The response, e.g., x-ray intensity, peak height, or any other experimentally measured variable, is plotted vs. the added quantity of analyte, and extrapolating to zero response provides a measure of the unknown concentration of analyte.

However, it requires a set of specimens with precisely and accurately known properties, or standards. When the necessary standards are not available, the standard addition method may be employed. A sample is divided into pieces, and an increasing quantity of a known sample is added. The test results for quantity are plotted and fit to a line or curve, as shown in Figure 8.2. The intersection of the line with the response (vertical) axis gives the correct value of the property. For example, the amount of crystallinity in an unknown specimen can be evaluated by standard addition in calorimetry by adding successively larger quantities of a sample of identical material with a known percent of crystallinity, collecting crystallinity data at each addition.

Zero repeatability is the amount of variation introduced to a measurement by normal equipment variation in the measured values without a sample present. As the measurement variables increase, the differences often increase. The same is true of baseline repeatability. Baseline repeatability is measured with a calibration standard. Baseline repeatability determines the test-induced uncertainty in the measurement. Finally, there may be noise in the measurement, which induces additional uncertainty in the measurement. All of these sources of variation should be measured for a given piece of equipment on a regular basis, if true precision of quantitative measurements is desired. Increases in variability may also indicate a need for machine maintenance.

Linearity and Time Constants

Measurements usually depend on the test to provide measurements assuming a systematic change in the reported values. This region is called the linear response range. The response may not be linear. Any known response shape is acceptable, so long as an artifact created by the test protocol does not artificially change the results. In the linear response range, a change in test value will result in a well-behaved response, which can be evaluated according to the theory governing the test. Outside of this region, the test method will introduce artifacts (Figure 8.3). The values obtained will not be consistent with testing conducted in the linear range. Tests should be conducted in the linear region of the method.

Linearity can also be a feature of the limitations of the equipment. For example, at the fastest heating rates, a DSC cannot keep up with the heating rate of the controller due to thermal lag, resulting in changes that can vary from test to test. Alternatively, at the slowest heating rates, heat can be lost to the environment, reducing the measurement. Therefore, a range of heating rates should be used to evaluate a calibration material. The value of the area, the onset, and the end of a thermal event should be plot-

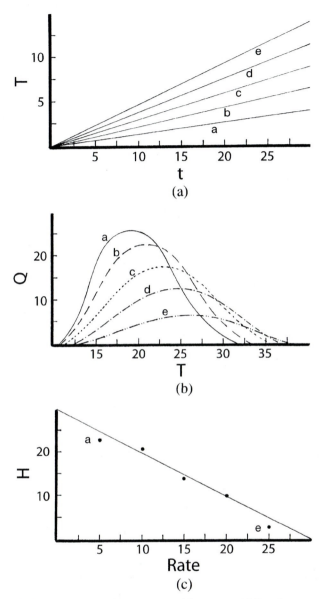

FIGURE 8.3 Tests such as calorimetry are sensitive to rate. (a) Rate increases from a to e. (b) The heat flow during melting, the response variable, drops in peak value and spreads out from a to e. (c) A plot of the area under the peak, the enthalpy of melting, vs. heating rate is a straight line. In this example, the first point, a, and the last point, e, fall below the line. At the slowest rate, heat can be lost to the machine, and the peak area will be artificially low. At the fastest rate, the time constant of the machine will restrict the response rate, again resulting in an artificially low peak area. In between these points, the peak height will vary linearly with the rate of the temperature ramp.

ted as a function of heating and/or cooling rate. The minimum and maximum heating rates can be determined from the point at which the curve is no longer linear. Any heating rates in the linear range can be safely used to extrapolate accurate thermal data.

The shapes of response curves are modified by the time constant of a test. For a machine, the time constant is determined by the response time of the electronics and physical elements of the equipment. For a chemical test, the time constant is determined by the reaction rate. A first-order property should manifest as an infinitely sharp peak and a second-order property should be a discontinuous step change, respectively. Tests cannot respond in this idealized fashion, causing a smearing of an infinitely sharp peak into a measurable, finite peak. The sharp step of a second-order change is smeared into a gradual step change. The time constant is also proportional to the experiment rate, and increases with increased testing rates. The time constant can be determined using the positive control and evaluating the distortion of the measured result from the ideal response. An x-ray spectrophotometer, for example, is limited by the response time of the x-ray detector.

Signal to Noise

Identifying the presence or absence of a peak, or when two peaks are separated enough to evaluate separately, is also an issue in experiment analysis. There are two common standards for setting the separation. One uses a decision limit, and requires that the peaks be separated by 6× the standard deviation of the signal. This restricts error of both type I and type II to 0.0013, or 0.13%. The detection limit is defined by a peak separation of 3× the standard deviation of the signal. This gives a 6.7% chance of both false positives and negatives.

Design of Experiments and Optimization

Putting together a good experiment plan is a guided art. If there is one control variable of interest and one response variable, the design is simple: change the variable, conduct the test, and record the results. The issue is more complicated when multiple control variables are of interest. One problem is identifying how to sample control variables simultaneously, in order to optimize the response variable or variables. Usually, a scientist or engineer exploring the control variables will test each control variable over a range of values independently. This will work well if the control variables do not interact, or if comparative information is desired. If the absolute best value of some response, such as the best

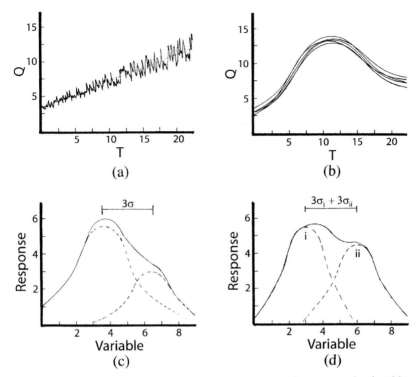

FIGURE 8.4 Noise in test signals: (a) A noisy background. On average, the signal is a straight line, but there is variation both above and below the line. (b) Noise in the signal can result from slight differences in the peak with each test, including variation in the point where the peak starts, the width of the peak, and the height of the peak. (c) If the peaks of two signals are separated by 3 standard deviations, the signals are characterized as being separately identifiable. (d) In order for the two signal peaks to be characterized as separable, they must be separated by 6 standard deviations.

modulus, is desired, or if the control variables are interdependent, then a simple testing strategy will not provide sufficient information.

An approach to multivariate testing is the factorial design. A 2^n (2-factorial) design, for example, uses a high and low value of a control variable as the test conditions. This yields a set of test pairs. This provides the 2. Each test variable adds to the exponential: 3 control variables require 8 experiments; 4 variables require 16 experiments. The experiment grid is set up to organize tests ranging from all high values to all low values, with every combination in between, as shown in Figure 8.5. Choosing the high and low values for the control variables requires some sense of the effects of the variables, some experience with choosing variable values

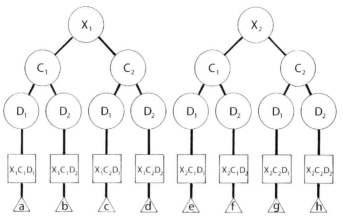

FIGURE 8.5 A 2^n plan for evaluating the interactions of three variables on a property of a composite or nanocomposite. An example of three variables might be the size of the particle, X, the concentration of the particle, C, and the dispersion of the particles, D. In order to evaluate the contributions to modulus, for example, high and low values for each variable are chosen. This leads to 2^3, or 8 experiments.

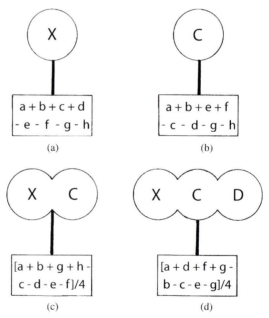

FIGURE 8.6 Calculating the effect of responses for the 2^n plan. (a) The effect of X can be evaluated by comparing the value of the modulus for all the experiments with high X to all those with low X. (b) If the effect of C is desired, all experiments with high C are compared to all those with low C. (c) The interaction effects between pairs of values, or (d) all three variables, can also be evaluated using the experiment results.

342

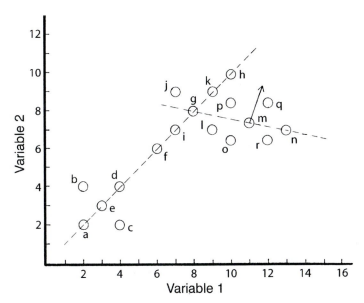

FIGURE 8.7 The method of steepest gradient is a technique for quickly zeroing in on an optimal solution. A group of 5 tests, (a)–(e) is conducted. A line is drawn along the line showing the greatest improvement in response. Tests are conducted along this line, (f)–(h), until the response ceases to improve. A new group of tests is created around the best response, (g) in this example, and a new gradient is established. After Haswell.

that will cause measurable changes in the response variables. From comparisons of the means of the response variable resulting from the given test conditions, the importance and interdependence of the control variables can be determined.

The factorial designs identify the importance of control variables and variable interactions. Exploring the response surface to pursue an optimum response is another task. One approach is the method of steepest gradient. In this approach, a rosette of experiments is performed around a starting point. A two control variable case is illustrated in Figure 8.7. The central point and the point with the best improvement in response are used to define the changes in variables for the next experiments. Tests are conducted along this slope until the response stops increasing. The best value of this series of tests becomes the center of a new rosette of experiments for determining the next gradient to follow. This proceeds until the best response is achieved within the limits of interest. This technique guides a more limited number of experiments, instead of requiring a complete, randomized set of tests across all of the available test space. Other optimization techniques are available.

Testing should generally be "blocked" to reduce the effects of systematic effects. For example, conducting all of the tests for one variable on Monday, a second variable on Tuesday, and a third variable on Wednesday can lead to obscured results if there are test parameters, such as cooling water temperature, equipment availability, time effects on standards, or so on, that vary systematically day to day. Once a test is designed, the order of the tests can be randomized to randomize systematic errors of this type. Also, tests should be run on different days, on different samples, and with repeat measurements, to randomize error as much as possible.

SAMPLE PREPARATION

In order to test composites accurately, a representative piece of material of a desired characteristic must be prepared. This preparation has specific requirements based on test design, eliminating artifacts, separating behaviors, and so on.

Sample Collection

Sample collection is an important art. The usual purpose of collecting a sample from a larger collection of specimens is to determine the properties of the entire group. Choosing samples for testing is a primary concern for getting representative results. In order to assess the population of the specimen, the sample gathered must be sufficient to represent the entire population. The simplest answer is to choose the entire population as the sample size, but this may leave insufficient material for later work, may be onerous and boring, or may even be dangerous if the material being tested is toxic or otherwise hazardous. Often, sampling is necessary. Any source of material for characterization presents specific challenges. Is the sample small? How can enough material be tested to provide a meaningful answer? Is the sample large? Well, then, how can one be certain that a small piece is representative of the whole?

The most common question is "how much is enough?" The size of an individual specimen for testing is often dictated by the test geometry. The sample size, on the other hand, must be sufficient to provide a desired level of precision. The normal distribution, given in Equation (8.1), depends on the sample size, n, and the standard deviation of the sample, so there is no firm answer to the question. A sequential sampling strategy can be adopted. Here, a new specimen is tested, and the value is added to the measurement pool. The results are evaluated against a decision tool, such as a null hypothesis, a desired degree of precision, or so on. If the results are not conclusive, measurements are taken until the requirements for a decision, positive or negative, are met.

Alternatively, tests can be repeated until the answer stops changing: that is, when the mean and standard deviation remain the same despite the collection of new test points. Either of these approaches is a "correct" way to determine how many specimens to include in a test. There are national standards that are used to remove the guesswork from choosing a sample size. Military specifications and ANSI standards have guidelines for determining an adequate size based on the population size. These standards use best generic algorithms to set standards that should work

---

Do It at Home 8.1

Playing Cards as a Model for Sampling

Experiment 1

From a standard deck of playing cards, choose 7 cards. Pick a card as a favorite and place the cards in any order, but make sure the favorite is located on an end. Have someone pick a card from your hand. Repeat the experiment multiple times. How many times is the "favorite" card drawn by the picker? What about the other cards? Try slightly extending the "favorite" card and repeating the experiment. How does the selection rate change?

Experiment 2

Stack a standard deck of cards according to suit and ascending rank. Select 7 cards from the top of the deck and record the numbers and colors. Select 7 cards at random from throughout the deck and record the numbers and colors. Shuffle the deck thoroughly and repeat the experiments. Is there any real difference between sequential and random sampling in the shuffled deck?

Experiment 3

Select 20 sets of 5 cards from a randomized, standard deck of cards. Record the poker hands collected. Do your results agree exactly with the statistics of large numbers used to calculate the probabilities of collecting specific hands, such as pairs, three of a kind, and so on?

well enough for many purposes and remove the guesswork and exploratory analysis necessary to set individual decisions. They are usually planned to over-sample, in order to ensure product compliance.

Regardless of how many individual points are needed, sampling strategies are a challenge. Most sampling requires random choice of specimens to eliminate the chance of biasing collection toward some hidden flaw. Randomness is difficult for a human collector. Humans tend automatically to select specimens that are separate in space and time, that show something interesting to observe, that are easier to collect, and a host of additional factors. Collecting every nth sample, where n is some integer number, is one strategy that helps with randomness. If the samples are randomly arranged, then gathering every nth item will assure randomness. The only problem with such a strategy is if any systematic variation occurs in such a way as to be missed by this strategy, such as a failure that occurs every nth event, and begins at some unsampled point. Stratified sampling occurs when a family of specimens is previously grouped, by preference randomly. Testing can be conducted evenly on samples from each group. If the groups are not the same size, however, this can lead to over-representation of test results from one particular group.

There are other sampling strategies of value. There are several factors to keep in mind when designing a sampling strategy: avoid over-selecting or under-selecting from areas with especially interesting features; stick to a sampling plan and do not vary from the established pattern; make sure to sample closely, frequently, and persistently enough to catch all the features of interest. The easiest way to ensure randomness is to start with a random population. In this case, any sample will be representative.

In a typical, designed experiment, the sample testing size is often 100% of all available specimens. That is, an analyst will usually test the modulus of all of the dogbone test pieces cut for testing, and chemists will usually test the pH of the entire novel acidic polymer that was just invented at the bench. After all, the population was created to be tested! This should be considered in reporting.

Sample Reduction

The next problem that must usually be overcome is the dimensional size of the sampled specimen itself. Seldom when preparing for an experiment is the sample just the right size or geometry for the test. There may also be a variation in composition depending on position within a larger specimen. Often, a bulk specimen must be reduced, cut down to samples of the correct dimensions, or a sample collected from a point of interest in a larger part.

One of the simplest ways to prepare a nanocomposite for testing is to cast the piece in the correct configuration. Parts made using *in situ* polymerization can be poured into a mold of the appropriate dimensions. A mold for single or multiple uses can be made using rubber. The liquid latexes for sale in hobby stores are good for this purpose. The mold produced is flexible; after casting and curing, the mold may be flexed to dislodge the part. Latexes are often waterborne, and so there is no volatile organic component. There can be some toxicity or allergies associated with latex, however, and the additives may contaminate the specimen, to the detriment of chemical analyses or to the properties. The thickness of the molded part plays a role in determining whether the amount of contamination will hamper analysis.

Common alternatives for latex rubber are silane polymers, or silicone rubbers. These materials are often stable over different ranges of temperature, depending on the chemical formulation. Some are stable at high temperatures, some at low temperatures, and so on. They are also available in a range of mechanical performance. Many silicone rubbers are two-component systems, with a liquid resin and an initiator to cause the solidification. The silane polymers are 100% solids, and again they normally contain no volatile organics or other contaminants.

Casting molds is usually done from a positive casting. Cavity cutouts are cut from polished metal, hard plastics, or ceramics. The cutouts are then placed in a flat, level container. The molding resin is poured over the inserts and cured according to the instructions. Removing the mold from the positives, or removing parts from the resulting mold can be difficult. It usually helps if the positive mold can be twisted or deformed to facilitate removal. Common causes of sticking are penetration of the liquid resins into porosity, physical bonding, or chemical interactions with the molding surface. Using polished surfaces can minimize porosity in the positive mold. Degassing the rubber resin before pouring the mold reduces porosity in the mold. The liquid can be placed under vacuum to pull out dissolved gas and gas bubbles trapped in the viscous rubber precursor liquid.

Chemical interactions must be blocked. A common method for blocking chemical interactions in molds is to apply a mold release. Common mold releases include simple barrier materials such as petroleum jelly or wax (hydrophobic), or waterborne grease (hydrophilic). There are also spray releases specially formulated not to contaminate specific types of plastic compositions. Care must be taken when using spray mold releases. Aerosols, as discussed previously, can remain suspended in air for

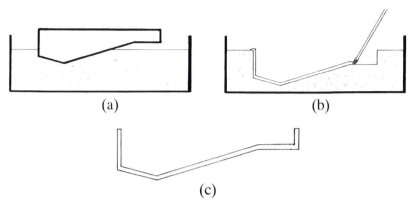

FIGURE 8.8 Making a flexible positive mold: (a) A positive casting is used to create a cavity. (b) The cavity is coated with a liquid rubber resin. (c) The flexible positive is removed.

long periods. The mold release can then contaminate any surface in the laboratory. In turn, this can interfere with chemical analysis, adhesion between reinforcement and matrix in composites, and many other activities common in an analysis laboratory or production facility.

If bulk material is all that is available, or a specific location in a test piece must be analyzed, then often a sample can be collected by cutting, grinding or breaking. A common way to create a proper sample is by cutting a larger piece to the proper dimensions. Metals have a host of appropriate, specialized equipment for machining. Metal lathes, saws, drills, grinders and so on are often used to good effect. Machining ceramics can be a special challenge, due to their brittleness. There are saws with abrasive blades that wear away material, allowing proper cutting. Ceramic saws often move more slowly than woodworking or metals saws, in order to reduce the risk of a catastrophic shock to the sample that will cause it to fracture. Softer polymers can often be cut using the same techniques that a carpenter uses to cut wood: drills, saws, lathes, grinders and sandpaper. If woodworking approaches don't work, the metals and ceramics processing techniques will often serve.

If cutting to a specific geometry, e.g. thickness or shape, is unnecessary, the material may be reduced more generally by breaking or grinding. Breaking, or fracturing, a sample does not introduce the saw and cutting marks common to other techniques. A particular feature of reduction by fracturing is failure along the weakest point in the material, such as grain boundaries, fault lines and micro-cracks, along the boundaries between reinforcement and matrix, and at joins. Milling, such as ball

TABLE 8.2 Some polymer-solvent pairs.

Polymer	Solvent	Non-Solvent
atactic Polypropylene	benzene, chlorinated hydrocarbons, cyclohexane, diethyl ether, hydrocarbons, toluene, isoamyl acetate	
isotactic Polypropylene	elevated temperature: 1,2,4 trichlorobenzene, decalin, halogenated hydrocarbons, hydrocarbons, xylene	
polymethylmethacrylate	acetic acid, acetone, benzene, chlorobenzene, chloroform, dioxane, hot isobutanol, MEK, methylene chloride, xylene, ethanol/water	carbon tetrachloride, diethyl ether, absolute ethanol, turpentine, linseed oil
polytetrafluoroethylene	perfluorokerosene	
polystyrene	benzene, carbon disulfide, chloroform, hot cyclohexane, MEK, ethyl acetate, methylene chloride, phenol/acetone, THF	acetic acid, acetone, alcohols, diethyl ether, diols, phenol, saturated hydrocarbons
polyethylene teraphthalate	hot DMSO, phenol, chloral hydrate, chlorophenol	ethers, hydrocarbons, ketones, chlorinated hydrocarbons
nylon 6	acetic acid, m-cresol, phosphoric acid, sulfuric acid	alcohols, ethers, esters, chloroform, hydrocarbons, ketones
nylon 11	higher primary alcohols, DMF, DMSO	

milling, is a sample fracturing technique that works well with brittle materials. Softer materials must often be cooled significantly before reduction by fracture predominates above mechanical deformation under the impact of the milling media. This may be an advantage or disadvantage. Using grinding, the fracture occurs by mixed-mode forces, usually shear, to abrade material away from a surface. Again, because mechanical force is applied to create the fracturing force, there may be implications to the appearance of the particles produced using grinding.

An alternative to *in situ* polymerization, cutting or grinding is dissolution. Many thermoplastic polymers can be dissolved in a solvent. The mixture of solvent and dissolved polymer can be poured out to form a film. After the solvent is removed, a film of any desired thickness is left behind, controlled by the concentration of the film and the quantity of solution used. An advantage of dissolution for sample preparation is that many additives are not soluble in the same solvent, and are thereby removed from the test specimen. Another is that very thin films suitable to analysis in transmission mode can be created. A disadvantage shared by many available techniques is a lack of a common process history with the original material.

Thermoplastic materials can also be melt-pressed. After determining the melting point of the polymer, powders, or pellets of the material can be placed in a hot press. Using pressure, films or pieces with simple geometry can be created. Subsequent machining to create the proper dimensions will result in a proper specimen. In order to effectively melt press a film, a temperature in excess of the T_m, usually by about 50°C, is necessary to fully melt the polymer and lower the viscosity sufficiently to allow flow. A negative mold of the desired dimensions is used to insure proper control of thickness. Based on the density of the material, an about 10%vol excess of powder should be added to ensure that the mold is completely filled with slight flashing. A possible disadvantage is changes to the crystal structure during processing.

Surface Preparation

After cutting or milling, an unnatural surface is left behind. There are scratches and possibly chemical contaminants at the newly created surface. Further, surface irregularities may contribute to unstable behavior. A clear example of this is ceramic mechanical testing. Brittle ceramics are particularly sensitive to the surface cracks left behind after machining. Grinding, polishing, and milling may further prepare the surface.

Grinding is a familiar technique. Using an abrasive surface, large scratches are scraped away by introducing smaller scratches. Grinding

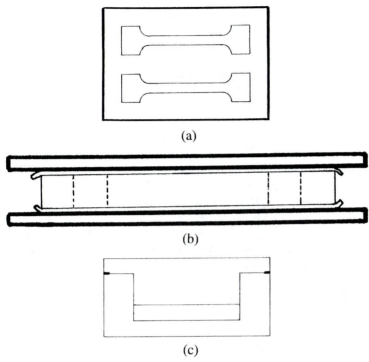

FIGURE 8.9   A negative mold for melt pressing films, plates and other parts. (a) A metal plate is cut with holes of the target part shape. (b) This mold plate is filled with polymer, heated and compressed to form parts. Release sheets may be used between the platens and the mold to facilitate removing the mold. (c) A fitted mold.

works by shearing, tearing and knocking loose pieces of material off the bulk to create a surface with features of a scale similar to the roughness of the grinding surface. Abrasive surfaces include emery paper, sand paper, abrasive plastic pads such as the Scotch-Brite™ range of polishing pads, grinding stones, or the crosscut surfaces of a file. Grinding can be applied in successive steps. A large grit or coarse surface can be used to remove the largest defects introduced by the sample cutting technique. A smaller grit surface can be used to remove the defects caused by the coarser grit. A still finer grit surface can grind off the scratches from the larger grit, and so on, until the finest available grinding surface has been used. The grinding surface can become clogged by material removed from the surface of the specimen. This will slow or stop material removal. Removing the top layer of abrasive can regenerate some clogged grinding surfaces, such as grinding wheels. Sometimes, lubrication can prevent the accumulation of debris. Sometimes, the grinding surface must be replaced frequently to assure that effective grinding continues.

Polishing is a less obvious process. Unlike grinding, polishing uses loose particulates to remove material from the surface of the specimen. Common consumer polishing compounds include cleanser, toothpaste, and jeweler's rouge. Polishing appears to work by a melting phenomenon. A material can only accomplish polishing with a higher melting temperature than that of the specimen to be polished. High melting materials such as carbides, alumina, and diamond, just happen also to have high hardness. The hard materials are not necessarily needed, as is demonstrated by jeweler's rouge, which is composed of very fine iron oxide (rust) particles. Particulates are removed from the specimen that are of the same size as the polishing particles. The particles are removed from the highest points remaining on the specimen. They often end up deposited in nearby scratches. They usually only physically are caught there, not trapped. Polishing media are available down to approximately 100 nm in diameter. They cannot produce a surface appropriate for examination below this level.

If a smoother surface than can be obtained by polishing is desired, a number of specialized processes can carry the process further. If the specimen is composed of a soluble material, then applying a solvent to the surface can dissolve and re-precipitate a molecularly smooth surface. The solvent cannot be applied with a cloth. The solvent will lower the glass transition of the polymer, and the cloth, no matter how soft, can then scratch the surface. Etching is another alternative. Very small particles are used to abrade the surface atomically. One example is ion etching, where applying an electrical potential creates particles. These

---

Do It at Home 8.2

Gather gravel, cut pieces of polypropylene from a bottle, sand, and kitchen cleanser. Take a waste CD from some source. Rub the surface with gravel. This will create large scratches. Rub the disk with sand, next. The scratches should be reduced in size to the scale of the sand particles. Wash the disk in water after this step. Does the result change? Continue to polish the disk with the kitchen cleanser. If the original scratches were not so deep as to hit the information layer, and if you polish radially rather than along the arc of the disk, the CD may even still be playable. Try scratching the disk with a soft polymer, with a lower melting temperature than the polycarbonate of the CD.

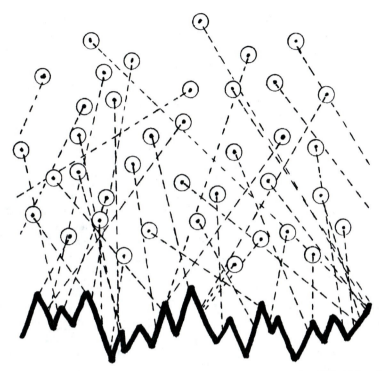

FIGURE 8.10 Erosion occurs as moving particles impact on a surface. The high points are more exposed, are hit more often, and therefore tend to wear away more rapidly.

particles are then accelerated at the target using an accelerating voltage. The impact causes atomic scale damage to the surface. This type of etching is only effective at removing very small damage. Controlled chemical corrosion or erosion is another technique. The surface is put in a chemically damaging environment. The high points of the surface are more exposed to the environment. The low points are less affected, and the high points wear away the fastest.

Fixation

Some specimens require fixation. Samples that contain water or solvents, materials that are soft, specimens with fragile structure, or all of the above, are easily damaged during sample handling or analysis. For example, polymers are often very soft and difficult to cut without crushing. Fixation applies any technique to make a specimen more stable for whatever analytical technique will be later applied. Fixation will make the material easier to handle and analyze without damage. There are both chemical and physical methods for stabilizing soft materials.

Soft materials, such as polymers or biomaterials, can be cross-linked to make the material stiffer and fix microstructure. Highly reactive oxidizing chemicals are one way of introducing cross-links. Aldehydes, such as glutaraldehyde and formaldehyde, and ethylene oxide, are small molecules that react with a variety of common chemical functional groups, such as alcohols, ethers, esters, and so on. Permanent chemical bonds between adjacent molecules form during reactions between the chemicals and a material containing these reactive groups. Small aldehydes are also volatile and easily removed after treatment. The chemicals are introduced in excess, allowed to penetrate the specimen and to react completely. The excess solvent and chemical are removed, and the sample can then be subjected to further preparation techniques with reduced risk of introducing artifacts. Chemical reactants must be able to permeate the specimen, and there must be suitable reactive sites in the material. An alternative route to cross-linking is high-energy sources, such as x-rays, electron beams, or gamma reactors. Under radiation bombardment, free radicals form at random points on a molecule. These radicals may react with a neighboring site, resulting in a permanent covalent bond. Even non-functional materials, for example polyethylene, will react under the influence of the high-energy particles. Chemical bonding may induce sample shrinking. Chemical changes also occur that may affect the results of analysis.

Physical fixation techniques introduce reversible, temporary, or otherwise non-transforming changes. Freezing a material will lock any structure in place during subsequent handling, such as cutting and fracturing. Freezing the material minimizes the difference in mechanical properties between a brittle reinforcement and a ductile polymer matrix.

If water is present, the water crystals expand during cooling and can destroy any microstructure. However, if freezing is very rapid, crystal growth is suppressed by nucleation. The crystals will conform more exactly to the original shape of any structure and minimize displacement, for example, of nanoparticles. As an alternative to freezing with water present, the specimen may be dehydrated. Just drying the material will usually result in deformation and distortion from the high surface tension of the retreating water. If the specimen is placed under sufficient surrounding atmospheric pressure, dehydration can occur at or above the critical point without a volumetric liquid-to-gas transition. Sublimation also minimizes the distortion caused by the phase change from solid to liquid. Finally, the water may be extracted with a suitable solvent. Pure ethyl alcohol is mixed with the specimen, replacing much of the water

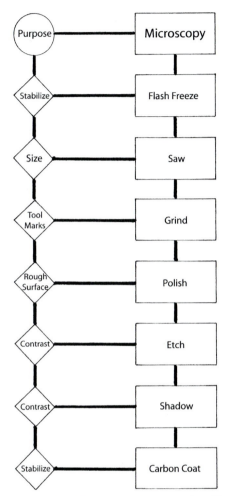

FIGURE 8.11 An example of the flow of sample preparation. The process begins by identifying the purpose of the specimen. The next steps may be undertaken to stabilize the matrix material, reduce the specimen size, clean the surface, or enhance the contrast.

with alcohol. Of course, the first extraction will not remove all of the water. By successive extractions, the water can be replaced with alcohol to any desired purity. The advantage of solvent replacement is that alcohol crystallization is non-expanding, and alcohol has a lower surface energy than water.

Finally, embedding can create stable specimen for testing. With embedding, a liquid material is introduced to the material, is allowed time to

TABLE 8.3 *Some nanocomposites preparations.*

SEM Microscopy of Montmorillonite clay distribution in polypropylene		X-ray analysis of crystallinity in nylon 11/silica composite		Wear resistance of a carbon-black containing rubber	
Step	Purpose	Step	Purpose	Step	Purpose
Cut with saw	Reduce size	Cut thick film	Fit to sample holder, select region	Cut with knife	Reduce size
Grind w/200 grit	Remove saw marks	or:		Ensure even thickness	Eliminate need to compensate for slope
Grind w/successive papers	Reduce surface flaw size	Freeze	Allow even grinding through specimen	Grind w/successive papers	Eliminate extraneous scratches
Polish w/successive grits	Reduce surface flaw size	Grind to powder	Reduce size, ensure even distribution of crystal orientation	Polish	Seldom needed
Erode w/N_2 Plasma	Remove last flaws, reveal clay at surface			Apply scratch at well defined speed	Scratch depth often depends on weight and rate
Gold decorate	Enhance edge contrast			Test within a defined test time	Scratch may recover with time in a polymer
Carbon coat	Reduce charging without changing MW contrast				
Mount on stub with glue, paint sides of specimen with carbon or silver paint	Prepare for analysis				

permeate the specimen and fix all features of interest, and then is solidified. Liquids used for this technique include epoxy, polyester resins that can be cured to high or low hardness, and wax. The embedded specimen can be machined or sliced to remove very thin sections.

Artifacts

When preparing a specimen for testing, artifacts may be introduced. An artifact is any element introduced by the preparation technique that will result in an inaccurate interpretation of the subsequent analysis data. For example, cutting with a saw puts scratches on a newly formed surface. Visualizing the surface using any of the microscopic techniques discussed in Chapter 9 will uncover these artificially induced scratches. A typical surface structure may be completely obscured by these scratches. Another example is contamination. Machinists often use a lubricating milling oil when machining metals. Subsequent chemical analysis of the metal part may reflect more about the milling oil than about the composition of the metal. Designing a proper sample preparation technique that minimizes defects, optimizes preparation time, and uncovers interesting features is an art developed with practice and study.

Composite materials, usually by design, incorporate materials with different hardnesses, and other mechanical properties. This can lead to

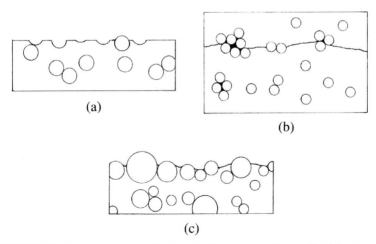

FIGURE 8.12 There are several artifacts particular to composites: (a) Particles may pull out of the surface during polishing. (b) Fracture may occur along a line of anomalously high particle concentrations. (c) A free surface may have more or fewer particles than is representative of the bulk.

significant artifacts that form during sample preparation. The difference in hardness can sometimes be overcome by freezing the specimen to a temperature low enough that both materials are very brittle. The failure lines introduced by saw, fracture, or shear then would propagate rapidly through either phase and travel without regard to composition.

Sample Conditioning and Accelerated Testing

Material testing is normally planned for immediately after sample preparation. This provides measures of the initial properties of the material, which are typically the properties of the greatest interest. However, in service the material will age during exposure to heat, light, oxygen, thermal cycling, and other environmental factors. This may cause the properties to change with time. If these changes are significant, there may be design limits to how the material may be used. There are some techniques for conditioning and aging materials to improve predictability in the composite behavior.

Conditioning, Standard Temperature, and Humidity

Some properties can be strongly affected by the local temperature and humidity. Adhesives, thin films or hydrogels, and textile fibers made with natural fibers are examples of materials with properties that will be influenced by the environment even at short time scales. These materials must often be tested under controlled temperature and humidity. Some testing facilities have CTH (controlled temperature and humidity) rooms. Samples must be stored in these rooms for a period sufficient to equilibrate them to the standard conditions. The required time can be determined by testing samples at different intervals until the required interval is determined. In practice, the conditioning time is set at a standard value, above the true required time, for convenience.

Other samples may need different types of conditioning. Hydrogels may require pre-soaking to completely saturate the material. Damp fibers may require drying to provide consistent test results. The required conditioning should always be considered during test design.

Artificial Aging

During use, many materials will experience changes in properties resulting from natural processes, such as exposure to oxygen or light. If the product is going to be buried for 100 years or used to build a bridge that will stand in the sun and rain for 50 years or more, the impact of these changes may be important. During the aging process, some degradation of the yield modulus may occur, for example. Artificial aging protocols are developed to help engineers choose stable materials for products with a long lifecycle.

One method of artificial aging is to soak the part in an oven at elevated temperature. The temperature may be above glass transition to accelerate physical and chemical changes. The rationale is that reactions that occur slowly at room temperature will occur more rapidly at higher temperatures.

A related approach is to put the part through cycles. Weathering chambers are available that expose parts to cyclic sources of ultraviolet light, mist, salt water, lightning and all manner of abuse designed to simulate and accelerate natural weathering processes.

Surfactants have been found to cause cracking in some plastics. Long-term exposure to various solvents, cleaning solutions, weak acids or bases, and so on, may accelerate the apparent age of a material very effectively. These are sometimes found in accelerated test protocols.

Such aging techniques do accelerate natural changes, but may also cause atypical performance variation. The amount of acceleration does not necessary correlate well with specific environmental factors. Still, withstanding a rigorous weathering schedule for six months provides more confidence that a material will be durable. The investigator generally cannot guarantee a lifetime performance unless a material is tested in service.

IMAGING

One of the most common goals of characterization is to visualize the structure and behavior of composites and nanocomposites. Whether as pictorial information ranging from magnified microscopic images, through abstract diffraction or absorption data, to a plot of the mechanical behavior changes as a function of process changes, visualization is part of the analysis process. Images also provide a parallel source of information to illustrate new knowledge. While a slight deviation from the usual approach to nanocomposite study, consideration of how the investigator and the audience see, interpret, and reproduce images can aid the analysis of nanocomposite test data.

Photographic images are a regular feature of structure analysis and method illustration. Pictures may be simple depictions of the experimental equipment, or may be highly magnified electron microscope views that show the shapes, spacing, and interactions between agglomerated nanoparticles. These pictures at their most valuable contain experimental information about features of interest in the material. Interpreting the image may require understanding how the data in the image is seen through the eye or the computer, how the image is stored and represented, and how to break the image up into measurable data. Seeing is related to the performance of the eye or the computer. Storage and representation of images depend on the type of information present: black and white, color, and specific content. Image analysis usually consists of thresholding, segmentation, and measurement.

Visualization

Photographic images are not the only type of data that benefits from visual representation. Several important recent microscopic techniques provide visualizations, rather than magnified images. For example, orientational imaging microscopy (OIM) provides a spatial map of the

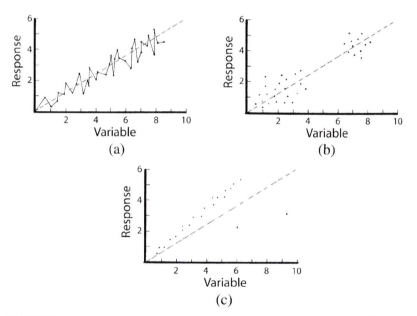

FIGURE 8.13 Scatter plots: (a) A noisy, linear response. Many students draw a line connecting the test points. If the experiment is conducted properly, the points should not be related, and connecting them is inappropriate. (b) A scatter plot with the corresponding linear fit. The distribution of points is bimodal. (c) A scatter plot with corresponding linear fit. Two anomalous results draw the fit artificially away from a more appropriate location.

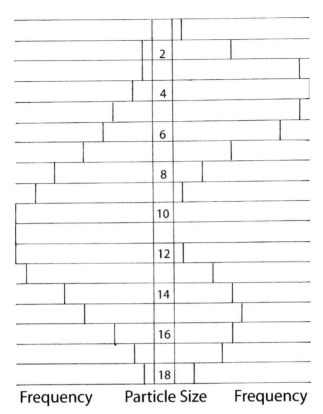

Frequency   Particle Size   Frequency

FIGURE 8.14 An example of a stem and leaf representation comparing two particle-size distributions. This format allows rapid, visual comparison.

orientation of crystals in a specimen; atomic force microscopy (AFM) can provide a spatial map of the attractive chemical force exerted between the microscope probe and the sample surface. The results are presented as an image, where visual intensity or color information is substituted for direction or chemical information, respectively.

Graphs are also an example of data visualization. Graphs provide an easy way to compare the relationship between two variables. As such, data plots should be considered with some of the same tools as traditional visual information. A particularly useful form of data visualization is the scatter plot. Scatter plots should include all the data, minus any outliers. However, the data should not be connected with lines unless the lines are appropriately fitted, and should reflect the certainty of the individual measurement. The data is a direct, spatial representation in the phase space of a controlled and response variable, and represent only single estimates.

Histograms and bar charts represent another type of visualization. Histograms report the number of observations made at a certain level. A stem and leaf diagram is a related type of representation. These diagrams provide a shape for the response space, providing information about symmetry of the distribution of responses, and so on.

Graphic representations taken together form a group of what is considered non-parametric data evaluation. There are no set parameters assumed to describe the data. Evaluations are drawn directly from the distribution of the data in a relevant variable space. Some parameters can be developed based on the graphs. A fitted line can help visualize the relationship the data may have. Fitting the data is usually accomplished in a plotting program using ordinary least squares regression (OLS). OLS works by minimizing the errors in the predicted response variable (y) given an input variable with no error (x). This may or may not be achieved in designing an experiment. The accuracy of the fit is assessed by the amount of deviation from the curve, which is given by a correlation coefficient. Presenting a fitted line without the surrounding data field can be misleading, because a single correlation coefficient can describe a fit to many data sets. Some fitting programs will also allow curve fitting without calculating parameters. One non-parametric line-fitting example is provided by the average of the slopes between any two data points in the data set. Splines are another visualization technique for drawing lines that do not assume any relationship between data points.

Another useful form of information can be obtained from Fourier space. Fourier space is the result of converting real space to frequency space using the Fourier transform. Fourier space has a real analogue in the reciprocal space formed at the diffraction plane of a microscope. The image data can be subject to various manipulations in Fourier space. Noise, such as diffuse scatter, or repeated patterns introduced by unintended diffraction, can be filtered out by removing the information in the appropriate regions of the Fourier transform before transforming back to image space.

Seeing

In tasks ranging from interpreting images to constructing figures for presentations, the elements of vision are an important consideration. The human eye is a remarkable organ. Many people are familiar with the basic components of the eye, as shown in the sketch in Figure 8.15, including the lens, or cornea, which focuses the light from an image onto the

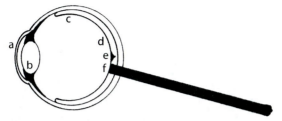

FIGURE 8.15 The human eye: (a) the cornea, (b) the lens, (c) the rod-rich outer reaches of the retina, (d) the central part of the retina containing the cones, (e) the fovea, the most sensitive part of the retina, (f) where the optic nerve leaves the retina, the eye has a blind spot.

retina. The retina has a region that is rich in the cones required for color vision, called the fovea, which occupies about 1.4° of the visual field, while the rest of the retina is rich in the rods that provide black-and-white perception. The remainder of the eye provides the mechanical system for adjusting focal length and for maintaining eye health. This is, however, a simplified overview of the eye. The process of seeing is complex, and results from the interconnections between the cellular elements of the vision process.

No element of the visual process is simply connected, with a

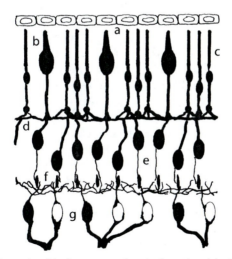

FIGURE 8.16 Schematic of the interconnections in the retina: (a) a thin layer of protective cells affects the sensitivity, (b) there are two types of cone and (c) a large number of rods, (d) The cones and rods are highly interconnected, (e)–(g) the signals from the cones and rods are combined and processed prior to entering the optic nerve.

FIGURE 8.17 Optical illusions are created by a combination of eye structure and expectations. In this diagram, the proximity of a dark edge next to a light edge makes the light edge appear artificially lighter.

one-to-one correspondence between element and nervous signal. The eye has three types of cone with sensitivities of 430, 540, and 575 nm, corresponding with red, green and blue. However, these signals are combined in the eye into blue-yellow and red-green signals before being sent to the brain. The rods, especially those farthest from the fovea, connect multiply to a single ganglion. The advantages of the cross connection are improved low-light sensitivity and the screening out of visual errors. The disadvantage is a loss in spatial resolution. There are also cross-connections that result in signals to enhance edges, enhance contrast, sum spatial information, and average out visual noise. The results of these cross connections lead to a pre-processed signal to the brain. Visual experience is complex both to interpret and to standardize.

Vision is also subject to visual tricks such as optical illusions. Perception is unique to the individual, but there are common features that can lead or mislead an observer to form similar observations from an image.

Since the eye is such a complex organ, interpreting the physiological responses is not simple. This makes acquiring, storing and reproducing images that directly represent the image seen by the eye challenging. There are empirical standards for image and video processing and storage that are important.

Image Capture

Film-based collection of information captures the intensity in the accumulation of a pigment based on the intensity of exposed light. The resolution of the film is affected by the fineness of the salt emulsion grain. Coarser emulsions are common for simple images; fine-grain film is more commonly used to acquire scientific images. Microscope images were traditionally taken on cut film, with negative sizes of approximately 4 × 5 inches. Thirty-five mm film has a resolution suitable for enlarged images, and has replaced cut film based on improved convenience. Polaroid film provides immediate feedback on image quality, and can furnish a negative as well.

Continuing improvements in the convenience and quality of digital imagery have driven the replacement of analog film, even in scientific testing. Data taken on film is an analog representation, and cannot be directly compared with digital photography. Digital image data is acquired using a brute force approach. The image's intensity is recorded based on points within a field of view. The resolution is determined by the maximum number of sampling points, usually reported as pixels per inch, ppi, or as total pixel count for the entire image, or "megapixels." These two measures are easily converted using the standard image aspect ratio of 4:3. The aspect ratio of digital images also does not correlate directly with the dimensions of 4 × 5 cut film or 1.5 × 1 35 mm film.

Images that are captured on film can be digitized with an optical scanner. High-resolution film negatives are regarded as fully sampled using a 4000 dpi scanner. A complication for scanning is the Nyquist frequency rule. This rule states that the maximum resolution of the digital data is 1/2 of the sampling rate. Image train flaws further reduce the maximum resolution. Thus, a scan must be taken at least twice the resolution as desired in the final image. Higher resolution scanning has been accomplished. Doubling the scan resolution to 8000 dpi also increases the resolution of the emulsion grain, which confuses the data somewhat, but does increase the amount of data retained from the original image by up to 10%. 4000 dpi captures most of the available data from high-resolution images. Lower resolutions can be satisfactory for less critical images or lower resolution emulsions.

The required resolution to achieve adequate image representation is a common question. The answer depends on the presentation medium. Computer screens have adjustable resolution. A "standard" 1024 × 768 pixel setting on a 17-inch monitor, corresponding to a 12 inch horizontal, would display an image with a quality of approximately 85 ppi. Magazine-quality, or near photo quality, standard printing is centered around 300 dpi. Here, dpi is appropriate because digital images use dots to create an image. To print a 5 × 7 image at 300 dpi requires an uncompressed file size of approximately 10 MB, a digital field of approximately 3.5

megapixels, or a scan of the original 5 × 7 image at 1000 ppi. Any increase in image size could be accomplished only by cutting the resolution of the image or by adding pixels by interpolation between existing pixels. The extrapolation cannot add image information. Any image data lost during digitization is lost. If the original image is a 35 mm negative, the scan rate required would be 2400 ppi or greater to generate the same 5 × 7 image.

Black and white images are the simplest form of visual data, containing only grayscale, or intensity, data. If the intensity is stored as a single 8-bit number, the gray scale can be recorded as a value of between 0 and 255. If the intensity is stored as a single 16-bit number, the gray scale can be recorded as a value of between 0 and 65,535. 8-bit encoding is the traditional form; 16-bit is only used in cases of extreme brightness range. Color images contain multiple frequencies of light, so storage is more complicated. One of the earliest processes for converting color image

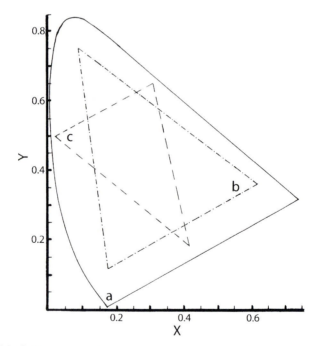

FIGURE 8.18 Illustrations of common color spaces: (a) the CIE standard, (b) the RGB space typical of a computer monitor, (c) the CMY space typical of a 3-color printer. Notice that neither the RGB space nor the CMY space completely covers the CIE color space. Also, there are colors available on a computer monitor that will not print on a color printer.

data to a storable, reproducible medium is CIE, named after the commission that designed the standard. CIE assigns colors based on the intensity of the combination of three colored lights required to reproduce, in the best judgment of the commission, a given color. This is similar to RGB color, which occupies a slightly reduced subspace of CIE color (Figure 8.18). Each color traditionally has an encoding range from 0 to 255.

The difference between light, with additive colors, and printing with pigments creates two different maps of color space. The standard for four color printing is CMYK. This is cyan, magenta, yellow and black. The black color is required because mixtures of CMY inks do not generate a strong black color and so decrease the available contrast range. Adding black ink allows a true black, but complicates the conversion from RGB to the corresponding CMYK colors. A method for further improving the color output from a printer is to add additional colors that are outside of those available with standard CMY ink mixtures. Mapping RGB colors that lie within mutually obtainable color space to CMY is a straightforward color mapping process. Mapping RGB to four or more color processes is challenging, since some of the original colors must be removed and replaced with the alternative colors. There are also regions of the two color spaces that are mutually unobtainable, leading to difficulties in, for example, reproducing the color seen on a video monitor.

The storage for digital images requires one byte for each pixel in grayscale or 3 bytes for each pixel of color data. This is the origin of large digital image file sizes. TIFF and RGB formats are common uncompressed image data files. Compression is possible by both lossless and lossy techniques. Lossless techniques use patterns in the data to allow saving less data without losing information from the image. An example is the run length encoding used in black and white facsimile transmission. The starting point of a color line and the length of the line is sent, resulting in a potential compression of storage length. Other algorithms are available. If some amount of loss can be tolerated from the original image, lossy techniques can be used. These storage algorithms, such as JPEG, can result in significant compression. The stored image can be recalled with usually acceptable results. Successively saving the image will result in further deterioration of the image. So, any image manipulation or measurements by preference should be made on an image stored in a lossless manner. Once an image has been stored using a lossy technique, data may be lost, and if so the loss is unrecoverable. Any further manipulation of an image stored in a lossy manner can result in more damage to the image and can introduce artifacts.

Printing reproduces an image on paper using one of several approaches. A very common technique is half toning. A 300 dpi printer can

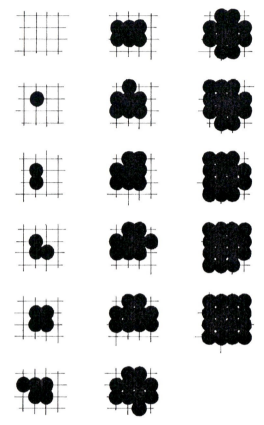

FIGURE 8.19 A half tone series with a 4 × 4 cell size.

print 300 dots in one inch in each direction. If these dots are broken into cells of 4 dots, then the resolution of the half-toned image becomes 75 dpi. Four dot cells are capable of rendering 17 individual gray levels. This may seem low, but is on a level with the number of grays available on a computer screen or from a Polaroid® print, although the film is considerably better than the print. New printers can achieve resolutions of 600 dpi of symmetric resolution. Since the printer places dots that overlap slightly, the image often appears darker than desired. Artifacts such as banding, aliasing and posterizing also can be introduced. These artifacts can be eliminated using a technique called "dithering" that adds random or patterned noise to disrupt the artificial edges created by half-toning. The side effect is a loss of sharp edge detail that may not be appropriate to scientific analysis.

Image Manipulation

Image flaws include noise, non-uniform illumination, and an undesirable viewpoint. The best approach is to collect an image correctly in the first place. However, after an image has been gathered digitally, some modifications can be made.

Collecting multiple images and averaging them enhances the signal and can reduce the noise. Alternatively, filters can be used to eliminate noise. Neighborhood averaging is a common approach. The intensity values of all pixels in a small neighborhood are averaged. The value of the central pixel is replaced by the average. Any pixel of noise is thereby reduced in importance. Unfortunately, this often causes displacement of boundaries. A neighborhood ranking, or median, method replaces the central pixel with the median of the ranked intensities of all neighbors in a defined region. This technique will not shift boundaries as much. All noise removal will soften edges. An edge is similar to noise in that there is a sharp change in the intensity at boundaries. Top hat and rolling-ball filters use a filter with a defined region. The algorithm looks for pixels that are too different from the surrounding pixels, too high for top hat or too low for rolling ball, and replaces them with the maximum allowed value.

If contrast or brightness is a problem, these can be expanded. If the total number of gray levels is less than 256, then the grays can be separated and expanded to occupy the entire range available. This often has the effect of improving the contrast. Other adjustments to the distribution of gray levels, or the histogram, can result in improvements to the image. Sometimes images may be non-uniformly lit. Features in dark areas of the image, or washed out features, can be obscured by this sort of flaw. There are algorithms that can be used to correct for this type of problem. Another type of manipulation is edge enhancement. The Laplacian operator, or unsharp masking, locates regions of greatest difference. This locates edges well. The information can be added back to the image to enhance edges. Derivative methods are even more effective at locating edges.

Image manipulation is a complex field. Adding artifacts or displacing features is a common result of misapplying the techniques. If an image is digitally manipulated, all transformations should be documented in any report of analysis.

Measuring

Once the image has been satisfactorily digitized, the final step is to

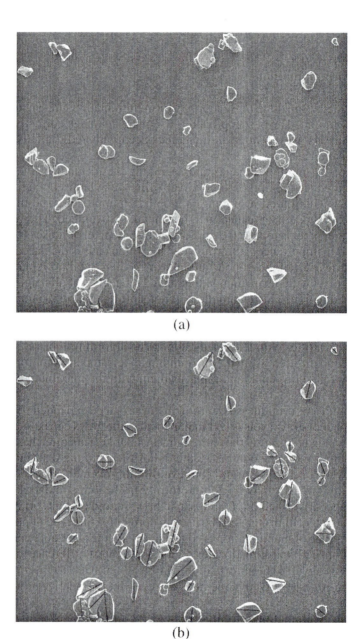

(a)

(b)

FIGURE 8.20 Artifacts in measuring particles: (a) An image containing multiple particles. (b) The image with lines measuring the particle diameter. However, the analyst
picked the longest axis for all measurements, and an artificially large particle size measurement will result. (c) The same image, evaluated correctly for the Feret's diameter. (d)
Measurement sampling lines that miss the particles entirely. Using these lines would artificially lower estimates of the diameter.

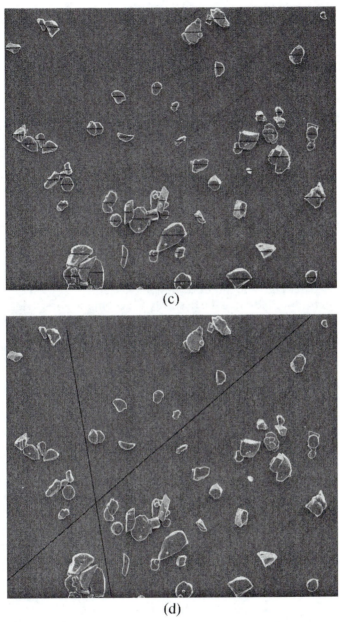

(c)

(d)

FIGURE 8.20 *(continued)* Artifacts in measuring particles: (a) An image containing multiple particles. (b) The image with lines measuring the particle diameter. However, the analyst picked the longest axis for all measurements, and an artificially large particle size measurement will result. (c) The same image, evaluated correctly for the Feret's diameter. (d) Measurement sampling lines that miss the particles entirely. Using these lines would artificially lower estimates of the diameter.

measure the features of interest. A trained operator can reduce an image to data in a number of ways, whether the image has been digitized or not. As shown in the examples in Figure 8.20, there are many ways to create significant errors in making measurements. Experience and preparation are important to avoid these errors.

Computer automated measurements are increasingly common. While automatic face recognition, infinite resolution of images, and other amazing video and image analysis techniques are shown on television, most of this technology does not exist. The eye and the computer "see" in profoundly different ways, and what is easy for us may be hard to do with a computer, and vice versa.

In order for a computer to interpret an image, the image must be divided into foreground objects and background noise (Figure 8.21). A trained human operator is usually the gold standard for identifying the edges of objects in an image. However, an observer is subject to variance in training, fatigue, and slight deviations in applying decisions. Automated algorithms are often the best way to accomplish image identification without operator bias. The process of identifying foreground and background is segmentation, dividing an image into regions that correspond to objects. Thresholding is one technique for identifying objects: a gray-level range is selected that contains the objects of interest, hopefully completely and exclusively. This range is then accorded a specific gray level identifying the status as object, often black. Other gray levels are assigned a different gray level, often white, corresponding to background. If the image field is flat, the background illumination is even, and all objects are approximately the same size or the background is uniform, then a single threshold range can be effective. Figure 8.22 shows some of the difficulties with picking a single value for threshold scanning electron microscopy images. For complex images, there are specialized techniques for choosing local thresholds to properly identify pixels as foreground or background, or to divide the field into foreground and background using other approaches entirely. Once the image is broken into objects, the boundaries of the objects must be uniquely identified. Figure 8.22 shows the mechanical process of the Pavlidis algorithm, a technique for automated object outlining. The measurement of object features, such as size, shape, separation and so on, depends on this boundary. A particular concern with segmentation is accurately defining the boundary. A difficulty can occur where objects overlap in gray level, leading to some misclassification of boundary pixels during thresholding. Objects with a large range of gray level gradation also may result in distortion of the boundaries. Inclusions may create holes in the object where none really exist. All of these features create uncertainty in

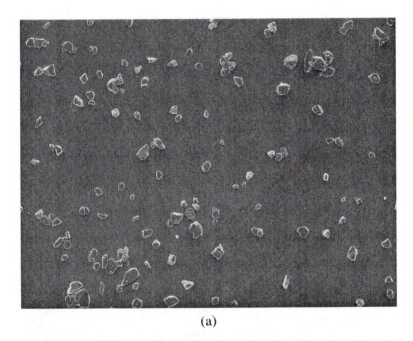

(a)

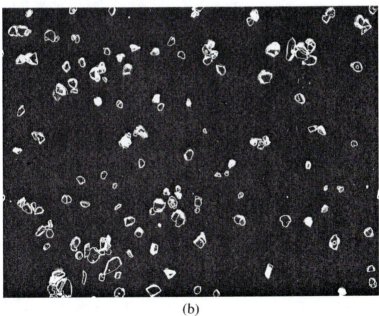

(b)

FIGURE 8.21 Segmentation of an image of particles in a composite matrix: (a) The original image. (b) A threshold can be picked that selects the borders of the particles. (c) The insides of the particles are identified and designated as particle. (d) The particles are replaced with equivalent ellipses.

373

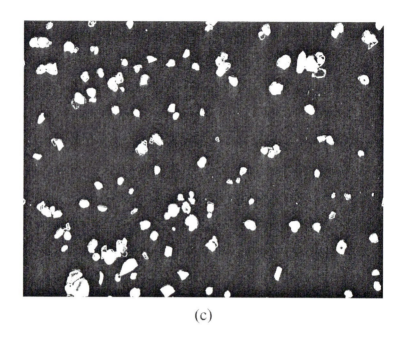

(c)

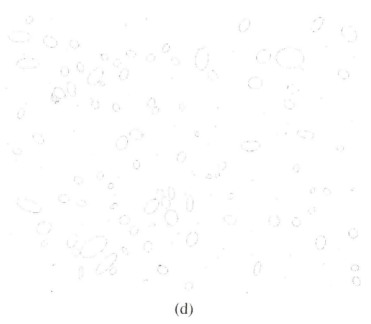

(d)

FIGURE 8.21 *(continued)* Segmentation of an image of particles in a composite matrix: (a) The original image. (b) A threshold can be picked that selects the borders of the particles. (c) The insides of the particles are identified and designated as particle. (d) The particles are replaced with equivalent ellipses.

374

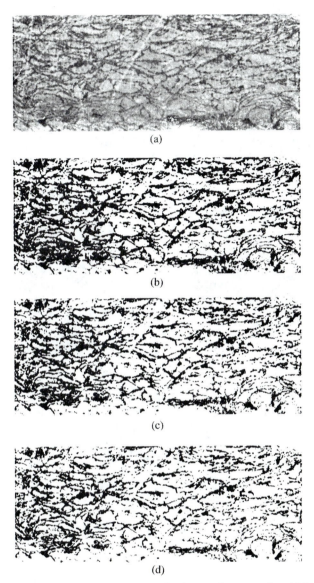

(a)

(b)

(c)

(d)

FIGURE 8.22 The steps in analyzing a complex image of a composite with laminar features: (a) The original image. (b)–(d) The image with thresholds set 5 gray levels apart. Slight changes in threshold selection may have a significant impact on characterization. (e) Two thresholds, the first is set at the peak in the gray scale distribution, the second is set at 3 standard deviations from the maximum. (f)–(h) The Pavlidis algorithm is an example of segmentation, or quantifying the outlines of the features, based on a threshold. (f) A point is chosen, and the rules of the algorithm are followed to find the boundary of the feature. (g) The algorithm will then follow and quantify the boundary. (h) The algorithm can find internal boundaries, but the process depends on the starting point.

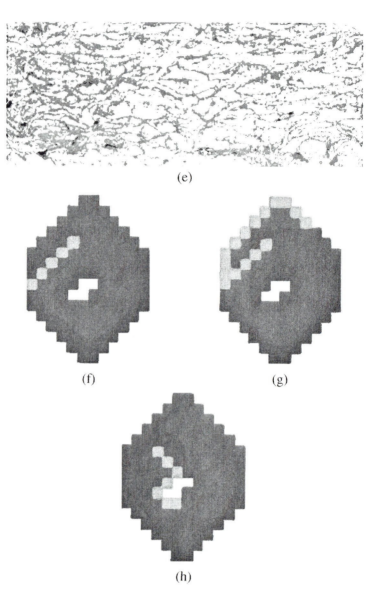

(e)

(f)   (g)

(h)

FIGURE 8.22 (continued) The steps in analyzing a complex image of a composite with laminar features: (a) The original image. (b)–(d) The image with thresholds set 5 gray levels apart. Slight changes in threshold selection may have a significant impact on characterization. (e) Two thresholds, the first is set at the peak in the gray scale distribution, the second is set at 3 standard deviations from the maximum. (f)–(h) The Pavlidis algorithm is an example of segmentation, or quantifying the outlines of the features, based on a threshold. (f) A point is chosen, and the rules of the algorithm are followed to find the boundary of the feature. (g) The algorithm will then follow and quantify the boundary. (h) The algorithm can find internal boundaries, but the process depends on the starting point.

376

the location of the boundary. Dilation and contraction are image processes that can be used to assist with properly capturing the outlines of objects. The science behind thresholding and segmentation is rich and informative. If the student is often in a position of analyzing image data for nanocomposite analysis, study of these techniques may be required.

Measurements are made on images, directly through human measurement, or automated with thresholded, segmented, binary images using a range of techniques. One of the most common approaches on native or computerized images is stereology. Stereology is the science of making estimations of length, surface area, and volume for 3-D objects from 2-D images. The mathematics for justifying the image measurements is extensive, but the conclusions are straightforward. Some common sampling techniques for stereology are shown in Figure 8.23. The estimate is based on the data gathered by the casting technique.

For point analysis, random points are sampled in an image. The simplest, unbiased sample is generated using a regular grid. The grid spacing should be narrow enough to ensure that many of the points are on an object. If the grid is too widely spaced, then the grid will have to be cast on the image many times in order to achieve a good measure. However, if the grid is too narrowly spaced each object will be multiply counted, which is inefficient. Each grid point is assigned as background or foreground. A sufficient sample set of points is collected to provide an accurate representation of the image. The number of points required to do this is subject to sampling requirements.

The same general approach may be taken with lines. A number of lines are generated, and the length of the line inside an object, or the number of intersections as a function of line length, is collected. This approach also allows the simple measurement of the average size of objects in the image, which is only possible using a probe with length information. Likewise, surface area information can only be gathered using a probe with measurable area information.

A commonly used area probe is the counting frame. The counting frame is a rectangle with two edges that continue to infinity. The counting frame must be able to sample any portion of the image, or the entire image, while counting or measuring any object once and only once. A typical image shows a problem at the edges, where objects cannot be completely measured, or may re-enter the frame of reference and be counted twice. In order to combat this, an asymmetric frame is placed with sufficient border to identify the behavior of objects at the edge. To prevent double counting, objects that are partly within the image, but touching in any degree the left or bottom "half" of the frame are not counted, while objects that touch the top and right "half" of the frame are

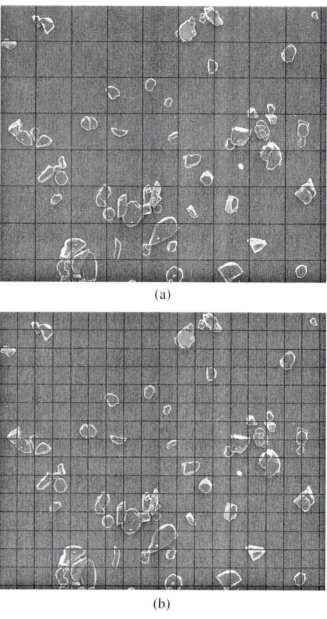

(a)

(b)

FIGURE 8.23 Counting and measuring techniques: (a) A point net that is too coarse, since many objects are completely missed. (b) An appropriate point net. (c) A point net that is too narrow will over-sample the objects, adding unnecessary calculation time to measurements. (d) Radiating measurement lines. (e) Randomly drawn test lines. (f) A counting frame. Particles that touch the left stair-step line within the frame are not counted. Everything in the frame or touching the right stair-step line are counted.

378

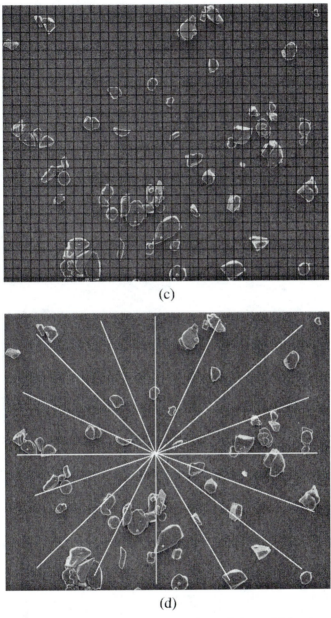

(c)

(d)

FIGURE 8.23 (continued) Counting and measuring techniques: (a) A point net that is too coarse, since many objects are completely missed. (b) An appropriate point net. (c) A point net that is too narrow will over-sample the objects, adding unnecessary calculation time to measurements. (d) Radiating measurement lines. (e) Randomly drawn test lines. (f) A counting frame. Particles that touch the left stair-step line within the frame are not counted. Everything in the frame or touching the right stair-step line are counted.

379

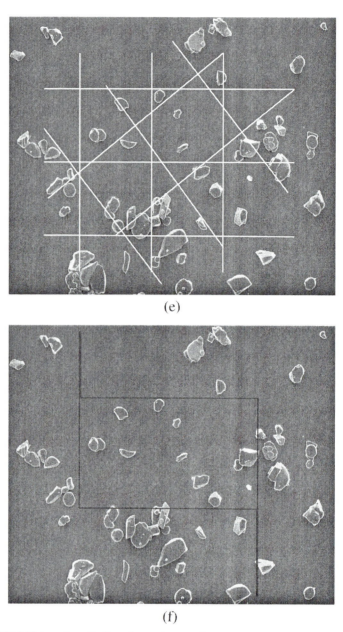

(e)

(f)

FIGURE 8.23 (continued) Counting and measuring techniques: (a) A point net that is too coarse, since many objects are completely missed. (b) An appropriate point net. (c) A point net that is too narrow will over-sample the objects, adding unnecessary calculation time to measurements. (d) Radiating measurement lines. (e) Randomly drawn test lines. (f) A counting frame. Particles that touch the left stair-step line within the frame are not counted. Everything in the frame or touching the right stair-step line are counted.

TABLE 8.4 Some measurements that can be accomplished using point, line and area calculations.

Unbiased Measures

Feature Dimension	Probe	Counting Frame
Number	Disector	
Length	Volume	$\dfrac{\text{Length}}{\text{Volume}} = 2\,\dfrac{\Sigma\,\#\ \text{Intersections}}{h\Sigma\text{Reference Points}}$
Area, Surface	Line	$\dfrac{\text{Area}}{\text{Volume}} = 2\,\dfrac{\Sigma\,\#\ \text{Intersections}}{\Sigma\text{Test Line Length}}$
Volume	Points	$\dfrac{\text{Intersecting Volume}}{\text{Reference Volume}} = \dfrac{\text{Intersecting Length}}{\text{Reference Length}} = \dfrac{\text{Intersecting Points}}{\text{Reference Points}}$

Specific Measurements

Type	Examples	Type of feature
Object Information	Counting	Number
	Neighbor Relationships	Length or Area
	Alignment	Area or Volume
	Orientation	Area or Volume
	Centroid	Area or Volume
Caliper Dimensions	Length	Length or Area
	Width (breadth)	Area
	Perimeter	Area
	Area	Area

Table from Unbiased Stereology and Image Processing Handbook.

measured. If the frame were displaced to the left by one frame width, the formerly excluded objects would become countable. Therefore, this probe has some geometric features in common with a unit cell.

Measurement descriptors are a part of quantifying image data. Image level information about relationships between objects is gathered without regard for individual objects. Image level measurements include object counting, nearest neighbor, and object alignment. At the object level, traditional, caliper-type measurements are often recorded. A calibrated reference line or grating is used to set an accurate scale. Shape data is presented as unitless ratios of object measurements. The data to be extracted from the image depends on the location of the periphery of objects in the image field. If the periphery runs out of the field of interest, the object is not measured.

Object measurements are plentiful, once a perimeter has been defined. The maximum and minimum Feret's diameter, or the major and minor

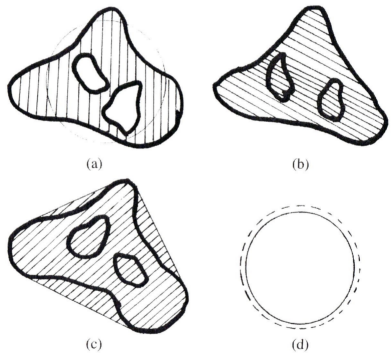

FIGURE 8.24 Particle assessment: (a) The net area and the equivalent circle. (b) The filled area. (c) The convex area. (d) The equivalent circle changes in diameter for these three measures.

axes, of a particle are a typical object measure. The center of gravity is the average center of an object. For simple objects, this is the midpoint of the object. However, if the perimeter is very irregular, then using the average chord length may place the center in the wrong place and other measurements may be more robust. Orientation is measured as the angle formed by the major axis of the object relative to a fiducial edge.

Box 8.1 The Laboratory Notebook

The laboratory notebook is the permanent record of research and characterization. A laboratory notebook should: record the exact steps and measures during an experiment; document the rationale for conducting an experiment; and provide certifiable proof of work. The notebook should contain everything that a scientist or engineer does. Because of this, a notebook always accomplishes multiple goals.

The laboratory book is a document of all the work done. The thoughts about a project should be recorded. Will ideas change? Will some be discarded without being explored? Certainly, but an early idea left initially unexplored may later lead to solving problems. Often, the advice that "the first answer is usually right" proves as true in science as in test taking. References read can be documented in the notebook. This provides an excellent, private forum for critical evaluation of the literature in the field. Of course, all details of every experiment must be recorded in the lab book. Reproducing the results, at a later date, is always part of the invention and testing process. Other investigators may be interested in reproducing the results, as well. Finally, all analysis can be recorded in the lab book. The fresh, immediate response to test results recorded in the laboratory notebook may provide insight that can easily be lost later due to familiarity with the experiments, materials and results from repeated work.

The laboratory notebook is a legal document. Why is this important at all levels? Even at the academic level, society is often looking to place blame somewhere. Was a classmate harmed during an experiment? Has someone accused your laboratory of "fudging" results? The notebook is a legal way to refute any accusation of wrongdoing. Did something remarkable happen when a simple, traditional experiment went slightly wrong? Have long hours of research finally resulted in a nanocomposite clay with true melt intercalation between gallery layers? The notebook is a legal document for establishing precedence on invention and experiment.

Keeping a notebook is not difficult. The basic principle is that notes should always be taken in ink, in a bound book with sequentially numbered, blank pages. Note taking should start at the beginning of the first page, and should proceed sequentially through the book. The ink insures that data will not be erased. Any errors should be crossed out with a single line, a note made of the reasons for striking the information made, and initials and date appended. The author should sign each page. Ideally, someone that can understand the contents of the page should witness each page. The progress of a single experiment occupying several pages should begin each page with "continued from page" and end with "continued on page" if needed.

Box 8.1 *(continued)* The Laboratory Notebook

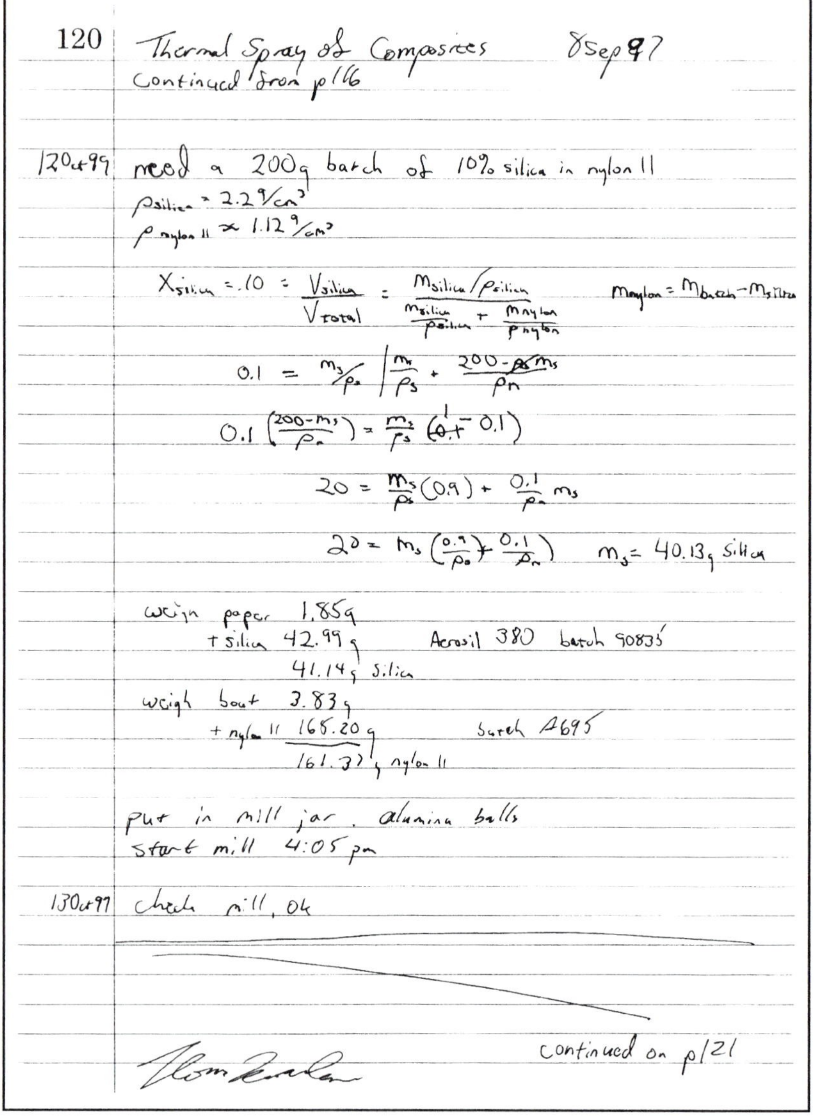

120 | Thermal Spray of Composites   8 Sep 97
Continued from p 116

12 Oct 97 need a 200g batch of 10% silica in nylon 11

$\rho_{silica} = 2.2 \text{ g/cm}^3$

$\rho_{nylon\,11} \approx 1.12 \text{ g/cm}^3$

$$X_{silica} = .10 = \frac{V_{silica}}{V_{total}} = \frac{m_{silica}/\rho_{silica}}{\frac{m_{silica}}{\rho_{silica}} + \frac{m_{nylon}}{\rho_{nylon}}} \qquad m_{nylon} = m_{batch} - m_{silica}$$

$$0.1 = \frac{m_s}{\rho_s} \Big/ \left[\frac{m_s}{\rho_s} + \frac{200 - m_s}{\rho_n} \right]$$

$$0.1 \left(\frac{200 - m_s}{\rho_n} \right) = \frac{m_s}{\rho_s} (1 - 0.1)$$

$$20 = \frac{m_s}{\rho_n} (0.9) + \frac{0.1}{\rho_n} m_s$$

$$20 = m_s \left(\frac{0.9}{\rho_s} + \frac{0.1}{\rho_n} \right) \qquad m_s = 40.13 \text{ g silica}$$

weigh paper   1.85 g
  + silica   42.99 g   Aerosil 380 batch 90835
       41.14 g silica

weigh bout 3.83 g
  + nylon 11 165.20 g   batch A695
       161.37 g nylon 11

Put in mill jar, alumina balls
Start mill 4:05 pm

13 Oct 97 check mill, ok

continued on p 121

FIGURE 8.25 A sample notebook page: All entries should have dates. Errors should be crossed out with a single line. If the reason for the correction is significant or not obvious, the correction should be initialed and dated. If the page is not to be completed, a line should be drawn canceling the rest of the page. Every page should be signed. Information about where the next entry for a project will begin may be of use.

TABLE 8.5 Dimensionless shape descriptors.

Descriptor	Formula	Variables
Roundness	$\text{Roundness} = \dfrac{4A}{\pi d_{max}^2}$	A = Area, d_{max} = maximum diameter
Aspect Ratio	$\text{Aspect Ratio} = \dfrac{d_{max}}{d_{min}}$	d_{min} = minimum diameter
Compactness	$\text{Compactness} = \dfrac{\sqrt{\dfrac{4}{\pi}A}}{d_{max}}$	
Extent	$\text{Extent} = \dfrac{A}{A_b}$	A_b = Area of the bounding rectangle
Curl	$\text{Curl} = \dfrac{D}{L}$	D = diameter (Feret's, Martin's, etc.) L = real object length
Form Factor	$\text{Form Factor} = \dfrac{4\pi A}{P^2}$	P = perimeter

After Russ.

There are several area measurements available. The net area, the total area of pixels in the object, is the obvious choice. However, if there are flaws that create the appearance of a hole in the object, than the filled area, including the net area and any extra area wholly inside the object, may provide a better measure of the object. A highly complex shape may have a small net area, but the convex area, identified by an irregular polygon connecting the outermost features, may provide a measure that more accurately predicts property measurements such as light scattering. Measuring the perimeter also seems to be a simple matter of counting the pixels along the edge. However, the convex perimeter may provide more physically relevant data. Further, a strict counting of pixels may be confounded by uncertainty in drawing the perimeter.

All of the measures described so far have values with units, and are specific to the objects. Some shape descriptors are unitless ratios of these measurements, and may provide information that is independent of object size. In this way, for example, a circle is identifiable as a circle and a square as a square, regardless of the size of the object in the image. Table 8.5 shows the equations for various dimensionless shape descriptors with application to simple objects, including form factor, roundness, aspect ratio, convexity, solidity, compactness, modification ratio and extent, and to fibers, including elongation, and curl.

So far we have introduced some of the general features of analyzing composites and nanocomposites. In the next two chapters, more techniques will be investigated in detail. Good experiments were designed, data gathered, and visually represented data yielded unique insight. The final step is presenting these results in paper and presentation form. Falling down at this stage would be a shame. Thus, notes on presentation close out the general discussion of analysis.

The scientific report is the primary technique for disseminating new results. Presenting results begins with a good abstract. The solid abstract conveys in a sentence or two apiece: what you did, why you did it, and what you found. The first two sentences should tell a reader or listener exactly the goal of the testing and why it was important to do the work. The remainder of the abstract should contain brief, solid statements specifying the results. An audience will be busy, and the abstract may be the only opportunity to generate any interest in a complete reading of your work.

The report is a linear and detailed statement of why the work is important (introduction and the background to substantiate the claims in the introduction), how you did the work (methods), and the results (results, discussion and conclusion). This is the generally accepted format for reporting scientific results.

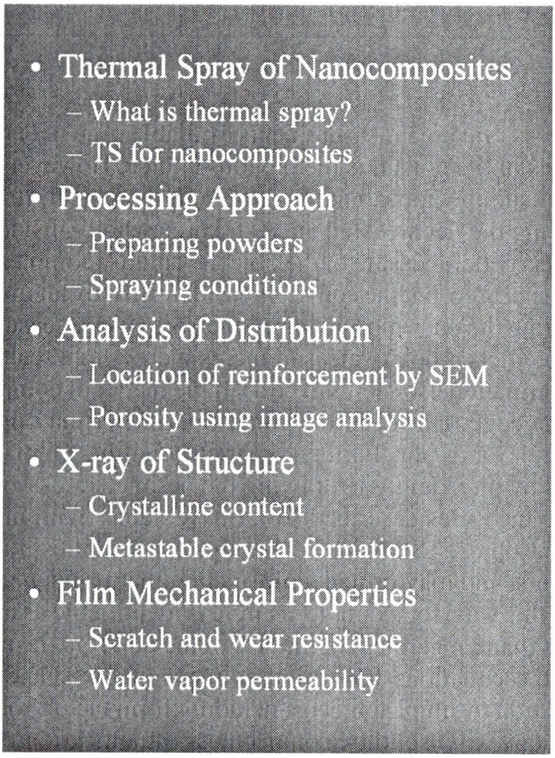

- **Thermal Spray of Nanocomposites**
  - What is thermal spray?
  - TS for nanocomposites
- **Processing Approach**
  - Preparing powders
  - Spraying conditions
- **Analysis of Distribution**
  - Location of reinforcement by SEM
  - Porosity using image analysis
- **X-ray of Structure**
  - Crystalline content
  - Metastable crystal formation
- **Film Mechanical Properties**
  - Scratch and wear resistance
  - Water vapor permeability

FIGURE 8.26 An example of an active outline for a nanocomposites presentation.

Box 8.2 *(continued)* Presentation of Analytical Results

There are two common errors in reporting made by beginners. The report is not a mystery story. The experiment process does uncover properties and performance previously unknown, but the results, once discovered, are the point of the story. The task becomes convincing the audience that the testing was done well. Mystery stories are unlikely to succeed in reaching this goal. The report is also not a journal. While the mistakes made during the investigation were instructive in guiding the investigator to a new and useful conclusion, and while all of the calibration is necessary to assure that the results are good, these steps are already assumed by an audience. The results are always the focus of the story, and the details of the process are restricted to the procedure.

For oral presentations, also, there are a few simple rules to follow. Words like "introduction," "outline," "background" and so on should be avoided. Instead of presenting an outline with simple descriptors, active descriptors should be used to provide the audience with the second presentation of the talk content, the first having been the abstract. In the body of the talk, images should be simple: text should be limited to no more than seven lines, containing no more than seven words; there should be no more than one figure per image. The story must always lead to the results. Any background slides should support the results. Discussion of methodology should lead to confidence that the results are trustworthy. The conclusion need not be labeled "conclusion," the content, in restating the major points of the results and discussion, should be adequate to alert the audience that the end has been reached.

The final piece of presentation advice is that multiple data streams are an investigator's strongest tool. During testing, a good investigator will always collect duplicate tests, and will make a measurement using two different techniques where possible. In presenting, the abstract, the written results and discussion, and the charts and graphs, are all multiple data streams for conveying information to the audience. Pay attention to each data stream: the audience is more likely to understand the message presented in several different ways at several points through the presentation.

REFERENCES

Haswell, S., ed., Practical Guide to Chemometrics, Marcel Dekker, New York, 1992.

Shinya Inoue, Video Microscopy, Plenum, New York, 1986.

Mathot, V., ed., Calorimetry and Thermal Analysis of Polymers, Hanser, New York, 1994.

Russ, J., The Image Processing Handbook, 2nd ed., CRC Press, Boca Raton, 1995.

Scheirs, J., Compositional and Failure Analysis of Polymers, John Wiley & Sons, New York, 2000.

CHAPTER 9

Structure Characterization

The structure of a nanocomposite exists at multiple scales, including primary chemical structure, tertiary crystalline structure, and large-scale distribution. One of the most important considerations is the distribution of the nanoreinforcement. A well-distributed reinforcement will generally be more effective. A poorly distributed reinforcement may arise from agglomerating forces native to the nanoparticles or from matrix incompatibility. Further, the reinforcement or matrix may have separate substructures important to the properties. Substructures may have a texture. Texture in materials' parlance relates to crystal orientation. Here, texture refers more broadly to large-scale patterns of order, originating from diverse sources including localized particle distribution, differences in crystal habit, or from specific conformations of the polymer chains. This texture definition is more consistent with image analysis.

The distribution, particle structure, matrix structure, and texture are all elements of the physical manifestation of the overall composite structure. This structure determines, in turn, the physical properties of the composite. A nanocomposite specimen may have a set of properties resulting from a given set of processing conditions. Manipulating these properties usually requires an understanding of the different levels of structure induced by processing, and how those structures are influenced by changes in processing. Determining structure and structural changes can be accomplished using a range of characterization techniques.

Composites' structure can be visualized, quantified, and classified. Visualization occurs through microscopic imaging, via optical or electron microscopy. Classification and quantification of crystallinity is usually accomplished using diffraction, by X-ray, electron, or neutron scatter-

389

ing, or using physicochemical techniques, including GPC and probe microscopy.

SCALES IN NANOCOMPOSITES

There are many important analysis scales in composites and nanocomposites. The measurement scales range from as large as the millimeter to as small as the nanometer scale. Each scale or feature may have an impact on the properties of the material worth considering. The chemical structure is measured on the angstrom to nanometer scale. The physical structure, amorphous or crystalline, layered or exfoliated, sphere or tube, is measured on the nanometer to micrometer scale. The dispersion of the particles is assessed on the micrometer to the millimeter scale. Distribution and texture describe the relative relation of structural features and may be present at any size scale. Different characterization techniques are appropriate and effective at different scales.

An example of the features in a particulate nanocomposite is shown in Figure 9.1. Large quantities of silica are present at the edges of areas containing little silica. This distribution is a consequence of the thermokinetic processing technique. Concentrated silica particle layers surround regions of nearly pure matrix. Higher magnification shows that there is considerable agglomeration in the silica regions, but that the particles in fact penetrate for some distance into the matrix.

Controlling the reinforcement distribution pattern is a key process goal. In preparing composites, a reinforcing material must be brought into intimate contact with the matrix. The process begins with contact between a reinforcement-rich and a matrix-rich phase. Under shear, the phases will mix. Ideally, components will be mixed thoroughly and homogeneously. The quality of the dispersion may depend on the scale of examination, the sizes of the domains of the reinforcing component, and the distribution of these domains. The particles may be well distributed on the bulk scale, and yet show considerable local agglomeration. The particle distribution is usually characterized by microscopy, but x-ray techniques can sometimes provide additional clues to the interactions between particles. Measurements to assess the effectiveness of processing include determining the spacing between two particles and any trapped porosity.

Particle structure is a second potential influence on composite properties. Many nanoparticles have no observable structure. Carbon black and fumed silica, two of the most common nanoparticles, are naturally amorphous. At the single-nanometer level, there are only a few atoms along

FIGURE 9.1 A thermally sprayed nanocomposite of nylon 11 and fumed silica. The feed material is composite powders, with nylon 11 coated with silica prior to deposition. The silica remains concentrated at the boundaries of the particles.

any axis of choice, rendering crystallinity irrelevant, even in ordinarily crystalline metals and ceramics. At tens or hundreds of nanometers, particles can be composed of single crystalline solids. Some important nanomaterials, such as mica and montmorillonite clay, also have a distinctive layered structure. The layers can be in the native state, intercalated with ions, semi-separated by external pressures and partial matrix infiltration, or exfoliated. The layers usually must be cleaved to provide the best reinforcement in a polymer. Thus, characterizing the degree of exfoliation provides important structure-process information. The particle crystallinity may be quantified using x-ray diffraction. Visualizing the separation of the layers usually is the province of transmission electron microscopy.

The matrix structure, primary and tertiary, also contribute to overall composite performance. Many nanocomposite matrices are amorphous. The *in situ* polymerization of liquid resins, including epoxies, acrylics, polyolefins and polyesters, usually leads to an amorphous matrix. In such materials, reinforcement occurs only due to particle distribution and volume fraction. On the other hand, semi-crystalline polymers have a structure that may dominate the performance of the composite. The primary chemical structure can be characterized by spectroscopy, including infrared or nuclear magnetic resonance spectroscopy. Characterizing crystal structure is accomplished with X-ray or electron diffraction. Visualizing the texture of crystals may be accomplished with scanning electron microscopy or backscattered electron diffraction imaging.

Particles or matrix features can be classified into domains with separate, quantifiable characteristics. These characteristics include both an individual domain size and the distribution of the domains throughout the material. One domain in the example material comprises the matrix-rich regions. Examples of structural features within these domains include the ratio of shell to core, the thickness of the ceramic-rich shell, the size of the unit including core and shell, the shape of the domain, and the shape of the core. Another domain of interest consists of the reinforcing particles. The particles within the domain may be agglomerated or well distributed. The particle domains also can be classified for size and shape. Other potentially important factors include the density of particles in a domain and the average number of neighbors for each particle. These shape factors are related to what is discussed here as texture.

TEXTURE

The texture provides an evaluation of the distribution uniformity in any region. If, for example, nanoparticles are agglomerated but the agglomerates are scattered evenly, the behavior of the material is expected to be different than if the particles are all well separated and also evenly distributed, or if the particles are well separated from each other and poorly distributed throughout the matrix. Each of these arrangements will exhibit a different texture, with associated distribution statistics. The difference in texture cannot be detected using only the average separation between particles. An example is shown in Figure 9.2.

Texture is something that may or may not be obvious to casual observation. Side by side, two images with the same first-order statistic and different second-order statistic are often obviously different. The quantitative difference in texture is easily discerned using the second and higher moments of the average separation. The second moment of separation distance emphasizes the particles that are farthest apart. The third moment probes subtler differences in particle arrangement. A physical technique for assessing the second moment is to throw a needle, or a line with a set length, onto an image of a test piece. A plot of the number of times that the needle lands with both ends inside a particle against the

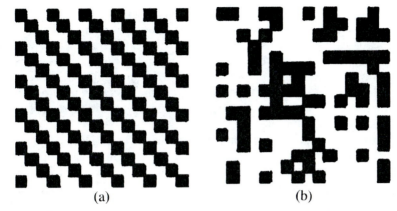

FIGURE 9.2 Two textures of distribution: (a) a perfectly regular distribution. (b) a random distribution.

number of throws provides the second moment of the particle distribution. The third moment may be evaluated by throwing a triangular counting device.

ELECTROMAGNETIC ENERGY

The structure of a material, the dispersion, morphology, primary, and secondary structure, can be examined in many ways. For the most part,

---

Do It at Home 9.1

Take a compass and a piece of graph paper. Draw a circle such that 10 intersections are encircled. Measure the distance between all of the intersections inside the circle and calculate the average {first moment}, distance average {second moment} and area average {third moment}. Draw a new circle and randomly place 10 points within the circle. Measure the distances and again calculate the various averages. Repeat this experiment by drawing the circle, and then closing your eyes and marking the 10 points blindly. Measure and calculate the average separation between points again. Put all ten points around the perimeter of the circle, or put nine points in one hemisphere of the circle and 1 point in the other hemisphere. Calculate the median of all of these samples. What information can you gain from the various measures of average?

---

studying the structure of a material requires instrument-aided examination of the material. The examination uses analytic instruments to augment or supplement ordinary vision: light microscopes, electron microscopes, magnetic resonance imaging, and infrared spectroscopy to name a few. Examining a material requires illuminating the specimen and then collecting a relevant data. Many of these techniques are unified by a common use of the electromagnetic spectrum to provide the energetic probe illumination.

The electromagnetic spectrum interacts with a material via scattering, diffraction, and absorption. Scattering is the basis for image formation, diffraction probes structural order, and absorption is the basis for spectroscopy. Thus, various types of structural information can be gained by examining a material using specific types of electromagnetic energy. The electromagnetic spectrum is illustrated in Figure 9.3.

Different regions of the spectrum are appropriate for different types of analysis. Infrared light interacts with the chemical structure, while x-rays interact with the patterns of a crystal structure. The primary characteristic of the photons in light is the energy, defined by the frequency, ν:

$$E = h\nu = \frac{hc}{\lambda} \tag{9.1}$$

where h is Plank's constant, c is the speed of light, and λ is the wavelength. Electromagnetic beams with energies ranging from electrons through infrared light can be used for image formation. A narrower set of beam energies, ranging from electrons through x-rays, provides diffraction patterns from molecular gratings.

Electromagnetic beams interact with the sample, and are changed in

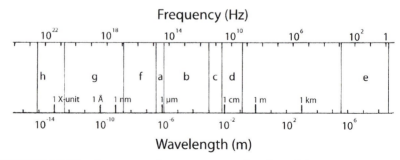

FIGURE 9.3 The electromagnetic spectrum, wavelengths and frequency: (a) visible light. (b) near infrared. (c) infrared. (d) far infrared. (e) microwave and radio wave (f) ultraviolet. (g) X-rays. (h) electrons.

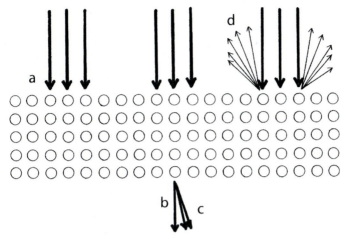

FIGURE 9.4 Interaction of light with a surface: (a) absorption. (b) transmission. (c) diffraction. (d) scatter.

the process. Studying the changes can provide information about the material.

Spectroscopy

Spectroscopy is often used to probe chemical structure. Spectroscopy is the science of studying the light absorbed or emitted by a material when specific types of electromagnetic energy are passed through the specimen. Absorption occurs when transmitted energy encounters some feature of the material that interacts with it, such as molecular vibrations. Ultraviolet light is absorbed by systems of long-range electron interaction, such as pi bonding systems and double bonds. Vibrations of the molecules in the solid absorb infrared light. Flipping parts of the molecule, including the nucleus, electrons, or a permanent dipole, will absorb microwave energy. The frequencies of the absorbed energy are determined by the basic chemical structure of the material.

Infrared

Infrared analysis (IR) can provide information useful for identifying a material, following a chemical reaction, providing information for determining the molecular structure, elucidating conformational information, or for a host of other purposes. The utility of IR spectroscopy comes from interactions of light with specific molecular vibrations. When three at-

oms form part of a chemically bonded structure, the atoms may participate in vibrations relative to each other. These vibrations resonate, that is,
absorb a maximum of energy, at specific frequencies of light characteristic of the material chemistry. These resonances occur in the IR region of
the EM spectrum. This is why IR is perceived as heat: IR is the energy of

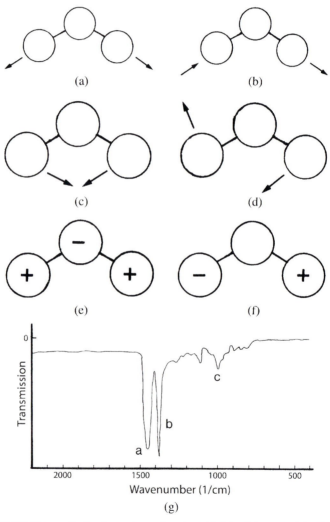

FIGURE 9.5 Types of vibrations shown by atom triplets: (a) symmetric stretch. (b)
asymmetric stretch. (c) scissoring. (d) rocking. (e) wagging. (f) twisting. (g) the infrared
spectrum of polypropylene: a. CH_2 scissors. b. CH_3 symmetric bend. c. CH_3 rock.

specific motions in molecules, and molecules that are more energetic have higher temperature then do slower ones.

The infrared spectrum of a material provides a fingerprint unique to that chemical structure. The infrared spectrum is gathered in the modern laboratory using an interferometer system, as shown in Figure 9.6. A single frequency of IR light, typically generated using an IR laser, passes through an interferometer, creating a narrow spectrum of light, which passes through a test sample. The interference pattern of the light passing through the specimen is measured, and transformed into the transmitted light using a Fourier Transform. There are several advantages to using an interferometer. First, multiple scans can be successively added to improve the resolution. Noise, being both positive and negative, will sum to zero in successive scans, while the signal will be reinforced with each additional collection. Second, the entire spectrum can be collected at once, rather than collecting information about the transmitted light intensity separately at each individual frequency.

The material can be tested in transmission or reflection. The sample of composite to be tested may be sliced into a representative thin film, or ground to a fine paste to be suspended in a non-absorbing matrix, such as NaCl. If the material cannot be prepared for transmission analysis, reflection may be successful. By bouncing the beam off the surface at the critical angle of refraction, a very small part of the energy is absorbed by vibrational groups. Multiple bounces will improve the resolution of reflected light analysis.

The Fourier Transform IR provides both the transmitted light spectrum and the absorbance spectrum. The transmission, T, spectrum is related to the absorbance, A, spectrum:

$$A = \log\left(\frac{T_0}{T}\right) \tag{9.2}$$

where T_0 is the intensity of the incident light. The absorbance can be used in quantification using the Beer Lambert law:

$$A = \varepsilon b C \tag{9.3}$$

where ε is the efficiency of an absorber, b is the thickness of the film layer, and C is the concentration of the absorbing species. The concentration is usually known. The thickness is also set by a sample thickness, if the test is conducted in transmission, but can be difficult to evaluate in reflection. The efficiency often must be determined through experimentation.

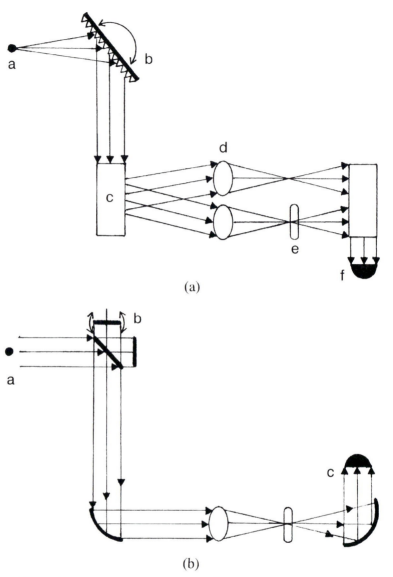

(a)

(b)

FIGURE 9.6 Infrared spectroscopy: (a) a traditional, dispersive spectrometer. a. a multifrequency IR source. b. a diffraction grating, which splits the light. The angle of the grating determines the frequency of the light entering the specimen chamber. c. Optics, including a splitter. d. One beam is used as a reference. e. Specimen location. f. Infrared detector, comparing the specimen and reference beam. (b) An interferometer spectrometer. a. The source is typically a monochromatic IR laser. b. A Michelson interferometer uses moving mirrors to generate a frequency spectrum. c. The detector records the interferogram. A Fourier transform applied to the interferogram converts it to the spectrum. Thus, the interferogram contains information about the entire absorption spectrum.

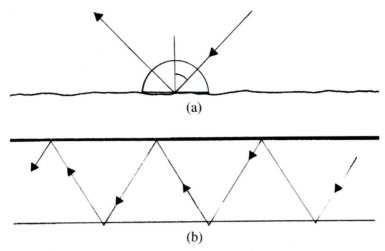

(a)

(b)

FIGURE 9.7 In reflection spectroscopy, the beam is transmitted through an IR transparent crystal, cut so that the beam is completely reflected at the surface of the crystal. When a specimen is in contact with the surface, a little of the light passes through the material. (a) an ATR cell, with a single bounce. (b) a multiple bounce cell.

The composition information gathered by infrared spectroscopy may be dependent on the structural environment of the atoms. Certain bands will be enhanced or suppressed relative to deformation. Some bands are characteristic of helices. IR can provide information about chain conformation and basic chemical bonds.

Only vibrations that distort the center of mass of a molecule can be detected by standard IR testing. Raman IR spectroscopy uses the light scattered from molecules due to vibrations that do not involve a change in the center of mass. This may provide information that is complementary to IR testing. Since many atomic, inorganic solids have vibrations of this type, Raman IR can be more useful for these solids. Standard IR has proven successful for normal analyses of organic materials.

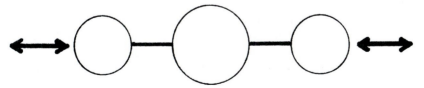

FIGURE 9.8 In Raman infrared active vibrations, the center of mass does not change as it does in infrared-detectable vibrations.

TABLE 9.1 Some characteristic IR bands.

Band	Approximate Frequency (cm^{-1})
NH stretch	3300
CH$_2$ asymmetric stretch	2900
CH$_2$ symmetric stretch	2850
C=O stretch	1750
C=C stretch	1650
NH bend	1590
CN stretch	1550
CH$_2$ bend (amorphous)	1450
CO stretch	1250
CF$_2$ asymmetric stretch	1250
NCO stretch	1200
CF$_2$ symmetric stretch	1200
CH$_2$ wag (crystalline)	1180
CH$_2$ twist (crystalline)	1050
CH$_2$ rock (amorphous)	880
Aromatic in-phase CH wag	730
CF$_2$ bend	550
Aromatic out-of-plane ring bend	540

Derived from Polymer Handbook.

NMR

Nuclear magnetic resonance, as well as magnetic resonance imaging, is also used to investigate molecular structure. The technique probes nuclei that have an odd number of particles, such as ^{1}H and ^{13}C, making it appropriate for studying organic molecules. Nuclei with an odd number of particles have a net spin. The spin can occupy one of two states, which have the same energy under ordinary conditions. In a magnetic field, the two states are separated. Radio frequency energy, applied as a stimulating force, will cause these spins to change state. When the spin drops back down, a photon is emitted. The frequency of this output energy is then measured.

The energy of the spin states in the nucleus depends on the atom bonding. A hydrogen atom bonded to a carbon atom will have a different energy than a hydrogen atom bonded to an oxygen atom. This causes a shift in the energy of the photon emitted. The absolute energy is not measured. Rather, the shift factor, δ, is measured—in ppm. The shift factor is measured relative to tetra methyl silane (TMS). A sketch of the hydrogen NMR spectrum for polypropylene is shown in Figure 9.10.

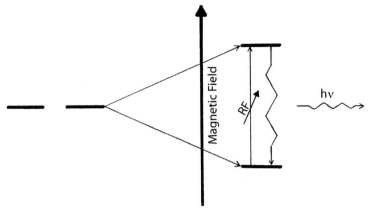

FIGURE 9.9 Nuclear resonance spectroscopy: (a) Ordinarily, the magnetic spin quanta in the nucleus are at the same energy. Under a magnetic field, the spin energies are split. Oscillating, radio frequency energy will excite the nucleus. When the nucleus returns to the low energy state, a microwave photon is emitted.

The total integrated area of a peak is proportional to the number of hydrogen atoms in that specific chemical environment. The peaks may have a substructure that is correlated with hydrogens on neighboring carbons.

^{13}C spectra are measured in a similar manner, but the information is about the chemical environment of the carbon in the specimen. However, while the natural abundance of ^{1}H is 99.98%, ^{13}C is only 1.11% of all carbon. Spectrum collection can take considerably longer. ^{13}C spectra are particularly useful for providing information on chemical structure. For polymers, this may include monomer sequences in copolymers.

Atomic Absorption Spectroscopy

Atomic absorption spectroscopy (AAS) and atomic emission spectroscopy (AES) are used to characterize the composition of an atomic solid. These techniques are particularly helpful in analyzing the composition of inorganic reinforcements. Using one of various techniques, a material is vaporized, in order to deliver its atoms to a controlled region, where the measurement is made. The material's various liberated atoms will absorb light of specific frequencies, so that the absorption spectrum of the gas can be used to determine the atoms present in the gas, and therefore in the solid. In atomic emission spectroscopy, the atoms are ionized to plasma. As the ionized atoms relax back to the ground state, a characteristic spectrum of photons is emitted. Identifying which atoms are present is then a straightforward analysis of the characteristic frequencies emitted. However, quantifying the amount of a specific atom present in the solid is more challenging. The Beer-Lambert law can be applied, but the effectiveness of liberating various atomic species during pyrolysis is not linear. Therefore, each element must have a separate calibration curve.

VISUALIZATION

Imaging

Electromagnetic light scattering is the basis for forming images. Any object of a size sufficient to interact with energy of a specific wavelength can scatter incident light. Familiar elements, such as lenses, mirrors and apertures, can be used to gather scattered light and focus this light into an image. The gathered light energy can be used for microscopy to create magnified visual data, visible light for optical microscopy, and electrons for electron microscopy. Appropriate lenses are available in each case; magnetic lenses for electrons and glass for visible light. Lenses for light

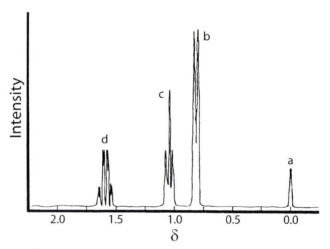

FIGURE 9.10 ^{1}H NMR of polypropylene: (a) Tetramethyl siloxane (TMS) peak, added as a reference. (b) The peak for the H's in CH_3 is split by the neighboring H in CH into a doublet. (c) The peak for the H's in CH_2 is split by each of two neighboring H's in CH's into a triplet. (d) The peak for the H in CH is split by the neighboring H's in CH_3's and the two neighboring H's in CH_2's into a complex substructure. The integrated areas under the peaks are proportional to the relative number of H's. The shift factor is unitless, and measured in parts per million.

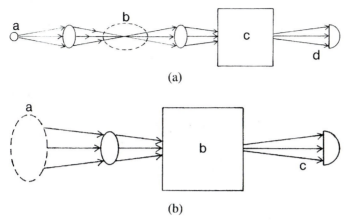

FIGURE 9.11 Atomic emission spectroscopic techniques: (a) In atomic absorption spectroscopy, a sample is vaporized, typically in a flame. Broad-spectrum light is passed through the vapor, and the absorption spectrum is recorded. a. light source. b. sample vapor. c. monochromator. iv. detector. (b) In atomic emission spectroscopy, the sample is vaporized and converted to plasma. The emitted output spectrum is characteristic of elements present in the sample. a. ionized sample plasma, b. Optic system. c. Detector.

are made from any transparent material with a high index of refraction; magnetic lenses operate on moving electrons to manipulate their path of travel. X-rays, lying in the intervening region of the energetic spectrum, are too energetic to be bent by magnets, and do not interact sufficiently with glass lenses, and so cannot be used to form images.

Basis of Image Formation and Resolution

Electromagnetic light striking an object will be reflected, refracted,

---

Do It at Home 9.2

Do you own a loupe? A loupe is a small magnifying lens, available at many camera stores. The lens is placed at the apex of a light-collecting device, usually a clear glass or plastic sphere. Most loupes provide an approximately 8× magnification, useful for examining photographs for fine detail, examining polished surfaces, and other common materials analysis tasks.

---

diffracted, or absorbed. An object can be defined as something as large as a barn or as small as a single plane in montmorillonite clay. The object must be on the order of, or larger than, the wavelength of the incident light in order to scatter the light. Any object that is much smaller than the resolution of the incident light will not be visible in that light. A well-dispersed nanoparticle in a clear matrix will not scatter visible light and will be invisible to many forms of illumination.

The nature of the interaction between the object and incident light depends on the composition and structure of the surface, the energy and character of the light, and on other features. Most materials absorb some of the incident radiation. Certain materials, e.g., metals or "white" ceramics such as gypsum, absorb very little incident energy. Light is more effectively scattered by these types of materials. For objects, light scattering translates to brightness of any image formed (albedo). Absorption

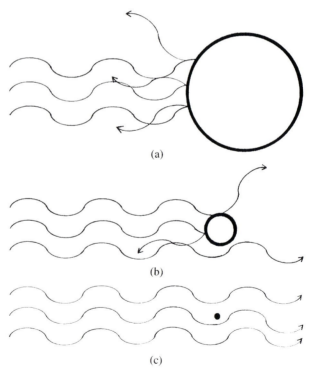

(a)

(b)

(c)

FIGURE 9.12 Interactions between light waves and particles is affected by particle size. (a) Large particles interfere with and scatter the light. (b) Intermediate size particles interfere with the light, resulting in diffraction as well as scatter. (c) Small enough particles do not interfere significantly with the light.

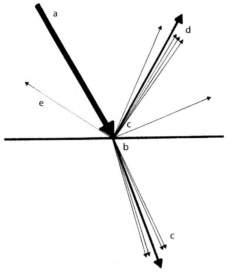

FIGURE 9.13 Light interactions with a surface. (a) Incident light. (b) Transmitted light, a primary component. (c) Diffraction occurs with part of the transmitted light if the solid has a regular crystalline structure. (d) Reflected light, another primary component. (e) Scattered light.

occurs based on specific interactions with the light such as described in spectroscopy. Silica glass, for example, will absorb most IR frequencies. The remainder of the light will be transmitted or reflected, based on the angle of incidence and the frequency of the light.

Figure 9.13 shows the various types of interactions. On striking a surface at an angle of incidence, some of the light is reflected, at the same angle as the incident light. Some of the light passes into the material. The angle of the transmitted light depends on the speed of light in the material, relative to the speed of light in the material, or the index of refraction:

$$n_1 \sin\theta_1 = n_2 \sin\theta_2 \qquad (9.4)$$

If light passes from a material with a higher index of refraction into one with a lower index of refraction, the light is bent toward the surface. If the angle of the incident light becomes shallower, there will come a point where all of the light will be reflected at the surface:

$$\sin\theta_1 = \frac{n_2}{n_1} \qquad (9.5)$$

Thus, if two materials have the same index of refraction, there will be no scatter at the interface of the two materials.

The reflected light can be collected to form an image of the surface. If the surface is not perfectly flat, or if the incident light is not parallel, the reflected light will be scattered in all directions. In order to reconstruct an image, the scattered light must be gathered by a lens and recombined.

The primary characteristic of a lens is its magnification, M. The magnification, however, is not the only important feature of a lens. The resolving power of the lens determines the smallest distance that can be separated, d. This is a function of wavelength, λ, index of refraction, n, and the angle of light that can be collected by the lens, α:

$$d = \frac{0.61\lambda}{n \sin \alpha} \tag{9.6}$$

The limit of resolution is the inverse of the resolving power. The denominator of the resolving power, $n\sin\alpha$, is also a lens characteristic called the Numerical Aperture (N.A.). A lens with a high index of refraction is capable of bending the light more than one with a low index of refraction. The numerical aperture is more important than the magnification, per se. Once the resolution limit has been reached, further magnification will be ineffective at revealing new detail. Two ways of increasing the effective magnification are to modify the Numerical Aperture, via, for example, oil immersion, to increase the index of refraction, or to change the wavelength of the light. Inserting the sample in the field of a magnetic lens can further reduce the numerical aperture in an electron microscope.

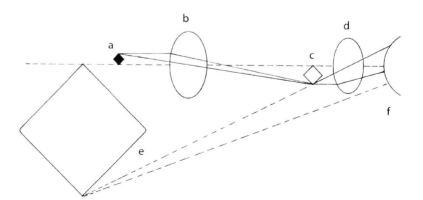

FIGURE 9.14 A magnifying system. (a) target object. (b) objective lens. (c) virtual image of the objective lens. (d) eyepiece. (e) the virtual image for the eyepiece. (f) eye.

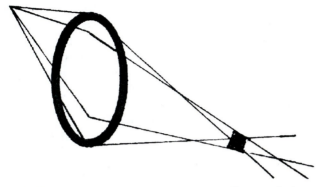

FIGURE 9.15 The imperfections of a lens create a zone of least confusion, over which the image does not get appreciably better regardless of changes in location.

The depth of field (L) of a lens is also important to the usefulness of the lens. The depth of field determines how much of the third dimension of an image, that is the hills, valleys and edges of an uneven surface, is in focus at the same time. The depth of field is inversely proportional to the square of the numerical aperture, so objectives with lower numerical apertures, that is, less resolution, have deeper depths of field:

$$L = \frac{\lambda n}{N.A.^2} + \frac{nd}{M \times N.A.} \tag{9.7}$$

where L is the depth of field and M is the magnification. The depth of field has an upper limit determined by the wave optics, which is the left-hand term, and a lower limit, the right-hand term, determined by the zone of least confusion of the focused light beam. Confusion in the focused beam is caused by light from the edges of the lens being focused in a slightly different place than light passing through the center of the lens. The zone of least confusion is the region where this effect is least pronounced. The zone of confusion is a lens defect arising primarily from spherical aberration. Another zone of confusion can be created by chromatic aberration in magnetic lenses. Lens limitations will be discussed next.

Lens Limitations

A number of defects are common to both glass and electronic lenses. Spherical aberration, coma, astigmatism, curvature, and distortion are all possible. These defects are due to the curvature required of a lens to focus the light. Spherical aberration occurs when the light at the periphery of a lens is refracted more or less sharply than at the middle of the lens. This aberration causes the light to focus slightly forward or to the rear of the focal plane, resulting in less clarity. Using an aperture to block the light from the periphery of the lens can reduce spherical aberration. This is effective, but obviously reduces the diameter of the lens, the *NA*, and therefore the resolving power of the lens.

Coma is a related aberration occurring with objects off the principal axis of the lens. Here, the light creates a smearing of the object image toward or away from the central axis in the image plane. The intensity fades as the image is smeared, creating a comet- or coma-shaped image. Proper lens alignment and an aperture can reduce this type of defect, as well.

Astigmatism arises, again, when an object is off the primary axis of the lens. The half of the lens on the 'same side' of the optical axis creates an image that is in focus on a different plane from the image created by the half of the lens on the other side of the axis. There is a zone of least confusion, analogous to that arising from the combination of aperture angle and working distance, in between these two images, which represents the sharpest image obtainable. Astigmatism also describes a familiar visual distortion, wherein an image is slightly out of true. Imperfect lenses can form images with this characteristic. Typically, astigmatism is limited in high-quality lenses. In magnetic lenses, astigmatism is corrected somewhat differently. Constructing perfect electromagnetic lenses is impractical. Therefore, small moveable permanent magnets are placed along the periphery of the lens. These are manipulated in order to optimize the roundness of the beam of illumination.

Another lens defect occurs when the lens curvature causes a change in focal length as the distance from the principle axis increases. This causes a flat object to project a curved image. This defect can also be removed by careful lens design.

The magnification of the lens also can change with distance from the principal axis, causing distortion of the image. Distortions are determined by the change in magnification, arising in either a pincushion or a barrel distortion. Distortion occurs if the principal ray, the one that would pass through the center of the lens, is blocked by a lens stop.

Chromatic aberration is not a factor of lens curvature. This type of aberration is the result of the difference in refraction of light with different frequencies. All materials have this energy dispersion. So, the blue component of a white light will be steeply refracted and focus nearer the lens, and the red component will be less refracted and will be in focus further from the lens, creating an image with a bluish center and a red nimbus. Combinations of lenses and lens materials, apochromats, can be crafted to limit chromatic aberrations over a defined color range. High-energy light is more sharply bent while low energy light is less steeply bent, arising in less sharpness at the focal plane. This effect can be limited by using a monochromatic light source. Some glass lenses, called apochromats, are specially shaped to limit chromatic aberration.

Using monochromatic light will also limit the effects of chromatic aberration. However, different lens materials can pass different wavelengths of light. Silica glass passes visible light well, but absorbs ultraviolet and infrared light. Pure quartz can pass ultraviolet well. Sodium chloride and potassium chloride lenses pass infrared light well.

Optical Microscopy

Optical microscopes are available for transmitted and reflected light. Transmitted light is most useful for objects in a medium transparent to the light source. In a transmission microscope, illumination is focused through the material. The light is scattered, transmitted, or absorbed to create contrast. An example of a nanocomposite best examined by trans-

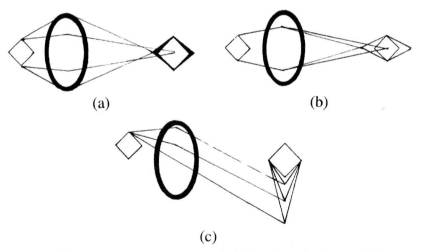

(a)   (b)

(c)

FIGURE 9.16 Lens aberrations: (a) In spherical aberration, affecting on-axis images, light passing through the edges of the lens are focused differently than light passing through the center of the lens. (b) In chromatic aberration, different frequencies of light are focused more or less effectively, resulting in distortion of the image. (c) In coma, affecting off-axis images, light passing through the outside of the lens is focused differently than light passing through the middle of the lens, resulting in image distortion. Even if spherical aberration is eliminated, coma may still occur.

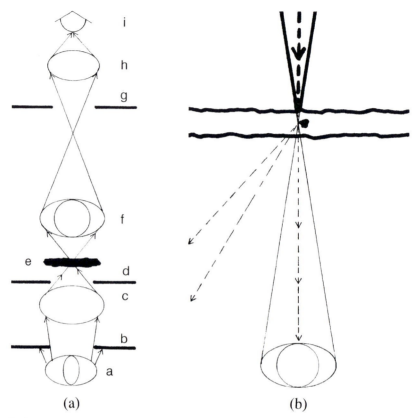

FIGURE 9.17 Transmission microscopy: (a) Light train. a. light source. b. iris 1, used to cut down stray light. c. condenser lens, to focus light on the specimen. d. iris 2, used to decrease spherical aberration. e. specimen. f. objective lens. g. iris 3. This aperture, by blocking the most highly scattered light, increases contrast. h. eyepiece. i. eye. (b) Contrast. Transmitted light striking an object in the specimen is highly scattered. This light is not recombined by the objective lens, creating contrast.

mission optical microscopy would include analysis of the quality of large-scale particle distribution of the reinforcing matrix. Another example is side-by-side comparison of two materials, one with a nanoreinforcement that absorbs some light and one that does not contain such reinforcement.

In reflected light microscopy, illumination is focused on the material where it is scattered by topography or absorbed to create contrast. Therefore, the rough edges of a nanocomposite fracture surface have a great deal of topological contrast. An area of plastic with a lot of ceramic load-

---

Box 9.1 Aligning a microscope

The first step in aligning the scope is to make sure the source is bright and focused on the sample. Remove the objective lens and adjust the condenser lens until the lamp filament is in sharpest focus. Adjust the condenser diaphragm so that the outline is just outside of the column. This reduces excess light from areas of the specimen that are not in the field of view. The objective iris should be used to limit the field to the region of interest. This removes outlying area, with the light passing through the outer edges of the lens with the largest lens errors.

---

ing will reflect light more than an area with lower ceramic loading, leading to contrast.

The magnifying power of a compound optical microscope is approximately the product of the two magnifications of the objective and eyepiece lenses. However, these are estimates based on an equation, and

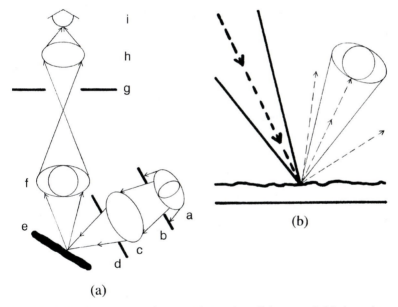

(a)

(b)

FIGURE 9.18 Reflection microscopy: a) Light train. a. light source b. iris 1, used to cut down stray light. c. condenser lens, to focus light on the specimen. d. iris 2, used to decrease spherical aberration. e. specimen. c. objective lens. a. iris 3. This aperture, by blocking the most highly scattered light, increases contrast. h. eyepiece. i. eye. b) Contrast. Reflected light striking asperities on the surface of the specimen is highly scattered. If the specimen is transparent, the light will be refracted at the surface, and is thus reflected and scattered by subsurface objects to be collected by the objective lens.

depend on the observer. Therefore, calibrating the magnification of a microscope requires using a calibrated grating or object, where the dimensions are known beforehand.

Contrast

In observing objects in a microscope, the contrast represents differences between the object of interest and the background (objects that are not interesting). The native sources of contrast in reflective optical microscopy are shadow and reflectance. Shadows are observed when the path of light scattered from one object blocks the light scattered by another. For example, a pockmark on the surface will have shadows at the edges that shift with the tilt of the surface relative to the incident light. An object such as a pin will cast a shadow on the background material. Reflectance also provides contrast at boundaries between materials in objects. A metallic object might reflect more light than the plastic medium in which it is embedded, leading to differences in the amount of light scattered.

In order to observe and make accurate measurements, the higher the contrast between objects, the better. There are many ways to improve the contrast within a sample. The most appropriate technique depends on what type of contrast is most effective. If edges between similar materials are important, such as grain boundaries, etching or shadowing techniques may be useful. If different crystal structures of similar properties are of interest, microscopic techniques such as dark field, phase contrast and polarized light may be the solution.

Chemical modification is often helpful in enhancing boundaries between objects. In etching, the surface of a sample is treated to cause one of the materials to erode away. This can be accomplished by immersing the sample in a chemical that will dissolve one of the materials on the surface. Alternatively, the etching can be done using energetic plasma. The ions in the plasma will attack the softest material at the surface and erode it away. Polymers will erode before metals or ceramics. Different polymers will erode away less quickly if a specimen is more crystalline, and edges will erode more quickly than broad planes, because the plasma can attack the edge from multiple angles.

Another technique for accentuating the difference between two materials is adding a contrast agent specifically to one component. Dyeing and staining are accomplished by adding a chemical that will react with the material and be differentially absorbed by the material. This chemical should have some feature that interacts with the electromagnetic energy

TABLE 9.2 Stains.

Stain	Attaches to:
Osmium tetroxide	Unsaturated bonds
Chlorosulphonic acid	Stain and stabilize amorphous olefinic bonds
Phosphotungstic acid	Surface functional groups
Ruthenium tetroxide	Aromatic rings and ether alcohols, surface staining
Silver salts	Biological materials

Derived from Scanning Electron Microscopy.

better than the unmodified material. For example, carbon-containing material is susceptible to absorbing osmium tetroxide. This makes the material more reflective to electromagnetic light and increases contrast.

Sometimes the most important features are the edges of crystals or other very small features with very little dimensional difference. In this type of sample, shadowing or decoration can be helpful. Decoration is a process in which a high-contrast material, such as a highly reflective material, is washed over the surface in a very dilute, low-concentration medium. For example, a dilute solution of colloidal metal can be used to decorate a hydrophobic polymer. Alternatively, a material such as gold can be evaporated on the surface. The metal will generally accumulate in depressions, making even small depressions more visible. A slightly different approach is to apply platinum at a fixed angle to the sample. The platinum will accumulate, analogously to snow in the macroscopic world, on one face of three-dimensional objects and be absent behind objects. This creates a high-contrast shadow image that can allow better visualization and the measurement of height using simple trigonometric relationships.

Electron Microscopy

Electron microscopy uses a beam of electrons as the illumination source. An electron beam is a negatively charged current, and magnets and magnetic lenses form an effective imaging system by manipulating the current. The original electron microscopes were often large and could simultaneously emit dangerous x-ray and radiation energy. Modern microscopes are much safer and can provide resolutions down to nanometers. The type of information and the ultimate resolution will be further discussed.

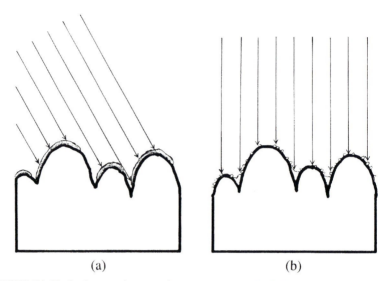

(a)   (b)

FIGURE 9.19 Surface coatings to enhance contrast: (a) In shadowing, platinum is evaporated from a target. The platinum rains on the specimen, and sticks where it lands. If the target is at an angle relative to the specimen, the coating builds up as shadows of surface texture. (b) In gold decoration, gold atoms are evaporated from a target. The gold does not stick on contact, and will accumulate at edges, such as surface texture and crystal strata.

Electrons in Microscopy

The wavelength of an electron beam is determined by the DeBroglie equation for particle waves:

$$\lambda = \frac{h}{\sqrt{2meE}} \tag{9.8}$$

where m is the mass of an electron and eE is the accelerating voltage. The wavelength of an electron is inversely dependent on the square root of the accelerating voltage. When an electron beam hits a specimen, the electron can interact in several different ways. Generally, the electron will penetrate the material to some depth before hitting an atom. The average distance traveled is the mean free path, which can be calculated assuming elastic collision probabilities as shown in Table 9.3. On hitting an atom, the electron can be scattered or generate energy during one of a number of types of inelastic events.

If the electron passes all the way through the sample, as can happen in very thin samples with large enough accelerating voltages, the electrons

TABLE 9.3 Depth of penetration based on elastic mean free path.

Atom	10 keV	20 keV
C	5.5 nm	22 nm
Al	11.8 nm	7.4 nm
Fe	0.3 nm	1.3 nm
Ag	0.15 nm	0.6 nm
U	0.05 nm	0.19 nm

Derived from Scanning Electron Microscopy.

can be used for transmission microscopy. The electrons may also diffract from a regular crystal structure. Some electrons will be absorbed, and must be conducted away or else they accumulate to cause a net negative charge that can interfere with the electron beam. Some electrons are knocked out of the material, creating low-energy secondary electrons, while some are eventually reflected, creating back-scattered electrons.

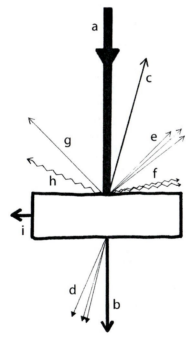

FIGURE 9.20 Electron beam interactions: (a) incident beam. (b) transmitted beam. (c) back-scattered electrons. (d) diffracted electrons. (e) secondary electrons. (f) characteristic x-rays. (g) auger electrons. (h) cathodoluminescence. (i) absorbed electrons.

Box 9.2 The replica technique

Polymers often cannot be examined directly in an electron beam without degrading. The replica technique creates a metallic analog of the surface of a polymer composite sample. A drop of a waterborne polymer, usually concentrated poly(acrylic acid), is put on a specimen. The drop is allowed to dry, and then removed. The replica is coated using evaporated gold or platinum, with or without shadowing. The plastic is floated on water in a crystallizing dish, filled to overflowing to prevent the replica from floating to the edge of the glass and sticking. The plastic dissolves in the water, leaving the metal replica floating on the surface to be picked up with a microscopy grid.

X-rays can be generated that have a characteristic wavelength useful for chemical analysis (EDAX or EDS), but the majority of the x-rays generated are white noise generated by the continuous deceleration of electrons being slowed down or absorbed.

Beam interactions can cause significant damage to the material through electron absorption or atomic displacement and structural damage. The size of the beam interaction volume depends on the number of electrons, or the beam intensity and energy, controlled by the accelerating voltage. Higher accelerating voltages create electron beams that interact less with the material, resulting in less degradation. Unfortunately, this improvement comes at the price of generating a greater amount of dangerous x-rays than standard accelerating voltages produce.

Metals can conduct away the charge accumulation from beam exposure. Therefore, ceramics and polymers are often coated with a very thin conductive layer, ideally no more than one or two molecules thick. The coating is accomplished by evaporating carbon or metal onto the surface or by sputter coating with a highly conductive metal. Metals and ceramics are also more tolerant of, and resistant to, beam damage than organic

---

Box 9.3 Orienting an electron microscope image

Orienting an electron microscope image so that the illumination appears to come from the upper left hand corner, with secondary preference for the top or left edges, is traditional. The micron bar should be a simple number, such as 1, 2, 5, 10, 50 or 100. The micron bar length should correspond to nearly the size of some feature of interest in the image, and should be between 1/4 to 1/2 of the width of the image. As the magnification changes with the focus, the magnification should be used to calculate the length of the micron bar after the machine magnification has been calibrated.

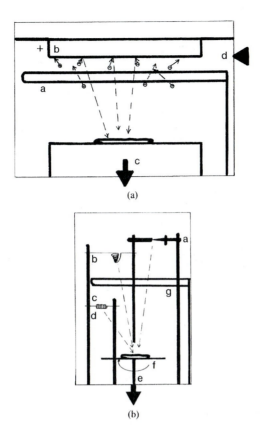

(a)

(b)

FIGURE 9.21 Devices for applying conductive coatings: (a) In a sputter coater, gold or gold-palladium is applied to the surface. a. Gas is ionized by radio frequency energy. b. The ions are accelerated toward a target, where they knock atoms of the target out with energy sufficient to adhere to the specimen surface on impact. c. Sputter coating requires a weak vacuum. d. Air, or a desired gas, is bled into the sputtering chamber to control the gas pressure, to maintain a stable plasma field. (b) In evaporation coating, metals or carbon are sublimated from a source, to solidify on all available surfaces including the specimen. An evaporator is typically set up to accomplish multiple, related tasks: a. Current passed through a sharpened carbon rod causes the carbon at the tip to heat and evaporate. b. Metal wire, such as platinum, can be placed in a tungsten basket. Current passing through the tungsten wire melts, and then evaporates, the platinum. c. Not all metals can be successfully evaporated using a tungsten basket, particularly gold and aluminum. A spring shape works better for these metals. d. Metal evaporated from an angle will cast a shadow on the specimen, enhancing contrast. e. Metal molecules will bounce off any atmospheric gas particles, and so will not travel any appreciable distance. A high vacuum is required, typically below 10^{-5} Torr. f. The specimen table may be enabled to rotate. This allows a specimen to be coated from any and all angles without breaking the vacuum. g. A mild vacuum environment is suitable for creating plasma. Therefore, the vacuum evaporator can be used, for example, to conduct plasma etching, platinum shadowing, and carbon overcoating without remounting the specimen or breaking vacuum.

417

materials. Therefore, one way to obtain a thin, stable electron microscopy specimen is the replica technique.

Calibrating magnification can be accomplished by examining a grid with a known size or latex beads of known diameter. Calibrating the rotation of the beam in transmission microscopy caused by precession of the moving electrons in the magnetic lenses is required when matching the images to the matching diffraction pattern. This calibration is usually accomplished using a single crystalline ceramic with a known diffraction pattern and crystal habit, typically molybdenum trioxide.

Transmission Electron Microscopy (TEM)

In a transmission electron microscope, the electron beam is generated in an electron gun. The beam is focused on the specimen using one or two condenser lenses to obtain a small, intense spot. The transmitted light is then gathered by the objective lens. One or two projector lenses can then further magnify the objective image. The image is observed on a fluorescent screen, which can be removed to allow the electron beam to expose film to record an image. Apertures are used to manipulate beam size, modify depth of field and eliminate defects including chromatic and spherical aberration.

TEM provides very high-resolution images. In general, fashioning reliable samples, using the microscope, acquiring good images and interpreting the results can be very challenging, requiring years of training and experience. The resolution is finite, restricted to features on the order of the diameter of the spot, as well as by limitations caused by lens defects. The total area of a specimen examined is also very small, so assuring that a representative sample is examined is important for drawing conclusions that can be generalized. However, TEM can provide useful images of nanocomposite materials, with achievable resolutions down to 0.5 nm.

TEM derives contrast from two basic sources. The easiest to understand is electron absorption. The incident energy is absorbed, transmitted or diffracted. Thus, the transmitted light intensity is decreased by increases in absorption or diffraction. Thicker areas of the specimen will absorb more electrons. Variations in material will result in more effective interaction with the beam, as well, based on the collision effectiveness. The collision effectiveness generally increases with atomic weight. Therefore, using metal shadowing or decoration on a carbon support replica will yield molecular weight contrast on a specimen.

The second source of contrast in TEM is diffraction. Crystals will diffract electrons, removing them from the transmitted beam. Features

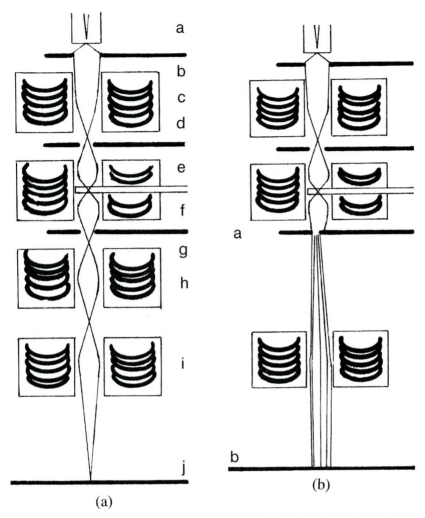

(a)

(b)

FIGURE 9.22 Transmission electron microscopy: (a) light train for image formation. a. electron gun (source). b. anode, to accelerate the electrons. c. condenser lens, to focus the beam, reducing beam size. d. condenser aperture, used to decrease spherical aberration. There may be more than one condenser lens. e. objective lens. f. The specimen is placed *inside* the objective lens. There is also an objective aperture to limit spherical aberration. g. intermediate aperture, which enhances contrast by blocking the most highly scattered light or diffracted rays. h. Intermediate lens provides increased magnification. i. Projector lens adjusts the image focus. j. image plane, which can be a florescent screen, or a photographic plate. (b) light train for diffraction. a. The intermediate aperture and lens are removed to allow the diffracted beams to travel to the projector lens. b. Instead, an aperture may be used to select a limited area of the specimen. This is useful, for example, in analyzing single crystals within a larger, polycrystalline film.

419

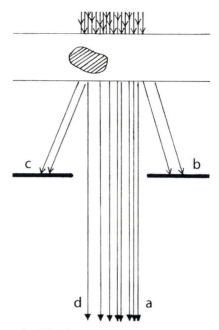

FIGURE 9.23 Contrast in a TEM image: (a) Some of the incident electrons pass through the film unhindered. (b) Some of the electrons are diffracted or scattered. The aperture blocks scattered electrons. (c) Inclusions composed of heavier nuclei are more highly scattered. (d) Therefore, less electron density continues in the transmitted beam.

causing more or less diffraction, such as edges and defects, will result in beam contrast. The microscope also can be realigned to send a diffracted beam through the projector lens system, or an aperture can be used to block the transmitted beam and pass a diffracted beam, called dark field imaging. In this case, the only visible features will be those that cause the diffracted electron beam.

Scanning Electron Microscopy (SEM)

Scanning electron microscopy uses the interaction of the beam with a thick specimen. A small electron beam interacts with the sample, generating electrons and x-rays. An amplifying detector gathers the electrons. The number of electrons detected determines the intensity of the resulting signal. The beam is moved across the sample by a raster circuit. This raster is matched to the raster of a cathode ray tube to reconstruct an image of the surface. Unlike transmission electron microscopy or optical microscopy, the image is not directly constructed from light scattered by

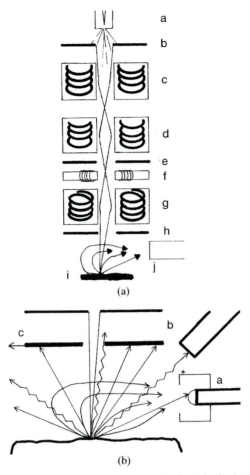

(a)

(b)

FIGURE 9.24 Scanning electron microscopy: (a) Light train for image formation. a. electron gun. b. anode. c. condenser lens 1. d. An optional, additional condenser lens may be used to make the beam still smaller and more focused. e. condenser aperture, to limit the spherical aberration. f. Scanning coils raster the electron beam across the specimen. The area swept by the electron beam controls the SEM magnification. gi. objective lens. This focuses the beam on the specimen. h. The objective aperture reduces spherical aberration. i. The specimen is the end of the electron beam. j. Various detectors are available. Shown is a secondary electron detector. (b) The various types of energy released by specimen-beam interactions are potential information sources. a. The typical secondary electron detector uses a cage with an adjustable, positive charge to attract low-energy electrons. These electrons stimulate a detector system. The larger the number of secondary electrons, the stronger the signal will be. b. The characteristic x-rays emitted by component nuclei can be detected, a technique called EDAX. c. A separate detector may be used for the high energy, backscattered electrons. One type of these detectors provides a sensitive electrode to collect only the electrons that bounce nearly straight back from the specimen, and are not attracted by the secondary electron detector.

the specimen. The magnification of the image is the ratio of the size of the area scanned on the surface to the area of the cathode ray tube (CRT). This magnification should be calibrated, since it depends on beam current, working distance, and other factors.

The interaction of the beam with the specimen is similar to that for transmission, but since the sample is thick, the electron will continue penetrating the specimen until an interaction occurs. The interactions occur within a pear-shaped region for a low atomic weight solid or a hemisphere for a higher atomic weight material (Figure 9.25). The collisions within the material can result in backscattered electrons (BSE). BS elec-

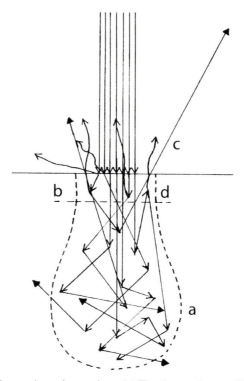

FIGURE 9.25 Electron beam interactions: (a) The electron beam will penetrate and interact with the specimen in a characteristic, pear-shaped interaction volume. This volume depends on atomic number. (b) Electrons may bounce off of multiple particles. At each interaction with an atom, a secondary electron may be knocked out of the specimen. (c) Electrons that hit an atom and are reflected back at a steep enough angle can be detected as backscattered electrons. These electrons can come from almost anywhere in the interaction volume. (d) Because of their low energy, only secondary electrons emitted from a shallow depth can leave the specimen.

trons are relatively high energy, and return to the detector traveling in a straight line. The collision can also generate secondary electrons with low energies, knocked loose from the material. These must be drawn to the detector. The backscattered or secondary electrons can be selectively gathered by a choice of detectors. The beam interactions also generate characteristic x-rays, often used for chemical analysis of metal and ceramic materials.

Contrast is provided by both surface topography and atomic number. Atomic number contrast is a result of the shallow penetration of the inter-action volume. More of the electrons created through collisions are re-leased from the surface, so compositions with high atomic number will appear brighter. Topography also influences how much of the interaction volume intersects the surface, as shown in Figure 9.26.

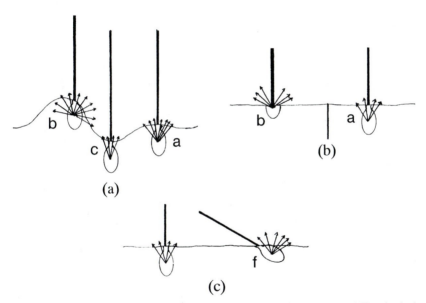

FIGURE 9.26 Sources of contrast in scanning electron microscopy: (a) Topological surface contrast. a. When the beam hits a flat surface, an average number of electrons are observed. b. On a hill, more of the secondary electrons may escape the surface, making high points brighter. c. In a valley, fewer of the secondary electrons escape. (b) Atomic number contrast. a. In a low-molecular-weight substance, the electron penetrates deeply. Most of the secondary electrons have insufficient leaving energy. b. In a high-molecu-lar-weight substance, the electrons penetrate less deeply into the specimen. The resulting interaction volume intersects more of the surface, and will appear brighter. (c) If the beam strikes the specimen at an angle f., more of the interaction volume will intersect the surface. This will result in shallower penetration of the specimen.

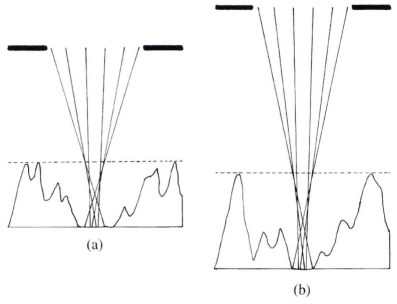

FIGURE 9.27 The working distance for a focused electron beam is created by the zone of least confusion. This zone is created by beam focusing, and depends on the lens angle. (a) If the angle of the beam is large, the working distance is smaller. (b) If the angle is small, the working distance, and the corresponding depth of field, will be larger. However, the maximum focus will be reduced.

The working distance determines the depth of field. As with all lenses, the resolution improves with the diameter of the lens. In SEM, resolution may be sacrificed to improve the depth of field by increasing the working distance, as shown in Figure 9.27.

The magnification of an SEM is limited by the spot size, the minimum controllable area of the raster, working distance, and the size of an image element on the CRT. If the image element is smaller than the beam size, there will be overlap in the image elements. At this point, further magnification will reveal no new information (hollow magnification). Unfortunately, as the spot size is decreased, the beam current is reduced, and so a balance between beam size and image intensity usually limits magnification. Higher intensity beams generated by LaB_6 (lanthanum hexaboride) or field emission electron guns significantly improve the beam current at smaller beam diameters, allowing better resolution.

Since the image is mapped, the potential defects in the image are a little different for SEM than other forms of microscopy. There can be projection distortions, such as pincushion and barrel distortion, image foreshortening for specimens tilted relative to perpendicular, edge distortions as the beam is diverted farther and farther, and beam distortion because the intersection between a cylinder and a tilted plane is not a circle. In addition, the raster scan has a pattern, and if the periodicity of the raster is too similar to a periodic structure in the specimen, a Moiré pattern will be generated. Finally, when the sample is tilted, the entire surface may not be in focus, if the depth of field is approached.

Electron Dispersive Spectroscopy (EDS), Auger Spectroscopy, X-ray Photoelectron Spectroscopy (XPS), and Electron Back Scattered Diffraction (EBSD)

The beam of an electron microscope generates characteristic electrons and x-rays that depend on the atomic composition of the surface. The energy level and intensity of these emissions can be detected and quantified. If the sample is crystalline, some of these electrons will also be diffracted by the crystal structure to form a characteristic pattern. These various ancillary sources of contrast create an opportunity for alternative imaging techniques.

When the electron beam in a microscope interacts with the specimen, electrons can be ejected from the material. The energy of these electrons, called Auger electrons, is very specific to the type of material. A detector capable of detecting the energy of these electrons can be used to determine the atomic composition of the material under the beam. By mapping the intensity of the electrons of different energies, a map of composition can be formed.

Auger electron microscopy is effective but somewhat expensive. A simpler technique is to monitor the characteristic x-rays that are emitted when the electrons from outer shells fall to lower energy levels to replace the Auger electrons. These x-rays are also specific to atomic composition. As the beam passes over the sample, x-rays are emitted that are detected, again providing a map of the atomic composition of the surface. This is the electron dispersion spectroscopy, or EDS, technique. Many scanning electron microscopes come with EDS detectors.

X-rays can also be used to irradiate a material surface. This stimulates the loss of characteristic electrons. This technique, x-ray photoelectron microscopy or XPS, probes chemical structure. The x-ray beam is not very narrow, however, so XPS does not generally provide structural or distribution data.

The electrons diffracted from a crystalline solid create a diffraction pattern in the chamber of an SEM. A suitable detector can monitor the pattern, providing information about the orientation of the crystal that

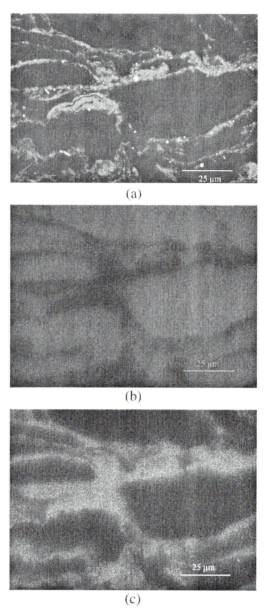

(a)

(b)

(c)

FIGURE 9.28 An x-ray detector can be used to analyze the characteristic x-rays from an element under the electron beam. Therefore, the composition as a function of position can be recorded. Each x-ray creates a dot. This is illustrated here with silica nanoparticles in a nylon 11 matrix. (a) A secondary electron image. (b) The location of sites generating characteristic carbon x-rays indicates that the dark areas in the original image are carbon. (c) The dot map for silicon shows clearly that the silica is concentrated in lamina surrounding the carbon-rich regions.

426

TABLE 9.4 Energy velocity and wavelength of neutrons.

Velocity (m/s)	Wavelength (nm)
200	2
400	1
1000	0.4
2000	0.2
4000	0.1
10000	0.04

the beam is striking. In this way, a map of the orientation of the surface crystallinity can be obtained in a technique called electron back scattered diffraction or orientational imaging.

Scattering

Like electrons, high-energy neutrons have a wavelike character. The mass of a neutron is much higher than the mass of an electron, so the wavelength is correspondingly shorter (Table 9.4). There is no characteristic variation to neutron scattering across the periodic table. If the nucleus has a non-zero spin (hydrogen and carbon 13), the neutron can interact on collision. On collision, the energy splits into coherent and incoherent components. Spinless nuclei collisions (deuterium, carbon 14), however, generate only coherent scatter, and therefore information. The wavelength of the interaction provides information on the order of 10s of nanometers. Neutron scattering has been used to measure the average size of single, deuterated polymer chains in a matrix of normal polymer. This technique is beginning to be effective in nanocomposite testing.

Diffraction

Unlike imaging, which produces a magnified image that reproduces the surface of a sample of material, diffraction provides structural information that is a little more abstract. Analyzing a diffraction pattern involves studying the energy pattern that is created by the systematic interference pattern of coherent light that is scattered from a grating. When light strikes a grating, the majority of the incident light is transmitted. Some light is randomly scattered from each plane of points, and this scattered light interacts constructively or destructively depending on the path length traveled by the light. Therefore, coherently scattered, diffracted light constitutes only a small fraction of the intensity of the total incident light, especially for very energetic light, such as x-rays and electrons.

Coherent light has only one component wavelength. Coherent energy scatters from each object within a solid. If the objects form a regular, crystal packing structure, the scattered, coherent light will form a pattern of reinforcement and interference called a diffraction pattern. Diffraction is caused by any grating created by any regular arrangement of objects that are of a size capable of interacting with energy of that wavelength. Incoherent light also scatters from a grating, but the interference pattern of the light can overlap, making the scattering uniform and providing little discernable information about the grating. Regular arrangements of nanoparticles, crystal arrangements of atoms, and other

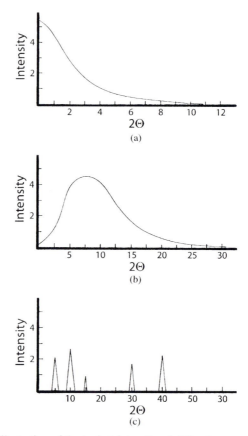

FIGURE 9.29 Illustration of the typical intensity of diffraction as a function of angle from: (a) gas, (b) liquid, and (c) a crystalline solid.

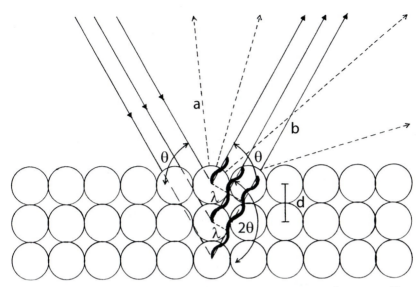

FIGURE 9.30 A beam of coherent light striking a regular grating, such as a crystalline, atomic solid, may cause diffraction. (a) Light striking the atoms is scattered randomly. (b) Scattered light creates patterns of constructive and destructive interference, and therefore generates bright beams in some directions. Constructive interference occurs when the incident beam is at the Bragg angle, θ. The Bragg angle depends on the wavelength of the light, λ, and the grating spacing, d. The angle of the specimen relative to the beam, 2θ, is easier to measure in a typical experimental setup.

regular structures with a repeat interval on the order of the wavelength of the incident light can diffract electromagnetic energy.

The pattern formed depends on the wavelength of the coherent light. Atomic crystals can diffract electrons and x-rays, with wavelengths in the sub-nanometer range. In diffraction, the pattern of energy scattering illuminates the structure of the grating and tells more about the ordering and orientation of crystals in the solid. The dispersion patterns of energy from a gas, liquid, and crystalline solid are shown in Figure 9.29.

Basis of Diffraction

Diffraction occurs if the incident light is all of identical wavelength, or monochromatic. Depending on the path traveled by the light, energy scattering from one point in the solid can add destructively to the energy from another point as shown in Figure 9.30. Bragg's law, below, defines the conditions for energy to diffract and create a beam with constructive interference:

$$n\lambda = 2d \sin\theta \qquad (9.9)$$

where n is the integer whole number of wavelengths, λ, d is the distance between planes of atoms and θ is the angle of the incident light. There is strong scattering, diffraction, in the directions predictable through the Bragg law, and no diffracted energy in any direction not corresponding to the Bragg conditions. The result is a spatially resolved distribution of light. If a projector screen were placed at some known distance from an irradiated material with a known crystal structure, then a predictable pattern would be visible. Contrarily, if a beam of energy is shone on a material with an unknown structure and a pattern is observed, the structure can be reconstructed using the same predictions in reverse.

According to the Bragg Law, light beams will occur intensely in only certain directions, called the characteristic reflections, and not at all in most other directions due to destructive interference. The directions of constructive interference are defined by the order of the diffraction, the wavelength of the light and the spacing of the grating. The minimum energy for diffraction is set by knowing that $\sin\theta$ cannot exceed unity, and that the smallest value for n is 1.

Libraries of x-ray spectra for a large number of materials are available. The basis of x-ray spectroscopy is comparing the spectrum of an unknown structure with a known atomic composition with known spectra of materials with those atomic compositions. The basis of x-ray analysis of structure is to evaluate the crystal reflections to uncover the source crystal structure. There are a limited number of possible unit cells. The unit cell determines the diffraction directions, and for each unit cell type, the rules for diffraction are known.

Note that only a fraction of the energy is scattered by Bragg diffraction. One difficulty in x-ray analysis is separating the diffracted beams

TABLE 9.5 Reflection rules.

	Reflections Allowed	Reflections Prohibited
Simple	All	
Base-Centered	h and k both even or both odd	h and k mixed
Body-Centered	h + k + l is even	h + k + l is odd
Face-Centered	h, k and l all even or odd	h, k and l mixed

Derived from Elements of X-ray Diffraction.

from the very bright, transmitted beam. The larger the camera distance, the greater is the separation between the transmitted beam and the diffracted rays. This is the primary differentiation between wide-angle and small-angle x-ray diffraction. Wide-angle diffraction (WAX) is useful for detecting the pattern from a polycrystalline material with small spacing between the crystal lattice points. Small-angle diffraction (SAX), which has a greater distance between the sample and the camera/detector, can be used to detect larger spacings, e.g., between particles and for visualizing single crystals, crystal preferred directions and other features.

Unfortunately, with high intensity x-rays shooting across the distance of a room, small angle x-ray analysis machines are unsuitable from a safety perspective. Small angle x-ray analysis activities are still carried out at high-energy physics centers. High-energy diffraction can also be carried out using electrons in a TEM, which produces a similar spectrum in a much safer piece of equipment, and has displaced SAX as the method of choice for spatially resolved diffraction patterns from ordered structures with larger grating spacing.

Paracrystallinity

Perfect diffraction occurs from perfect, infinite crystals under perfect radiation. Thus, the diffraction pattern is modified by typical crystal imperfections. Small crystals result in incomplete destructive interference. As the crystallites shrink in size, the diffraction spots will become broader and the number of orders of diffraction, or the diameter of the visible pattern, will become smaller. Therefore, all other conditions being known, the average size of the crystals can be determined by measuring the diameter of the diffraction spots. However, imperfections in the crystal can occur for two additional reasons. First, the atoms will vibrate, causing uncertainty in the location of the atom in the lattice, despite the average position of the atom being on center. A typical example of this disorder of the first type is thermal noise. Again, this will cause the size of the diffracted beams to broaden and the overall size of the pattern to shrink. Second, there can be variability in the separation of pairs of atoms. An example of this type 2 disorder is bond length variability. The primary effect of this type of error is a reduction in the number of discernible orders of diffraction.

Reciprocal Space

The diffraction pattern forms what has come to be known as reciprocal

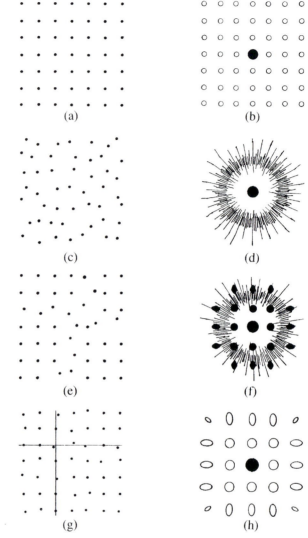

FIGURE 9.31 Paracrystallinity is a theory of how crystals and crystal defects affect x-ray patterns. (a) A regular grating, with long-range order, results in (b) a typical, regular diffraction pattern. (c) A random grating with no long-range order, typical of liquid, results in (d) a ring of light, called an amorphous halo. (e) A combination of small regions of crystalline order, in a field of random particles results in (f) a regular diffraction pattern with an amorphous halo superimposed. The smaller, non-infinite crystals result in larger diffraction spots and fewer ranks of spots, counting from the center beam. (g) If the regular crystal pattern is maintained but random, small displacements, such as caused by heat, are introduced, (h) the results include spot broadening and fewer ranks of spots counting from the center beam. So, the effects of decreasing crystal size and increasing temperature are indistinguishable.

432

space. In the diffraction pattern, a grid of closely spaced points will scatter the beam farther, and widely spaced points will deflect the beam less. A planar array of points creates a diffraction pattern consisting of a single spot. A single atom creates an infinite plane in the reciprocal space of diffraction. A line of atoms is visualized in diffraction space as a line of spots perpendicular to the actual objects. As a result, a finite crystal will yield a diffraction spot with some finite diameter inversely proportionate to the size of the crystal. Planes separated by a large spacing will create spots that are close together; planes separated by smaller distances will be visible as spots separated by a larger distances in reciprocal space. Making measurements for crystal structure in reciprocal space yields easier calculation of many unit cell parameters.

X-rays

Many x-ray analysis techniques have been developed for specific materials studies. There are far fewer in common use. The most common technique is wide-angle x-ray. Small-angle x-ray is also of interest as the precursor technique to electron scattering, having the same basic physics. High-energy x-ray sources at the National Laboratories are also still

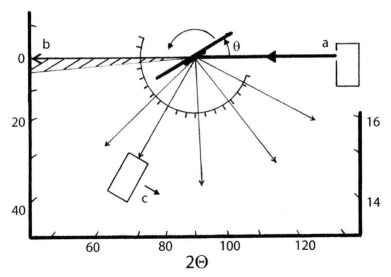

FIGURE 9.32 A wide-angle x-ray diffractometer. A polycrystalline specimen is mounted to a rotatable table. (a) A copper x-ray tube generates source x-rays. (b) The majority of the beam is transmitted through the specimen. So, the detector cannot be used too close to 0°. (c) The detector is moved to sense x-rays as a function of specimen angle.

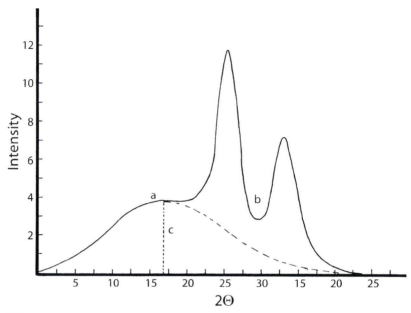

FIGURE 9.33 The X-ray spectrum can be analyzed to determine crystallinity. Prior to analysis, sources of noise, including thermal diffuse scatter, and Bremßtralung radiation are removed from the spectrum. The diffraction pattern of a semicrystalline polymer has (a) an amorphous halo and (b) one or more crystal peaks. In order to determine the intensity of x-rays in the crystal peaks, the amorphous halo must be eliminated. Crystal peaks often distort the amorphous halo. The amorphous halo is symmetric about a centerline (c), which may be used to assist in analyzing the amount of intensity in the halo and subtracting this contribution to leave the crystal peaks.

in use for small-angle diffraction studies. Valuable, bright x-rays are generated in particle accelerators when the particles are bent around a curve in the accelerator. The loss of momentum in one direction is accompanied by white x-ray radiation that can be filtered and used for small-angle x-ray analysis.

X-ray diffraction can be accomplished using a machine such as that shown in Figure 9.32. The path length can be maintained constant, and the angle of diffracted beams can be measured by rotating the crystal relative to a detector measuring the intensity of x-ray density. Measurable diffraction angles are limited by the diameter of the transmitted beam at small angles and by the sizes of the detector and x-ray source at large angles, but cover a wide arc.

In order for x-ray diffraction to be effective, all crystal orientations must be represented in the area under the x-ray illumination. Otherwise, the detector might not observe all diffraction beams. For example, if the material is a single crystal and the axis of x-ray detection does not cross the diffraction arc of the crystal, no intensity will be detected. Many samples are polycrystalline, with crystals that are small enough to ensure that all crystal orientations are represented in the area of the beam. Alternatively, a single crystalline material can be ground into a fine powder that is then analyzed. In polycrystalline samples, the diffraction pattern is observed as a series of rings centered on the beam and located at angles defined by the crystal lattice spacings and the Bragg law. The detector cuts a cross section through these rings starting near the beam and proceeding outward.

In addition to determining the unit cell and crystal structure of a material, x-ray analysis can provide information about the amount of crystallinity in a semi-crystalline polymer. In a scan of a polycrystalline, semi-crystalline polymer, there will be an amorphous halo due to the disordered, amorphous content of the material superimposed on one or more sharp diffraction peaks from the crystalline content. The ratio of the x-ray intensity in the amorphous halo to the intensity in the crystalline peaks is proportional to the percent crystallinity.

There are some limitations to the technique. The baseline for determining the intensity of each the amorphous halo and the crystal peaks is critical. This job is made even more difficult by the automatic corrections to the baseline made by modern diffractometers designed for metals. The diffractometer will automatically subtract thermal diffuse scattering and atmospheric scatter using computer-generated baselines. The computer-generated baseline may be drawn incorrectly, however, for a polymer having an amorphous halo exhibiting many of the characteristics of the thermal diffuse scattering phenomenon. In addition, finding a standard crystallinity for comparison is very difficult for polymers and for naturally occurring nanocomposites such as clay. Consider also that the gallery size in a nanoclay is large, and therefore the diffraction angle will be quite small. The limits of the detector at small angles will often interfere with WAX as a tool for analyzing the distances between particles in a nanocomposite.

The camera length, that is, the distance between the sample and a piece of film used to record the diffraction pattern, can be increased dramatically in small-angle x-ray diffraction (Figure 9.34). As a result, smaller divergences from the transmitted x-ray beam can be detected and measured. This may allow the measurement of ordered structures with greater separation between lattice points: for example, the separation between packed nanoparticles. Further, since the diffraction pattern is recorded as a piece of film, experiments are not limited to polycrystalline

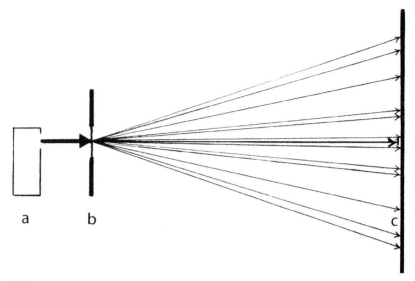

FIGURE 9.34 A small-angle x-ray diffractometer. (a) A copper x-ray tube generates source x-rays. (b) The x-rays are passed through the specimen. The transmitted beam travels some distance, the camera length, to be recorded using a detector or (c) film. The longer the camera length, the smaller the angle of the diffracted beam can be separated from the transmitted beam.

or powder samples, but can be used on single crystals or other types of material that have diffraction patterns that are not axisymmetric.

The shape of the diffraction pattern can yield information about texture and orientation. Texture in the crystals describes the long-range ordering of the unit cells. The crystals in a polycrystalline sample may have a preferred orientation. Large, small or imperfect crystals will also be observed in the size of the diffraction spots on a SAX pattern. If a crystal has plate-like, dendritic, or rod-like crystal habits, the diffraction spots will reflect these dimensions. If a preferred orientation is present, the diffraction pattern will be observed as arcs meridional to the axis of orientation instead of as complete rings. By measuring the length of the arcs and observing the orientation, both the direction and the percent alignment can be determined.

The wavelength of an electron with energy of 100 kV will be 0.004 nm. Electron diffraction is limited by all of the same factors as electron microscopy: the sample must be thin, and the scattering intensity decreases rapidly with angle. However, the scattering is very intense, and good diffraction patterns can be recorded in a very short time. This is especially important for beam-sensitive materials such as polymers.

PHYSICOCHEMICAL ANALYSIS

There are a many physical methods for assessing the structure of a material. Only the two most important will be discussed here. The molecular weight of a polymer has yet to be considered. Scanning probe microscopy and related techniques also have been very successful in analyses of nanomaterials.

Gel Permeation Chromatography (GPC)

GPC is one of the most widely applied techniques in common use for measuring the molecular weight of polymers. The technique generates information suitable for calculating any desired moment of the molecular weight. In GPC, the molecular weight is determined by passing a solvent containing the polymer through a porous media (Figure 9.35).

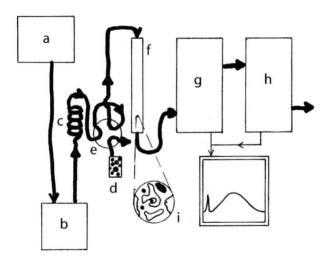

FIGURE 9.35 The components of gel permeation chromatography: (a) A reservoir of pure solvent. (b) A pump to push solvent and polymer through the column. (c) Pulses and irregularities in flow can be reduced using fixtures such as a coil. (d) A discrete sample of dissolved polymer is introduced at time t_0. (e) A distributor allows the sample to be introduced to the flow of solvent as a discrete event. (f) The sample passes through a column. The column contains a porous media. Larger molecules penetrate fewer of the pores and pass through the column rapidly. Smaller molecules penetrate more of the pores, hindering passage through the column. (g) The solvent stream passes through one or more detectors. Detector types include, for example, turbidity, flame, or conductivity. (h) The signal output of the detector is interpreted and visualized. (i) The separation column uses a porous separation medium.

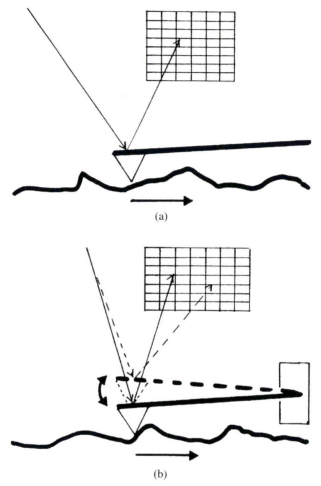

FIGURE 9.36 (a) Atomic force microscopy records the displacement of a contact probe as a function of displacement, similar to profilometry. The probe typically is an atomically sharp point. The tip-surface interactions are chemical or electrochemical. The force of contact and force to remove the tip are asymmetric. (b) In tapping mode, the probe oscillates, tapping along the surface. The tapping is included to eliminate artifacts caused by contact.

Entering the pore structure will slow the smaller molecules; longer chains will not interact with the media but pass rapidly through the column.

The polymer is separated by chain length. The delay, or residence time, of chains of known lengths is usually determined using calibrated molecular weight standards, typically polystyrene latexes with narrow molecular weight distributions. The molecular weight depends on the size of the molecule, which in turn can depend on the quality of the solvent for the specific polymer. Unfortunately, the unknown polymer and polystyrene seldom have identical interactions with a specific solvent, and so polystyrene standards will be offset from the unknown polymer. Therefore, measurements by GPC often provide estimates of the molecular weight biased by the difference in solvation between polystyrene standards and the unknown polymer. Comparison between different sets of the same polymer will be consistent with the molecular weight trend observed by GPC.

Scanning Probe Microscopy

The newest type of microscopy is scanning probe microscopy. The most common example of this family of techniques is Atomic Force Microscopy (AFM). A probe with a monatomic tip is dragged along or tapped along a sample. The displacement of the tip is monitored by interferometry. A laser is bounced off the back of the tip, allowing very precise measurements of displacement. The tip can also react with the surface, providing chemical information about the surface. Other probe types include scanning tunneling microscopy (STM) and magnetic probe microscopy.

REFERENCES

Cullity, B., Elements of X-Ray Diffraction, 2nd ed., Addison Wesley, Reading, Massachusetts, 1978.

Smallman, R. and K. Ashbee, Modern Metallography, Pergamon Press, Oxford, 1966.

Flewitt, P. and R. Wild, Physical Methods for Materials Characterization, 2nd ed., Institute of Physics, Bristol, 2003.

Goldstein, J., D. Newbury, P. Echlin, D. Joy, A. Romig, C. Lyman, C. Fiori, E. Lifshin, Scanning Electron Microscopy and X-Ray Microanalysis, 2nd ed., Plenum, New York, 1992.

Sibilia, J., Materials Characterization and Chemical Analysis, 2nd ed., VCH, New York, 1996.

Young, R. and P. Lovell, Introduction to Polymers, 2nd ed., Chapman Hall, London, 1991.

CHAPTER 10

Characterization of Physical Properties

This final chapter considers the characterization of the physical properties of nanocomposites. The goals of physical characterization are diverse. The goals of this chapter are to explain techniques for measuring physical properties, and to help develop an individualized analysis philosophy for reliably assessing performance. This chapter will cover traditional and non-traditional mechanical properties, tensile, shear and compression properties, viscoelastic properties, impact and hardness, and unusual mechanical issues. The measurement of surface mechanical properties of adhesion, wear, scratch, coefficient of friction, blocking, and tribological properties are considered. Thermal analysis will be discussed. Barrier properties, including water vapor and gas permeability, will be explored.

IDENTIFICATION

One of the most common analytical requirements is determining the identity of an unknown material. Identification is the first step in reverse engineering, where the material may be a complete unknown. Identification also is fundamental to quality control, where the basic assumption in developing assay methods is that the material was not made as intended. Identification tests conducted on a new material or a finished part are also a part of designing a synthesis or processing scheme.

The first tests to be considered are those suitable for basic identification. Identification techniques may be applied as screening to identify unknown materials or in quantitative manner to provide repeatable mea-

441

sures of characteristic parameters. Identity parameters include appearance, density, melting behavior, flammability and solubility. Identifying the composition of a nanocomposite can be particularly difficult. The reinforcement causes property changes that hinder identification by masking the performance of the individual components.

Much can be discovered about a material by simple examinations of certain physical properties. Figure 10.1 shows the process of determining the composition of a composite material. Notice that at each phase, progress is made by narrowing the available choices, rather than by directly determining the composition.

Providing identity information is often qualitative or semi-qualitative. Appearance factors, density, surface character, and other properties often require several complementary tests in order to provide enough information to faithfully represent quantitative distinctions, even for differences easily perceived using the senses. Quantifying the appearance of a specimen, however, can provide useful characterization metrics for material comparison or quality control.

General Phases

A few ID tests can be applied regardless of the basic nature of the material. Appearance can provide information regarding the matrix and the reinforcement. Density is usually a simple measurement for materials, and it provides comparative information. Melting point determination and the more rigorous thermal analysis techniques provide further general information about the material and important processing temperature ranges.

Appearance

Perhaps the very first thing to consider is the appearance of the material. Attributes of appearance include glossiness, haze or diffuse reflectance, transparency, color and texture. Glossiness and haze provide information about the appearance-based roughness of a material. Transparency relates to the ability to transmit or block light. Color can reveal the presence or absence of pigments, but also may indicate a type of material, a type of reinforcement, and other features. Texture describes a surface roughness that can affect the appearance. Composite surfaces often have haze due to surface irregularities imparted by the reinforcement. Composites are seldom transparent, and the reinforcement particles can contribute to perceived color. The particles can also contribute to surface roughness perceived as texture.

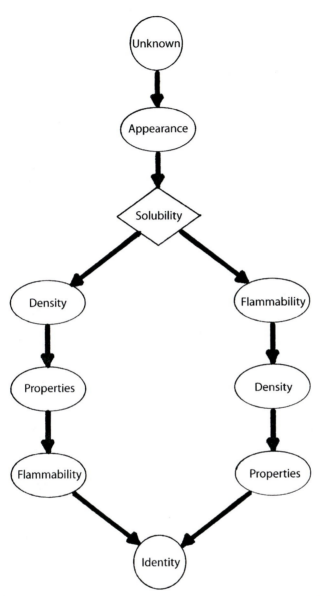

FIGURE 10.1 Flow of a sample through the logic of an identification matrix: Beginning with the sample, the first task is to evaluate the appearance. The second step is to evaluate the solubility. This is a decision point. Insoluble materials, such as thermosets, proceed to density testing. The next step is analysis of properties such as melting or softening points, hardness, and elasticity. The final step is analyzing the characteristics of flammability. If the sample is soluble in one or more solvents, this provides characteristic information about the material. This also allows the solid reinforcements to be removed from the sample and tested separately. The flammability, density, and properties are then analyzed.

TABLE 10.1 Plastics identification table.

Number	Standard Abbreviation (ISO 1043 ASTM 1600)	Name	Density unfilled g/cm³	Density filled g/cm³	transparent thin films	transparent, clear	hazy to opaque	usually contains fillers	leathery or rubbery, soft	flexible, resilient	hard	Sample slowly heated in pyrolysis tube (m = melts, d = decomposes)	Reaction of gas given off (al = alkaline, n = neutral, ac = acidic, sac = strongly acidic)	Ignition with small flame (0 = hardly ignitable, I = burns in flame, extinguishes in absence of flame, II = continues to burn after ignition, III = burns vigorously, fulminates)	Ignition with small flame	Odor of gasses given off on heating in pyrolysis tube or after ignition and extinction	Gasoline	Benzene	Methylene Chloride	Diethyl Ether	Acetone	Ethyl Acetate	Ethyl Alcohol	Water
1. Polyolefins																								
1.1	PE	Polyethylene - soft	>0.92		*		*			*		becomes clear. m. d. vapors are hardly visible	n	II	yellow with blue	slight	i	sw	i	i	i	i	i	i
		- hard	<0.96		*		*			*	*		n	II	center, burning droplets fall off	parafin-like color	sw	i	sw	sw	i	i	i	i
1.2	PP	polypropylene	0.905	1.3	*	*	*			*	*						sw	i	sw	i	sw	i	i	i
1.3	PB	polybutene-1	0.915		*	*				*							sw	sw	sw	i	sw	i	i	i
1.4	PIB	polyisobutylene	0.93	1.7			*	*	*			m. vaporizes, gases can be ignited	n	II	yellow, burns quickly	paraffin- and rubber-like	s	s	i	sw	s	i	i	i
1.5	PMP	poly-4-methylpentene-1	0.83			*					*	m. d. evaporates, white smoke	n	II	yellow with blue center, drips	slightly paraffin-like	sw	sw	i	sw	s	sw	i	i
2. Styrene Polymers																								
2.1	PS	polystyrene	1.05			*					*	m and evaporates	n	II	flickers, yellow, bright. very sooty	characteristic of city gas	sw	s	s	i	s	i	i	i
2.2	SB	high impact polystyrene w/polybutadiene	1.05				*				*	m, yellowish. d	n	II		like PS + rubber	sw	s	s	i	s	i	i	i
2.3	SAN	styrene acrylonitrile copolymer	1.08	1.4		*					*	m, yellow. d	al	II		similar to PS, irritating	i	s	s	s	s	i	i	i
2.4	ABS	acrylonitrile butadiene/styrene (copoly)	1.06				*			*	*	d. turns black	n (ac)	II		like PS + cinnamon	sw	s	s	s	s	s	sw	i
2.5	ASA	acrylonitrile strene (acrylate copolymer)	1.07				*				*	m. d. black residue	ac	II		like PS + pepper	sw	s	s	s	s	sw	sw	i

Solubility in cold solvents: s = soluble, sw = swellable, i = insoluble

444

3. Halogen-containing homopolymers

No.	Abbr.	Description	Density			behavior on heating	flame class	code	combustion / flame appearance	odor									
3.1	PVC	polyvinylchloride (55% Cl)	1.39	* *	*	softens, d. becomes brown-black	I	sac	yellow, sooty, lower edge of flame is slightly green	hydrochloric acid and also a burnt odor	i	i	i	i	i	i	i	i	i
		copolymer with VAC or similar	1.35				I	sac			i	i	sw	i	i	i	sw	i	i
3.2	PVCC	temperature resistent (60–67% Cl)	1.5	* (*)	*		I / II	sac			i	i	sw	i	sw	i	sw	i	i
3.3	PVC-III	impact resistent; made with EVAC	1.2–1.35	*	*			ac			sw	sw	s	sw	sw	sw	sw	sw	i
		made with PEC	1.3–1.35	*	*		I	sac			i	sw	sw	i	sw	i	i	i	i
3.31	PEC	chlorinated PE	1.1–1.3	*		m. becomes brown	I / II	sac	yellow, bright, sooty	HCl + paraffin	sw	between PE and PVC based on Cl-content							
3.4	PVC-P	plasticized (prop.s depend on plasticizer)	1.2–1.35 1.6	* *	*	similar to PVC	I / II	sac	bright (due to plasticizer)	HCl + plasticizer	i	sw	sw	sw	sw	sw	sw	i	i
3.5	PTFE	polytetrafluoroethylene	2.0–2.3	*		becomes clear, doesn't melt. d. at red heat	0	sac	doesn't burn, blue green edge on flame, doesn't char	at red heat stinging odor, HF	i	i	i	i	i	i	i	i	i
	PTEP		2.0–2.3	*			0	sac											
	PEA	PTFE-like molding materials	2.0–2.3	*			0	sac											
	ETFE		1.7	*			0	sac											
3.6	CTFE	polytrifluorochloroethylene	2.1	* *		m. d. at red heat	0	sac	like PTFE, sparks	CHl + HF	i	i	i	i	i	i	i	i	i
3.7	PVF2	polyvinylidine flouride	1.7–1.8	* (*)	*	m. d at high temp.	0 / I	sac	hardly flammable	stinging (HF)	i	sw	i	sw	sw	i	sw	i	i

4. Polyvinylacetate & derivitives, polymethylmethacrylate

No.	Abbr.	Description	Density			behavior on heating	flame class	code	flame appearance	odor									
4.1	PVAC	polyvinyl acetate	1.18	Dispersions	* *	m. brown, evaporates	II	ac	bright, sooty	acetic acid and additional odor	i	s	s	sw	s	s	s	i	i
4.11	PVAL	polyvinyl alcohol	1.2–1.3	*	* *	m. d. residue	II	n	bright	irritating	i	i	i	i	i	i	i	i	i
4.12	PVB	polyvinyl butyral	1.1–1.2	safety glass	*	m. d. foams	II	ac	blue w/yellow edge	rancid butter	i	sw	sw	sw	s	sw	s	sw	s
4.2		polyacrylates	1.1–1.2	Dispersions	* *	m. d. evaporates	II	n	bright, slightly sooty	typically sharp	i	s	s	s	i	s	s	i	i
4.3	PMMA	poly(methyl methacrylate)	1.18	*	*	softens, d. swells up and crackles, little residue	II	n	burns with crackling, drips, bright	typically fruity	i	i	i	i	i	s	s	i	i
4.4	AMMA	co methyl nethacrylate/acrylic nitrile	1.17	* yellow	*	brown, then, m. d. black	II	al	sooty, sparks slightly	first sharp, irritating	i	i	sw	i	sw	i	s	i	i

(continued)

445

Solubility in cold solvents: s = soluble; sw = swellable; i = insoluble

Sample slowly heated in pyrolysis tube: m = melts; d = decomposes

reaction of gas given off: al = alkaline; n = neutral; ac = acidic; sac = strongly acidic

Ignition with small flame: 0 = hardly ignitable; I = burns in flame, extinguishes in absence of flame; II = continues to burn after ignition; III = burns vigorously, fulminates

5. Heteropolymers

Number	Standard Abbreviation (ISO 1043 ASTM 1600)	Name	Density unfilled g/cm³	Density filled g/cm³	Usual Appearance	Elastic Behavior	Sample slowly heated in pyrolysis tube	reaction of gas given off	Ignition with small flame	Odor of gasses given off on heating in pyrolysis tube or after ignition and extinction	Gasoline	Benzene	Methylene Chloride	Diethyl Ether	Acetone	Ethyl Acetate	Ethyl Alcohol	Water
5.1	POM	polyoxymethylene	1.41	1.6	hazy to opaque *	hard *	m. d. evaporates	(ac)	II — blue, almost colorless	formaldehyde	i	i	i	i	i	i	i	i
5.2	PPO	polyphenyleneoxide (modified)	1.06	1.3	hazy to opaque *	hard *	becomes black. m. d. brown. vapors	al	II — difficult to ignite. then bright. sooty flame	first slight then phenol odor	i	s	s	i	i	i	i	i
5.3	PC	polycarbonate	1.2	1.4	transparent thin films *, transparent clear *	hard *	m. viscous, colorless. d. brown	(ac)	I — bright. crackly. drips, chars	first slight then phenol odor	i	sw	s	sw	sw	sw	i	i
5.4	PETE	polyethylene terepthalate	1.35	1.5	transparent thin films *, transparent clear *, hazy to opaque *	hard *	m. d. dark brown. white deposit above	ac	I / II — bright. crackly. drips. sooty	sweetish. irritating	i	i	sw	i	i	i	i	i
	PBTE	polybutylene terepthalate	1.41	1.5	transparent thin films *, transparent clear *, hazy to opaque *	flexible, resilient *, hard *		ac	I / II — difficult to ignite bluish yellow edge. crackly. drips. fibet forming	characteristic odor similar to burn horn	i	sw	sw	sw	sw	sw	i	i
5.5	PA	polyamides - crystalline [PA66 to	1.14	1.4	transparent thin films *, transparent clear *, hazy to opaque *	hard *	becomes clear. m	(a)	I / II	possibly phenol	i	i	i	i	i	i	i	i
		[PA12	1.02	1.4	transparent thin films *, hazy to opaque *	flexible, resilient *, hard *		(a)			i	i	i	i	i	i	i	i
		amorphous	1.12	1.4	transparent clear *	hard *	d. brown	(a)			i	sw	sw	i	sw	sw	i	i
5.6	PSP	polysulfone	1.24	1.5	transparent thin films (*)	hard *	m. bubbly. vapors invisible. brown	sac	II — difficult to ignite. yellow. sooty. chars	first slight amount. finally H₂S in tube	i	s	sw	i	sw	i	i	i
5.7	PI	polyimides	1.4		transparent thin films * yellow	hard *	doesn't m. brown on strong heating. glows	al	0 — glows	on strong heating phenol	i	i	i	i	i	sw	i	i

(continued)

5.8		Cellulose derivitives																	
5.81	CA	cellulose acetate	1.3	*	*		*	*	m. d. black	ac	II	m. drops yellow, green w/sparks	acetic acid + burnt paper	i	sw	i	sw	i	i
5.83	CP	cellulose propionate	1.2	*		*			m. d. black	ac	II	same as CAB	propionic acid, burnt paper	i	s	i	s	sw	i
5.84	CN	cellulose nitrate (celluloid)	1.35–1.4	*	*		*	*	d. violent!	sac	III	bright, violent, brown vapors	nitrous oxides (camphor)	i	i	sw	i	i	i
5.85	CMC	hydroxymethylcellulose	>1.29	adhesive raw material					m. chars	n	II	yellow bright	burnt paper	i	i	i	i	i	s
5.86		cellophane (regenerated cellulose)	1.45	*		*			d. chars	n	II	like paper	burnt paper	i	i	i	i	i	i
5.87	VF	vulcanized fiber	1.2–1.3			*	*	*	d. chars	n	I/II	burns slowly	burnt paper	i	i	i	i	i	i

6. Phenolic resins

PF - Phenol Formaldehyde: including cresol resin

6.1	PF	free of filler: uncured	1.25	industrial resins					m, d	n	I	difficult to ignite, bright, sooty	phenol, formaldehyde	i	i	i	s	s	(s)
	PF	molded or cast resins	1.3	(*)						n	I		phenol, formaldehyde	i	i	i	i	i	i
6.2.1	PF	mineral filled moldings	1.8–2		*			*	d. cracks	n (al)	0 / I	bright, sooty	phenol, formaldehyde, poss. ammonia	i	i	i	i	i	i
6.2.2	PF	organic filled moldings	1.4		*			*	d. cracks	n (al)	1/II	chars	as above + burnt paper	i	i	i	i	i	i
6.3.1	PF	paper based laminates	1.3–1.4		*			*	d. cracks	n (al)	II	bright, sooty	as above + burnt paper	i	i	i	i	i	i
6.3.2	PF	cotton based laminates	1.3–1.4		*			*	d. delamination	n	II	bright, sooty		i	i	i	i	i	i
6.3.3	PF	asbestos or glass fiber based laminates	1.8		*			*	d. cracks	n	0 / I	reinforcing structure remains	phenol, formaldehyde	i	i	i	i	i	i

7. Amino resins

UF - Urea Formaldehyde; MF - Melamine Formaldehyde

7.1	UFMF	uncured		glues					d. cracks, darkens swells up	al	0 / I	very difficult to ignite, flame slightly yellow, material chars with white edge	ammonia, amines disgusting fishy odor (esp. thiourea), formaldehyde	i	i	i	i	i	s
7.2.1	UFMI	organic filled moldings	1.5		*			*		al				i	i	i	i	i	i
7.2.2	MF	mineral filled moldings	2		*			*		al				i	i	i	i	i	i
7.2.3	MF+PF	organic filled moldings	1.5		*			*		al				i	i	i	i	i	i
7.3.1	MF	glass fiber fabric based laminates	1.8		*			*		al				i	i	i	i	i	i

(continued)

447

Plastic Materials (arranged in chemical groups)

Solubility key: s = soluble, sw = swellable, i = insoluble

Density: unfilled g/cm³, filled g/cm³

Sample slowly heated in pyrolysis tube: m = melts, d = decomposes

reaction of gas given off: al = alkaline, n = neutral, ac = acidic, sac = strongly acidic

Ignition with small flame: 0 = hardly ignitable; I = burns in flame, extinguishes in absence of flame; II = continues to burn after ignition; III = burns vigorously, fulminates

Number	Standard Abbreviation (ISO 1600 / ASTM 1043)	Name	Density unfilled g/cm³	Density filled g/cm³	Appearance	Elastic Behavior	Sample slowly heated in pyrolysis tube	reaction of gas given off	Ignition with small flame	Odor of gases given off on heating in pyrolysis tube or after ignition and extinction	Gasoline	Benzene	Methylene Chloride	Diethyl Ether	Acetone	Ethyl Acetate	Ethyl Alcohol	Water
8. Crosslinked reactive resins																		
UP - Unsaturated Polyester; EP - Epoxy Resins																		
8.1.1	UP	unfilled, cast resins (flame retarded)	1.2		transparent, clear	hard	darkens, m, cracks, d, possibly white deposit above	n (ac)	II (I)	styrene and sharp additional odor	i	i	sw	i	sw	sw	i	i
8.1.2	UP	moldings, laminates		1.4–2	hazy to opaque; usually contains fillers	hard												
8.2.1	EP	cast resins	1.2		transparent, clear	hard	darkens from edge, d, cracks, possibly white deposit above	n or al	III / II	depends on curing agent, ester-like or amines (sim. to PA), later phenol	i	i	sw	i	sw	sw	i	i
8.2.2	EP	moldings, laminates		1.7–2	hazy to opaque; usually contains fillers	hard	(uncrosslinked)	(ac)		difficult to ignite, burns with small flame, sooty								
9. Polyurethanes																		
	PUR	a) crosslinked	1.26		hazy to opaque	leathery or rubbery, soft; hard	m on strong heading, then d	al	II	typically unpleasant singing (isocyanate)	i	i	sw	i	sw	sw	i	i
		b) linear, rubberlike	1.17–1.2		transparent thin films	leathery or rubbery, soft		ac	II	difficult to ignite, yellow, bright; foams, then d	i	sw	sw	i	sw	sw	i	i
10. Silicones																		
	SI	mainly silicone rubber	1.25		usually contains fillers	leathery or rubbery, soft	d only on strong heating, white powder	n	0	glows in the flame; white smoke, finely divided white SiO2 residue	sw	sw	sw	i	i	i	i	i

Source Unknown.

448

Gloss, Haze and Diffuse Reflectance

A glossy material will appear shiny, even if there is no distinct image reflected. While the reinforcing component will often disrupt surface smoothness, the gloss will be affected by the size of the inclusions and the quality of the particle dispersion. Gloss is a quantitative assessment of the magnitude of specular reflection from the surface. Gloss is the angular selectivity of surface-reflected light resulting in highlights or images of objects to be superimposed on a surface. Put another way, if the image reflected from two points of a surface is merged, the material will appear to be less glossy than if the two images are not merged. This leads to testing the amount of light reflected from a surface at a specified angle. ASTM test D523-39 uses an angle of 60° to the normal. The gloss ranges from 0, low gloss, to 100, high gloss. The method can be refined to include measures at near normal angles, 20°, sometimes called sheen, and at high grazing angles, 85°. Since the angles are defined, simple ma-

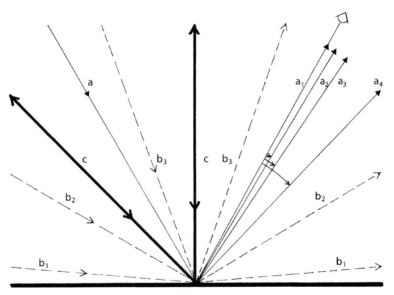

FIGURE 10.2 Gloss measurements: (a) Gloss measures are taken relative to a 30° incident beam, but various types of gloss are available. a1. The specular gloss is taken at 30°. a2. The distinctness of image (DOI) gloss is collected across 30° ± 0.3°. a3. Haze is measured relative to the gloss, with narrow-angle haze measured at 32° and a4. wide-angle haze measured at 34°. (b) Specular reflection may be measured at 20°, 60° and 85°. (c) Diffuse reflection is measured at 45°, either with a perpendicular incident beam or a beam at 45° measured perpendicular to the surface.

chines are available that can process systematic, comparable measures of the three data points. Capturing the reflectance of light at one or more angles provides an idea of how distinctly an image will reflect from the surface. Patent leather shoes and freshly waxed cars are examples of high-gloss finishes. Figure 10 .2 shows several gloss measurements.

Haze is another surface property, similar to gloss, that refers to how distinct an image will reflect from the surface. Haze measures the width of the specular lobe, based on a definitive angle relative to the 60° specular angle. Haze measures are defined in ASTM E 430. One measure is offset 0.3°, the distinctness of image, providing a measure reported as G_{doi}. A larger value of G_{doi} corresponds to a more distinct image. Another measure of haze uses the difference between the 20° and 60° measurements of gloss as the haze, H, where larger values of H indicate more haze. A well-polished spoon will often provide high gloss and high haze.

Reflection describes the ability of a surface to reflect incident light. Reflected light from a surface may form no coherent image. Diffuse reflectance occurs throughout the material, rather than exclusively at the surface. When light penetrates a material, the light colliding at any change in index of refraction can be partially scattered and partially transmitted. Diffuse reflectance may be an important component of appearance in nanocomposite coatings formed from transparent or translucent matrices, such as acrylic or polyethylene, respectively. This diffused light will also contribute to the appearance of the material. There are several standard reflectance measures. R_0 is the intensity of reflected light, or reflectance, of a unit thickness of material backed by a black body absorber. R_∞ is the reflectance of a unit thickness backed by an infinite thickness of the material. R_1 is the reflectance of the material backed by a perfect reflector. A definition of opacity used by the coatings industry is R_0/R_∞. Note that this definition of opacity differs from the traditional meaning discussed in the section below, on transparency. Measuring diffuse reflectance requires diffuse illumination. The light can be gathered from a sphere lined with a material that is close to perfectly reflective, such as barium sulfate or magnesium oxide. The intensity must be measured using a photometer.

Transparency

The transparency of the material is another indicator of composition and performance. If the sample is opaque, this implies heavy pigmentation, a semicrystalline matrix, or large reinforcement particles that block or scatter light. A clear material may indicate no crystallinity or very small crystallites, and no reinforcement, nanoparticles below the scattering limit of the incident light, or a matrix/particle system with a matching refractive index. Transparency is measured by turbidity, a measure of forward scattered light from a solution. The intensity of transmitted light, I, is measured relative to the incident light beam, I_0:

$$T = \frac{I}{I_0} \tag{10.1}$$

This is converted to transparency, or the turbidity, τ, a specific name for the absorbance per unit length of transit through the specimen, by the logarithm of the inverse:

$$\tau = \log\left(\frac{1}{T}\right) \tag{10.2}$$

The turbidity is an assessment of the opacity of the material. An opaque material transmits no light, a transparent material a large fraction of the incident light, and a translucent material an intermediate quantity.

Color

Reflected color is a complex phenomenon related to pigment color, as well as to intensity and brightness. The perceived color can be influenced by surface gloss and other parameters. A yellowed material may be indicative of matrix oxidation during processing or ultraviolet light degradation. The pigmentation or any other aspect of color may be identified using the CIE color scheme discussed in Chapter 8.

Additive colors change based on what pigments are present to reflect specific frequencies of light. Colors in pigmented, opaque materials are generally additive. Subtractive colors change light by absorbing light of a specific, limited visible frequency. For example, if the material absorbs red light, the reflected or transmitted light will appear blue, since there is less red light to balance the blue light. Colors in a transparent material are generally subtractive.

The CIE system characterizes colors by luminance and two color coordinates that specify the point on a complete spectral color space. Recall that the spectral color space is larger than the RGB color space of a computer monitor, and different from the CMY color space of printer pigments. The color coordinates, chromaticity, are calculated by measuring the light spectra at a series of frequencies. These coordinates are then multiplied by an integrated weighting factor, such as the distribution in Figure 10.3. There are two standard observer responses, slightly dependent on the equivalent angular aperture of the field of view. The most common is a 2° standard observer, but a 10° standard is also used.

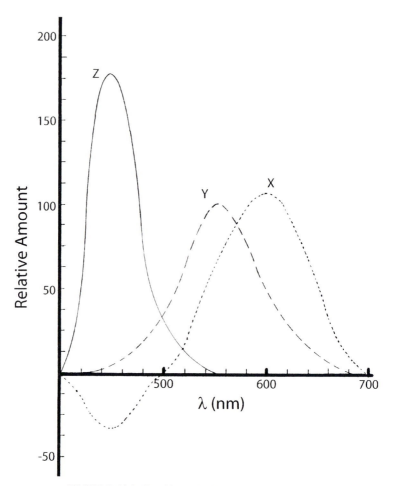

FIGURE 10.3 The 2° standard observer color spectra.

The center of the CIE range is achromatic light, that is, white light. However, this white is not a unique white. A black body emitter that generates white light provides the standard for true, pure white. The temperature of the most nearly equivalent blackbody radiator to a specific white can also be evaluated, providing a pseudo-"color temperature" for the source. The source may not be at this temperature at all. The color temperatures of some standard illumination sources are given in Table 10.2.

Color matching is another approach to assessing color, using the color wheel approach as a basis for applying a standardized subjective color judgment. Munsell color matching divides hue into 100 equal parts

TABLE 10.2 *Standard white light sources, equivalent color temperature and CIE coordinates.*

		CIE Coordinates	
Description	Color Temperature	x	y
Incandescent light	2854K	.1176	.4075
Direct sunlight	4874K	.3840	.3516
Indirect sunlight	6774K	.3101	.3162
Bright incandescent light	5000K	.3457	.3586
Natural daylight	6504K	.3127	.3297
Normalized reference	5500K	.3333	.3333

After Fortner.

around a circle. A saturation, or chroma, is assigned from zero to a number topping at 10 to 18, depending on the hue, distorting the circle. The number of perceptibly differences in hue defines the maximum value for saturation. A third value, a brightness "value" is assigned from zero, black, to 10, white. A color can be uniquely assigned by hue, value and chroma.

The Ostwald color system also matches colors to a set of standard samples. The standards are created from a dominant wavelength (hue), purity of wavelength (saturation), and luminance (brightness).

Texture

The feel of a material, slippery or waxy, sticky or inherently rough, blocking or soft, is its surface properties that are apparent to the touch. These properties are often characteristic of the matrix material. Certain of these properties may be quantifiable. The roughness of the surface can be quantified using a profilometer. The identity types of information for behaviors such as stickiness, slipperiness and softness, may be qualitatively evaluated using the coefficient of friction. The texture of a nanocomposite will be modified by the particles, especially particles at the surface of a part.

Profilometry

Among other sources of surface irregularity, composite particles are often more concentrated on the surface of a part in order to minimize the increased surface energy mismatch between particle and matrix. The profile created by these particles can be very characteristic of a material.

Further, surface profile may also provide an estimate of how much segregation of the reinforcement is occurring. Profilometry is a particularly useful technique for surface metrology. For years, the technique has been practiced using a sharp needle, lightly traced across a surface. The displacement of the tip, measured accurately, provides a measurable record of the peaks and valleys, the surface profile in a line across a sample. More recently, profile is recorded by measuring the distance traveled by a laser beam using interferometry. An example of a profilometry standard is ANSI B46.1. This provides definitions of roughness, waviness, and lay. Roughness provides information about the microscopic imperfections on the surface. Waviness describes larger frequency surface disturbances. The lay describes the relative amount and direction of orientation in surface features.

There are many different numbers reported as roughness, each with different applications or implications. Many, r_a, r_t and such, evaluate something related to the roughness. However, there are r-values that pro-

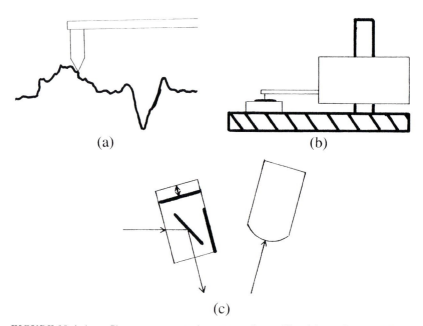

(a)

(b)

(c)

FIGURE 10.4 A profilometer uses a stylus to trace the profile of the surface. (a) The tip must be capable of conforming to the surface. The tip may not fit into some of the features. The tip must also be able to respond quickly to negative displacement. (b) A schematic of a profilometer. (c) Modern, non-contact profilometers use an interferometer to generate a complete interference pattern of reflectance from the surface.

TABLE 10.3 Profile parameters.

Description		Calculation	Discrete Approximation
R_a	Arithmetic mean roughness	$R_a = \dfrac{1}{L}\int_0^L z^2(l)dl$	$R_a = \dfrac{1}{n}\sum_{i=1}^{n}|z_i|$
R_q	Root mean square roughness	$R_q = \sqrt{\dfrac{1}{L}\int_0^L z^2(l)dl}$	$R_q = \sqrt{\dfrac{1}{n}\sum_{i=1}^{n}z_i^2}$
R_t	Distance between highest and lowest points on the surface		$R_t = z_{max} - z_{min}$
R_p	Distance between the highest point and the mean line		$R_p = z_{max} - R_a$
R_v	Distance between the lowest point and the mean line		$R_v = R_a - z_{min}$
R_{sk}	Skewness		$R_{sk} = \dfrac{1}{4R_q^3}\sum_{i=0}^{L}z_i^3$
R_{ku}	Kurtosis		$R_{ku} = \dfrac{1}{4R_q^4}\sum_{i=0}^{L}z_i^4$

From Hommelwerke Dektak Operation Manual.

vide information about waviness or about surface features. Table 10.4 provides roughness measurements. Figure 10.4 provides illustrations of these measurements. In surface roughness, as in many previous averaged calculations, the skewness and kurtosis of the averages can uncover distribution information about differences between two surfaces that have the same r_A.

Density

Density is a rapid and ubiquitous datum to collect from a material sample. The density is a function of the composition of the entire composite. The density of the matrix will not be determined independently of the reinforcement. The most common technique for obtaining density is the Archimedes principle. A block of material with precisely known mass is immersed completely in a liquid of precisely known volume. The increase in volume is measured, again precisely, and the total density can be measured as accurately as the precision of the measurements used:

$$\rho = \frac{\text{mass of sample}}{\text{volume of liquid displaced}} \qquad (10.3)$$

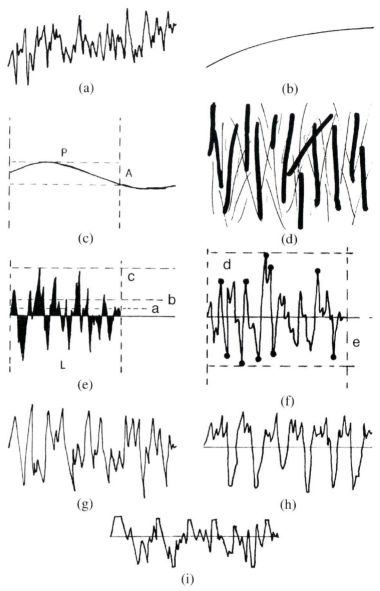

FIGURE 10.5 Sketches of various features of a rough surface: (a) An example surface. (b) An underlying slope. (c) Waviness is characterized by a period, P, measured from peak to trough, and amplitude, A. (d) Lay is characterized by a distinct orientation to surface roughness features. (e) and (f) Some of the characteristic roughness measurements. a. R_a, b. R_q, c. R_p, d. Average of the 5 highest and 5 lowest points. e. R_t. (g) An unusually rough surface and (h) a surface with an unbalanced surface, which would have similar R_a but different kurtosis and skew. (i) If the profilometer cannot conform with the surface, the peaks recorded may be distorted.

456

Polymers tend to be low density, but a ceramic reinforcement can increase the density significantly. Therefore, if the density is less than one, the material is either a pure polyolefin, or a polyolefin with very little reinforcement. For quality-control purposes, a density measurement can be a quick technique for assuring that the reinforcement has been properly incorporated while processing, for example, a polypropylene – nanoclay composite. In order to measure the density of the separate components, the two materials must be separated prior to taking the measurement.

There are potential difficulties with a simple density measurement of this type. If the material is porous, due for example to poor interaction between a matrix and reinforcement, then the holes introduce extra, and perhaps false, volume to the specimen. If the porosity is contiguous and open, then the liquid can penetrate into the sample. For complete penetration of the specimen porosity, the true molecular density is obtained from the measurement, after all. Thereafter, precise measures of the volume of the material without liquid penetration can provide measurement of the pore volume, or porosity. Preventing penetration can be accomplished by sealing the pores with wax or a thin lacquer coating, or by using a liquid such as mercury that will not penetrate small pores without force. Porosity will be discussed further below.

Melting and Thermal Transitions

Thermal analysis for composites is particularly useful, in that the transitions of each phase will usually be separate and can easily be distinguished, without resorting to complex techniques for mechanical separation of the phases.

The melting point is a relatively easy to measure thermal datum for a material. To conduct melting point analysis, a quantity of the material is heated while measuring the temperature. The precise temperature range for melting is recorded, and compared to known melting ranges. There are cheap, specialized units available in which a small, often powdered, quantity of the material is placed in a glass microtube. The thin-walled tube allows precise observation of sample melting, and an integrated thermocouple provides accurate temperature measurement. The melting temperature obtained is highly characteristic of many materials. In a composite, normally only the melting point of the matrix material is accessible at lower temperatures. Thermal transitions in metal or ceramic particles occur at higher temperatures, where the matrix will burn away in an oxygen environment, allowing separate analysis of the two components.

For simple atomic and molecular solids, the melting temperature is a thermodynamic, unchanging quantity. Several confounding factors are common. First, most ceramics and metals melt at temperatures well above those accessible to simple equipment. Another problem arises even for simple solids, where small quantities of impurities will elevate the melting temperature and depress the freezing temperature, creating hysteresis between these two temperatures. Most simple melting point equipment is unidirectional, designed for controlling increasing temperatures only.

Semicrystalline polymeric materials have melting ranges, rather than unique temperatures. As discussed in Chapter 2, the melting temperature of a polymer depends on backbone structure, crystal perfection, and the interactions between chains. During incorporation into crystals, polydispersity in polymer chain lengths will impose additional limits on crystal thickness. The chain ends create disorder, as they may displace farther than elements farther along the chain at the same temperature, which will lower any transition temperatures. The hysteresis between heating and cooling directions in transitions is also more dramatic in polymers. Visually determining the exact onset and completion of the event may be more difficult than for atomic solids, if the softening is subtle.

Amorphous materials do not have a melting temperature. A melting point test may provide ambiguous results. However, the equipment used for measuring simple melting point is not sensitive to the nature of transition type. Therefore, if the softening of an amorphous material at the T_g is sufficient to allow the material to flow under capillary force, then the T_g will manifest in the same way as a T_m would for a crystalline or semicrystalline material.

Combinations of materials often exhibit complex melting behavior. Immiscible materials, including composites, will usually exhibit multiple, defined melting points, one at each of the melting temperatures of the pure components. If the materials are miscible, the melting temperature is usually a compositional average of the T_ms of the components, according to the rule of mixtures:

$$T = m_1 T_1 + m_2 T_2 + \ldots + m_n T_n \qquad (10.4)$$

where m_n is the mass fraction and T_n is the melting temperature of a component. A harmonic average is more descriptive for some materials:

$$\frac{1}{T} = m_1 \frac{1}{T_1} + m_2 \frac{1}{T_2} + \ldots + m_n \frac{1}{T_n} \qquad (10.5)$$

This is common for random copolymers. These two averaging approaches for approximating transition temperatures provide bounds on the experimental temperature. The rule of mixtures provides an upper bound while the harmonic mean provides the lower bound.

Regardless of the challenges, even an approximate temperature range for melting can narrow the range of choices about a material very rapidly. Thermal analysis techniques can further refine the thermal data significantly.

Thermal analysis seeks detailed data about the transitions in a material. The most common techniques are dilatometry, differential scanning calorimetry (DSC), or differential thermal analysis (DTA). In dilatometry, volume changes with respect to temperature are recorded. In DSC, the difference in heat required to heat an unknown at the same temperature rate as a standard reference material is measured. In DTA, the difference in temperatures in an unknown and a reference are measured in a furnace heated at a constant rate. The two techniques provide similar data. DSC requires fewer theoretical assumptions for interpretation, but for practical reasons has a more limited temperature range than DTA. Other properties that are used in thermal analysis are oscillatory force—dynamic mechanical analysis (DMA), creep deformation—thermally stimulated creep (TSC), electrical response—thermally stimulated discharge current (TSDC), or mass loss—thermogravimetric analysis (TGA). In short, thermal analysis probes changes in properties of interest, e.g., mechanical, electrical or otherwise, as a function of temperature. These techniques can provide a connection between continuum measurements of properties to the molecular structure of the material.

In each technique, as the temperature increases, specific molecular movements cause measurable changes in the monitored property. Each technique is sensitive to particular types of movement. Mechanical tests such as DMA or TSC are particularly sensitive to motions resulting in changes to mechanical properties. DSC is sensitive to most transitions. The primary focus of this section will be DSC and DMA, the two most common thermal analysis techniques. Thermal analysis, using properly calibrated equipment, can provide information about the temperatures at which dramatic property transitions, such as glass transition, melting temperatures, and other molecular rearrangements occur. Quantitative information, including heat capacity and the amount of energy consumed in a transition, can also be generated with proper care.

Dilatometry

In a dilatometer, the volume change of a material is measured as a function of temperature. The volume change can be measured in several ways. Frequently, the length change in one dimension is monitored closely. The deformation can be measured without contact, using a laser displacement, or by contact displacement of, for example a rod. If displacement is used, the thermal expansion coefficient of the rod material must be calibrated. An isotropic length change of the material is typically assumed. Therefore, the volume change is simply the cube of the length change. Such an assumption is usually good for materials with an isotropic structure, for example, when the crystals assume polycrystalline or cubic structures. An asymmetric crystal or a composite with directions of anisotropy can lead to errors in the volume measurement.

Alternatively, the specimen can be placed in a known volume of a non-wetting liquid with a known thermal expansion coefficient. Further, the liquid must not be miscible with any phase of the specimen, and perhaps even including the gas phase. A common dilatometry fluid is mercury. As the temperature is raised, the total volume change minus the volume change expected from the mercury will provide the volume expansion of the specimen.

The larger the sample, the more accurate will be the volume change measurements. This makes dilatometry a technique better suited to testing composite pieces rather than for measuring the transitions of nanoparticles. The thermal expansion coefficient can be determined by the change in volume as a function of temperature in the absence of any thermal transition. Thermal transitions often cause step changes in volume, either positive or negative. In general, when a crystal melts, the volume will increase (water being a clear exception).

Dilatometry will allow measurement of the coefficient of thermal expansion in the solid. Volumetric changes due to phase changes and other transitions will also be reflected in the plot of volume vs. temperature, as seen in Figure 10.6.

Differential Scanning Calorimetry (DSC) and Differential Thermal Analysis (DTA)

In DTA, a furnace with very accurate electronic controls is heated at a constant rate. In this chamber, the difference in temperature between an unknown material specimen and a known reference material is measured. There will be a discrepancy between these two temperatures, caused by: (1) differences in heat capacity; (2) melting or crystallization in the material; (3) any other molecular motion that consumes or absorbs heat; and (4) thermal losses due to the machine. Differential scanning calorimetry measures heat flow instead of temperature difference.

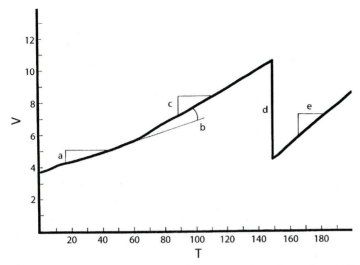

FIGURE 10.6 Dilatometry curve of a material with a negative volume of crystallization, plotting volume vs. temperature. (a) The coefficient of thermal expansion. (b) A second-order transition (such as the Curie temperature). (c) A new, higher coefficient of thermal expansion. (d) Melting, characterized by a discontinuous, non-integrable step change. (e) The density change of the molten material with temperature.

Thermal transitions are very dependent on both heating rate and thermal history. Increasing the heating rate will cause measured transitions to increase in observed temperature. Therefore, a lower heating rate will more accurately reflect the actual temperature of a transition, within the constraints of the equipment. The linear range for the heating rate should be determined for the equipment, and experimental heating rates should be chosen in this range.

A thermal history is introduced into any test material during the processing, post-processing and storage of a specimen. Therefore, the first, second and third scans of a material specimen can be very different. After the first scan, the sample is usually cooled in the calorimeter at a controlled, reproducible rate. This usually creates a repeatable thermal history, so the second and third scans should be identical. Thermal transition effects caused by processing are only detectable, or measurable, in the first scan. The idealized transitions are best determined by a second scan, usually conducted at both a heating and cooling rate that are in the linear range of the machine.

To measure the heat capacity of a material, a region of the thermal heat spectrum in which no thermal events occur is needed, that is, no heat consumption or generation resulting from molecular rearrangement. The first step of the analysis is to consider only steady-state heat changes, with no losses other than the thermal lag between the actual temperature and the measurement of the temperature. The thermal lag is assumed to introduce a constant of proportionality. The derivation based on thermodynamics yields an equation for the heat capacity:

$$C_s = C_r \left(1 - \frac{\Delta T - \Delta T_0}{\Delta T_r - \Delta T_0} \right) \tag{10.6}$$

where C_s is the unknown heat capacity and C_r is the heat capacity of a known reference material. ΔT is the change in heat over the region of interest, while ΔT_r is the change in heat of the reference over the same range, and ΔT_0 is the change in when there is no sample or reference in the test cells. This last is ideally zero, if both sample cells are perfectly balanced.

In order to use this equation, three measurements must be made. First, a measurement of the region of interest with no sample and no reference, ΔT_0, must be made. Second, a sample of known heat capacity is measured as a reference, typically a metal with low heat capacity, similar mass, and no transitions in the region of interest, providing ΔT_r. From this, the calibration constant for the machine can be determined. Finally, the sample is measured with the reference in the reference cell, producing the ΔT curve. Then, the heat capacity can be determined.

Thermal transitions are also very informative, allowing, for example, calculations of the amount of crystallinity present and quantification of the extent of an interaction between two materials. Thermal transitions are non-steady-state events. Explicit solutions require a differential heat flow.

Peaks or step changes characteristic of thermal events can be measured from a thermal scan. Second-order-type transitions such as a glass transition will be observed as a step change in thermal analysis. Melting or other first-order-type thermal events will occur as peaks. The temperature or time location along the horizontal axis and the shape of these events will change with the heating rate. Measurements from these shapes must be made with attention to the machine performance and the theoretical basis for thermal analysis. For example, in a peak there will be an onset, maximum, and trailing value. The onset is the temperature at which the first motions of the thermal event, such as the most imperfect crystals, etc., begin, with only the distortion of the time constant of the machine. The maximum occurs after half of the crystals have melted, while the trailing point is reached after all of the molecular motion is complete. Thus, the entire shape of a melting curve can provide a qualitative evaluation of crystal perfection and thermal processing effects. Properly reporting the temperatures can include the onset temperature, the maximum temperature, and the heating rate, depending on the specific transitions being analyzed.

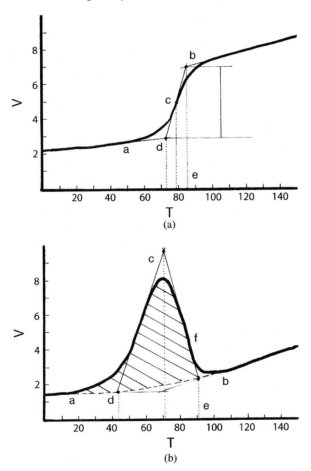

FIGURE 10.7 Interpreting thermal events: (a) A step change is characteristic of, for example, T_g. a. the lower baseline. b. the upper baseline. c. the midpoint temperature $T_{g,mid}$ of the transition. d. the onset temperature, $T_{g,onset}$, where the event begins. e. the trailing temperature, $T_{g,trail}$, where the event is complete. (b) A peak is characteristic of, for example, melting T_m. a. the lower baseline. b. the upper baseline. c. the midpoint temperature $T_{m,mid}$ of the transition. d. the onset temperature, $T_{m,onset}$, where the event begins. e. the trailing temperature, $T_{m,trail}$, where the event is complete. f. The area under the peak is characteristic of the enthalpy of melting.

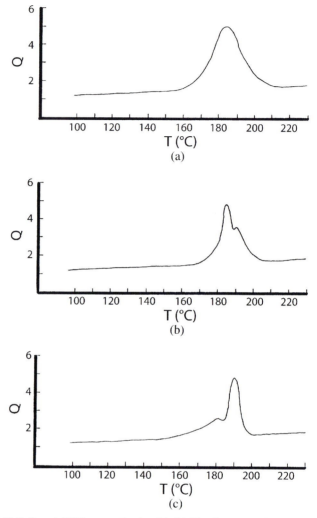

FIGURE 10.8 Sketch DSC scans of nylon 11: (a) The first scan of a commercial material, with a single, broad melting peak. (b) A rapidly cooled specimen has a larger amount of a less-stable crystal phase. (c) An annealed specimen will have less metastable crystal and more of the high-temperature, stable crystal phase.

A representation of the DSC of nylon 11 is shown in Figure 10.8. The thermal spectrum shows a glass transition at around 40 °C and dual melting transitions at around 190 °C, characteristic of the α and γ crystal structures. Processing conditions, including the presence of reinforcements, influence the crystal structure that forms. The amount of crystallinity can be measured by analyzing the melting peak.

Thermal Characterization of Melting

Semicrystalline or crystalline solids will exhibit a peak in DSC near or above the melting temperature. On cooling, a peak will manifest during crystallization near or below the freezing temperature. This is the result of making a measurement in finite time in a kinetically controlled experiment. Since melting and crystallization are first-order transitions, heat flow should yield an infinite discontinuity at the thermodynamic transition temperature. Even in a crystalline atomic solid, the melting event is smeared by a combination of the rate of response of the material and the time constant of the equipment. Thus, the peak is a representation of the melting event. The peak will be shifted by heating rate and machine response times, and can be affected by minor changes in crystal perfection as well. The more perfect a crystal is, the fewer flaws present, and the more difficult the introduction of motion in the lattice will be. Thus, real crystals will melt over a range of temperatures.

The melting event occurs over a range of temperatures depending on crystal perfection, crystal type, and other factors affecting molecular mobility. This includes reinforcement particles, increases in molecular weight, and thermal history. Characterizing any complex molecular solid, especially a polymer, by a single melting transition is usually misleading. The melting temperature should usually be reported by onset temperature (T_{mo}) and maximum temperature (T_m) to provide information not only regarding the location of the peak but also of the shape. The ideal, thermodynamic melting temperature can be difficult to obtain. There are tables of the best measurements in a variety of reference sources.

Calculating the amount of heat liberated during the melting event and comparing this to the value expected from a 100% crystalline material is a useful technique for measuring crystallinity. Several factors make these calculations problematic, however. The same factors affecting the location of the peak can influence the area of the transition event. Obviously, the heat of melting calculated from the peak area is influenced by the heating rate and the machine constants. This can be semi-qualitatively corrected for by analyzing a calibration material with a unique and well-behaved melting transition in the vicinity of the transition of interest in an unknown material or via a set of calibration curves depending on heating rate, sample volume, and temperature. However, the presence of multiple crystal forms and imperfect crystals can make quantitative anal-

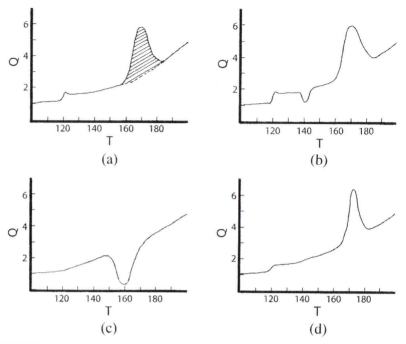

FIGURE 10.9 Crystallinity can be evaluated using DSC. (a) The peak area is measured by comparing the enthalpy of melting, estimated from the area under the peak, to the enthalpy of a perfect crystal. The process history influences the crystallinity measured by DSC. (b) The scan of a sample may show a recrystallization peak (the negative peak just above T_g). The recrystallized material will contribute to the measured enthalpy of melting. (c) On cooling, crystallization occurs at a temperature different from the melting temperature. (d) On a second heating, the melting peak is narrower, characteristic of more perfect crystals.

ysis of the melting peak more problematic, as well. The calculation is more accurate if a calibration sample of the material of interest is available. In the absence of appropriate calibration materials, the crystallinity calculated from the melting peak should be regarded as proportional to the real value of the crystallinity, providing a crystallinity number rather than the literal fraction of crystallinity.

Thermal Characteristics of a Glass Transition

The glass transition is a typical feature of amorphous and semicrystalline solids. The glass transition has the shape of a second-order transition, including a change in slope in first-differential properties such as volume, and a step change in heat flow. The location and magnitude of the glass transition, however, is highly modified by heating rate, and in most circumstances is observed around 50°C above any thermodynamic predictions.

The exact shape of the glass transition can depend on a variety of kinetic conditions. The basic methods for reporting an accurate glass transition include the onset (T_{go}) and the $1/2\Delta C_p$ (T_{gm}) temperatures of an idealized step change. However, several factors can displace the temperature and distort the shape of the transition, including stress, physical aging, molecular weight, cross-linking, blending, filling, crystallinity, and heating rate. The heat capacity of the liquid, above glass transition, is a thermodynamic property, while the heat capacity of the glass is not a strong function of temperature. Therefore, an alternate construction can be made. The heat capacities of the liquid and glass are extended. Then a vertical line is chosen through the step change, having one intersection with the glass C_p line, one with the curve and one with the liquid C_p line. The appropriate choice of this line is the line that makes the areas of the leading and trailing mismatches with the heat capacity lines equal to the mismatch with the step change. The temperature of this vertical line is a reasonably stable measurement of glass transition (T_{gh}), which follows the proper thermodynamic trends for a material. T_{go} and T_{gm} can show significant distortions due to thermal history.

Leading and trailing peaks to the glass transition step change can be introduced by changes in the internal free volume of the sample. In typical molding processes, including injection molding, pressure is applied. This has the effect of reducing the free space for molecular motion. Thus, the molecular motion will be kinetically suppressed until enough energy is absorbed to induce motion. This creates a characteristic shape similar to an infinite spike, wherein once the threshold energy is reached all of the suppressed motion begins to occur. Mechanical stress, such as internal stress, can cause peaks both above and below the glass transition. Typically, the rearrangement of molecules through storage or application under pressure, compression, or even through long-term storage at near the T_g where slow rearrangement can occur, will be accompanied by an observable peak above the T_g step change.

Various molecular features can affect the glass transition. The general concept of the glass transition is as a function of chemical composition. As polar interactions and chain stiffness increase along the chain, for example, the glass transition temperature will generally increase as well. As the flexibility and disorder of a chain increase, caused perhaps by randomly placed side groups, the T_g will decrease. Other molecular features will also have an effect. As molecular weight increases, the decreasing

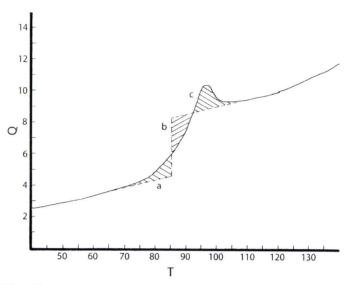

FIGURE 10.10 The glass transition may have features making interpretation difficult, such as a gradual departure from baseline and an overshoot. To interpret a complex transition event, the lower and upper baselines are extended as usual. A vertical line is drawn to divide the event into three areas, a, b, and c. The area of a + c should equal the area of b. This vertical line is T_g.

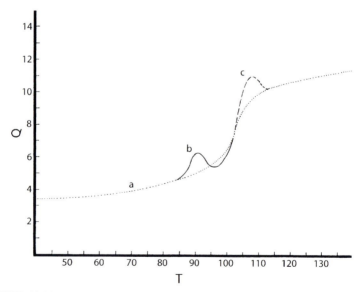

FIGURE 10.11 Process history can influence the T_g. (a) The dotted line is the base event. (b) Leading peaks, shown as a solid line, arise from thermal stresses caused by rapid cooling. (c) Trailing peaks, illustrated using a dashed line, are introduced by pressure and aging.

468

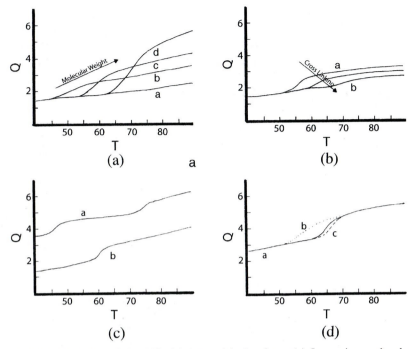

FIGURE 10.12 The T_g is shifted by material chemistry. (a) Increasing molecular weight, a–d, will shift T_g to a higher temperature. A non-polymeric molecule, including oligomers, will not show a glass transition. (b) Increased cross-linking, a–c, shifts the glass transition to a higher temperature and decreases the prominence of the event. a. is typical of a B-stage epoxy resin (pre-polymerized). b. is typical of a completely cross-linked thermoset, showing no T_g. (c) Blends of two polymers will show different behavior if they are miscible or immiscible. a. Immiscible polymers will show two T_g's, one for each polymer. b. Miscible polymers show one T_g, a weighted average of the individual T_g's. (d) Additives can also influence T_g. The basic event, a, is lowered by the addition of a diluent, b. The base T_g is increased slightly by the addition of reinforcement, c.

number of chain ends will decrease the T_g. This effect is strongest at lower molecular weights, and the T_g ceases to change at around 2,000 carbon units. If the chains are cross-linked, mobility will be decreased, requiring a relatively higher temperature to initiate the motions of T_g. The addition of fillers, diluents, or any other material can suppress or promote mobility, thus influencing the T_g.

The magnitude of the step change is proportional to the amount of amorphous material. One advantage, then, is that a 100% crystalline material, which is rare in semicrystalline solids, would show no glass transition. A quenched material, one that is cooled so quickly as to completely suppress the formation of crystals, will give rise to the largest step change possible. A100% crystalline material is seldom available, while a 100% amorphous material can often be crafted through rapid quenching of a polymer melt. Therefore, in some cases, measurements of crystallinity can be made more accurately using the magnitude of the step change in the glass transition temperature.

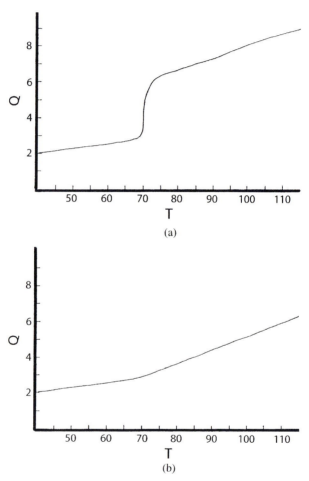

FIGURE 10.13 Crystallinity can be estimated using the T_g. (a) The thermal event shows a step change. The magnitude of the step change depends on the amount of amorphous material present. A quenched polymer, with zero crystallinity, would have the maximum possible step size. (b) A purely crystalline material would have no step change due to the glass transition.

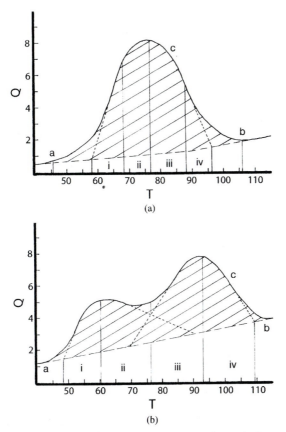

FIGURE 10.14 Dynamic DSC scanning provides a basis for analyzing a curing mono-mer. The curing event begins at a, and is complete at b. The area, c, determines the total heat of reaction, ΔH_R. i. shows the temperature range for the first quarter of the event, ii. for the first half of the event, iii. for three quarters of the event, and iv. completes the event. (a) a single peak for a single reaction. (b) a double peak, characteristic of a multiple reac-tion curing process.

Curing or Polymerization Rates by DSC

Thermal analysis is also useful for measuring the curing conditions and curing rates for polymers, used as nanocomposite matrices or as ad-hesives to join two composite parts. The basic approach is to use dy-namic scanning to determine the initial temperature of curing, the typical terminal temperature of reaction, and the total heat of reaction. The pro-cess is schematically shown in Figure 10.14.

After determining the broad reaction conditions and reaction energy, the kinetics as a function of temperature must be determined. A series of isothermal scans are conducted covering the range of temperatures determined using the dynamic scan.

Partial integrals of the heats of reaction as a function of time are mea-

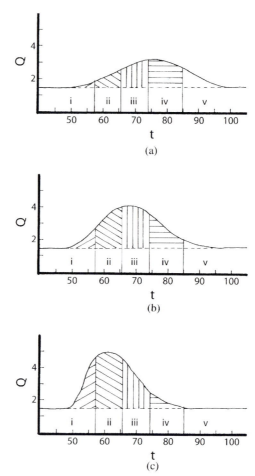

FIGURE 10.15 The kinetics are determined using isothermal scans. Shown are schematics for three temperatures, (a) lowest T to (c) highest T, divided into 5 periods. At the lowest temperature, almost no reaction occurs during period i. At the intermediate temperature, most of the reaction has occurred prior to period v, and by period iv at the highest temperature. The area under the peak up to the end of the period provides the enthalpy of reaction expended up until that elapsed time.

sured, as shown in Figure 10.15. These are used to determine the rate of reaction at each of the temperatures measured:

$$\frac{d\alpha}{dt} = k_R [1 - \alpha]^n \tag{10.7}$$

where α is the reaction position, the amount of the reaction that has occurred, which is related to the amount of polymer that has formed or the length of the polymer chains formed, for free radical or condensation polymerizations, respectively. This expression points out that the rate of polymerization is a function of the amount of unreacted groups remaining. The rate expression might not be a simple zero-, first- or second-order reaction. The rate expressions may have multiple reaction constants. Assuming each reaction generates a standard amount of heat, the reaction position is determined from the heat of reaction:

$$\alpha = \frac{\int_0^t Q(t)dt}{H_R} \tag{10.8}$$

where $Q(t)$ is the heat generated in an isothermal reaction as a function of time, and H_R is the total heat of reaction. The reaction constants, however, have the standard Boltzmann distribution:

$$k_R = Ae^{\left(-\frac{E_a}{RT}\right)} \tag{10.9}$$

where A and E_a are the front factor and activation energy for the reaction. The form of the reaction constant can be determined from the kinetic data gathered at several temperatures.

DMA

Dynamic mechanical analysis is really a series of techniques for applying an oscillating force to a test material and evaluating the mechanical response. Perfectly elastic solids, such as metals and ceramics, will respond to oscillating forces instantaneously, or nearly so. Viscoelastic polymers will therefore usually exhibit some lag between the applied force and the measured response. The nanoparticulate reinforcement will cause dramatic changes to the oscillatory response.

Since DMA applies and measures force, the technique is considered a

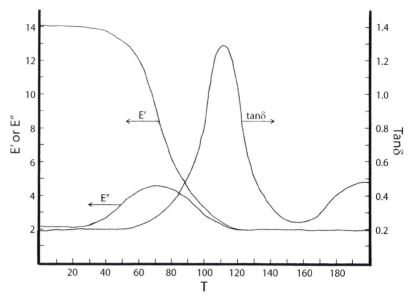

FIGURE 10.16 Sketch of a DMA curve. The T_g peak by E'' and tanδ are usually slightly different.

mechanical testing technique. The modulus can be determined, and only this technique can separate the real and imaginary parts of the modulus, or the storage and loss modulus. The storage modulus is the elastic response, and provides the elastic modulus. The loss modulus reflects the energy lost in viscous response. The phase angle between these two components of the complex modulus is the shift angle. The connection between the DMA response spectrum and the mechanical behavior in macroscopic, real applications is sometimes not obvious.

In a DMA test, an oscillating force with a period of ωt is applied:

$$\sigma = \sigma_0 \cos \omega t \qquad (10.10)$$

The stress is a function of the maximum stress, σ_0, and the frequency, ω, which is $2\pi\nu$. The strain is a response to the stress, determined by the compliance, as noted in the expression below:

$$\frac{\varepsilon}{\sigma_0} = J \frac{1}{E} \left(\frac{1}{1+\omega^2 \tau^2} \right) (\cos \omega t + \omega t \sin \omega t) \qquad (10.11)$$

$$J^* = J'(\omega) \cos(\omega t) + J''(\omega) \sin(\omega t) \qquad (10.12)$$

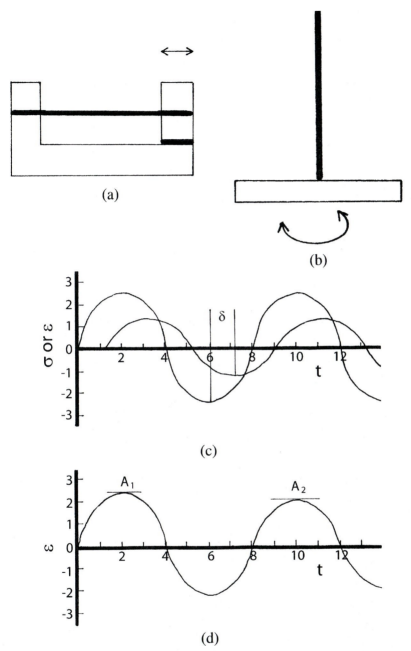

FIGURE 10.17 Sketch of dynamic mechanical analysis testing (DMA): (a) Forced oscillation. (b) Torsion pendulum. (c) The strain response lags the applied stress by a phase angle, δ, in the forced oscillation test. (d) The amplitude of oscillation decreases with every period.

where $J'(\omega)$ is the storage compliance, and $J'(\omega)'$ is the loss compliance. The storage measures the amount of reversible, elastic, in-phase deformation occurs, while the loss measures the amount of energy lost to heat to internal viscosity, 90° out of phase with the stress. This suggests that the compliance is related to the viscoelastic model parameters, E and η:

$$J' \sim \frac{\eta}{E^2} \tag{10.13}$$

$$J'' \sim \frac{1}{E} \tag{10.14}$$

A phase angle, δ, can be defined:

$$\tan\delta = \frac{\sin\delta}{\cos\delta} = \frac{J''(\omega)}{J'(\omega)} \tag{10.15}$$

The total mechanical strain will lag behind the applied stress according to the phase angle, δ. The mechanical response can be resolved into a real and an imaginary component:

$$J^* = J'\cos\omega t + J''\sin\omega t \tag{10.16}$$

This can be rewritten with the aid of the phase angle:

$$J^* = (J'^2 + J''^2)^{1/2}\cos(\omega t - \delta) \tag{10.17}$$

The phase shift response can be measured using a torsion pendulum as shown in Figure 10.15. After the initial oscillation begins, the amplitude will be damped. The decay in amplitude is related to the log decrement:

$$\Lambda = \ln\left(\frac{A_n}{A_{n+1}}\right) = 2\pi\frac{\lambda}{\omega} \tag{10.18}$$

Any relaxation, including the glass transition, the melting transition and any other molecular motion, will increase the amount of energy that can be dissipated by the material. Thus, as the temperature increases, or alternatively as the rate of oscillation increases, there will be a step drop in the storage modulus and peaks in the loss modulus and tan δ at each transition temperature. The three changes do not occur at a unique temperature. Relaxations from DMA are usually reported as the temperature of the maximum in the loss tangent (tan δ). Values determined in this manner are usually higher than those obtained from G'', because the temperature shift for (tan $\delta)_{max}$ is more sensitive to parameters such as cross-link density, filler content, or blend morphology. The changes in moduli or phase angle will be shifted in measured temperature by the frequency of the measurement at a single temperature, the temperature at a single frequency, the rate of heating, and other factors.

Chemical Tests

There used to be an array of chemical tests to identify or quantify materials. These techniques have been nearly entirely replaced by spectroscopic analysis to uncover the primary chemical composition, as described in Chapter 9. Atomic absorption or emission can quantify the composition of the reinforcement phase if it is metal or ceramic. Infrared and nuclear magnetic resonance spectroscopy are effective techniques for identifying an organic component.

Inorganic Phases

An important step in characterizing a composite is isolating and quantifying the reinforcement and any other inorganic components. There are two simple methods for determining these components: ashing and thermogravimetric analysis. Ash content is determined by burning off all organic material, leaving behind a, hopefully, unchanged inorganic component. Thermogravimetric analysis is accomplished by tracking mass as a function of increasing temperature. Ash content is the easier to obtain, but has limitations on precision and can be insensitive to organometallic compounds. TGA uses a specialized piece of equipment, but provides very precise measurements of mass changes as a function of temperature.

Ash Testing

In an ash test, any organic component of a sample of material is burned away. A known mass of material is placed into a crucible. The crucible can be heated using a flame source, such as a Bunsen burner, or in a high-temperature furnace. The crucible is heated until the organic component of the material begins to oxidize. The heating is complete when all of the organic material is driven off, usually determined by reaching a stable weight measurement. Typical test protocols use a furnace temperature of 900 °C. Organic materials are usually completely burned off by temperatures of approximately 400 °C. The difference in mass before

TABLE 10.4 Mass loss of some inorganic powders and pigments.

Powder	Composition	Mass Loss	Effect of Temperature
Titanium dioxide	TiO_2		No significant loss
Kaolin	$2SiO_2 \cdot Al_2O_3 \cdot 2H_2O$ or $Si_2Al_2O_5(OH)_4$	13.9%	OH-bound water loss begins below 500°C, complete dehydroxylation at 700°C
Calcium carbonate	$CaCO_3$	44%	Decomposition to CaO and CO_2 begins 500°C, complete at 900°C
Talc	$Mg_3Si_4O_{10}(OH)_2$	4.8%	OH-bound water loss begins 800–920°C
Muscovite	$KAl_3Si_3O_{10}(OH)_2$	4.9%	OH-bound water in range 500–850°C
Alumino silicate	$Na_2O \cdot Al_2O_3 \cdot 4SiO_2 \cdot 4\text{-}6H_2O$	7–20%	Water loss in range 120–400°C
Aluminum trihydrate	$Al(OH)_3$		Water loss starts at 180°C, stable as Al_2O_3 by 750C

From Thermal Stability of Paper Fillers.

and after the organic phase is burned off provides the fraction of inorganic component. TAPPI Test Method T 413 is one source for ash testing methodology.

There are several advantages to ash tests. The test is simple, fast, and inexpensive. The precision of the test can be improved by using a larger sample size, thereby increasing the amount of inorganic material remaining at the end of the test. The inorganic material remaining may be suitable for further analysis, such as density, hardness, or structural characterization by microscopy or spectroscopy. This further analysis can provide useful new knowledge about an inorganic reinforcement phase.

Two significant drawbacks to ash determination of inorganic materials are the amount of material required and any instability in the inorganic phase. The precision of the test measurement is dependent on the initial mass. Some samples may not be available in sufficient quantity, such as nanofibers and other experimental nanomaterials. The accuracy may be diminished if there is a metallic, or organometallic, component that can be forced to oxidize even at moderate temperatures. This will make analysis of the inorganic material subject to a systematic error. Table 10.4 shows inorganic materials that lose a significant proportion of their weight at various temperature thresholds. If temperature-sensitive inorganic materials are present in the matrix, longer heating at 525 °C is required to fully oxidize the organic phase while leaving the additive unchanged.

TGA

An alternative approach to ashing is thermogravimetric analysis (TGA). This technique uses a microgravimetric balance with resolution suitable to measure small mass changes and a furnace capable of producing a precisely controlled heating rate. The mass loss of a sample is recorded as a function of temperature. The heating may occur in an oxidizing environment, which is standard, or in other environments to probe, for example, anaerobic oxidation. A particular advantage of TGA is quantification of degradation steps, such as the separate loss of waters of hydration, followed by CO_2 loss in $CaCO_3 \cdot nH_2O$. Note that CO_2 loss is anaerobic, that is, requires no oxidation, while the loss of CO_2 from polyethylene ($-CH_2CH_2-$) requires the presence of oxygen. The mass loss as a percentage of total mass may provide information about the chemical changes occurring at each temperature transition. The technique also can generate kinetic information about degradation steps.

Organic Phases

Organic materials are susceptible to several factors that the inorganic components in a nanocomposite are not. Specifically, organic materials are degradable, and may absorb or dissolve in solvents. These properties may be useful in exploring the matrix material with less interference from the reinforcing phase.

Degradation

Flammability

Flammability, or rather combustibility, is characteristic of organic materials, including polymers. In flammability testing, the burning process of the material is observed. The color of the flame, the amount of smoke produced, and the tendency to continue burning in the absence of an ignition source, are indicators of the class of matrix material used in a composite. Polymers that degrade to monomer, such as nylon and PMMA, do not generate much smoke during burning, converting entirely to CO_2 and H_2O. Polyolefins tend to continue to burn, even after the source of the flame is removed. The results of burning require careful consideration for composites and industrial formulations, in which inorganic inclusions may interfere with the burning process, or may catalyze more rapid burning. Nanoparticles in a composite primarily act as inhibitors to flammability.

The flammability may be quantified for a variety of characteristics. Tests include sensitivity to ignition, vertical burning distance, horizontal burning distance, location and description of the pyrolysis process, extinction characteristics, and gas/smoke generation. For many materials, application-specific flammability standards are set by industries. ASTM has over 100 specific flammability standards for various substances and applications, ranging from clothing to architectural materials. For industrial applications, the conditions required to start burning, the extinction character of the material, and the type and quantity of smoke and fume produced are of the greatest import. In the standard tests, the results are usually recorded as ratings, such as Class A flammability, that correspond to the broad behavior expected in applications. For laboratory purposes, features such as flame temperature, luminosity and heat liberated may be of more interest and more characteristic of a particular formulation.

Ignition can vary with external conditions. A point source of ignition may consist of a steady source, such as a gas jet, or a temporary source, such as a burning methyleneamine pellet. Auto-ignition may occur without any source of ignition if there is sufficient heat. Ignition can only occur when a sufficient gaseous polymer fuel/air mix can form. The minimum heat for ignition is the critical heat flux. Some specific ignition information includes the contact time required to initiate burning, the intensity of heat required to generate ignition, or the ignition time as a function of the ambient temperature. For quality control or flammability rating purposes, a "pass" or a class rating may be set. This occurs, for example, when the material will resist ignition for a minimum amount of time. For discriminatory examination in lab work, the actual ignition conditions may prove more informative.

An example of an extinction method is to introduce a point source and measure the distance that the flame travels from the ignition source. This may be used as a quantitative test, with the results reported as a distance traveled, or as a quality-control test, with a "pass" issued if the flame does not spread more than a certain distance before extinguishing. Other parameters for evaluating extinguishing properties include: the time elapsed between removing the ignition flame and the extinction of the flame; the rate of flame progress from the ignition source; and the distance traveled if the material is held in a vertical or in a horizontal position.

Smoke and fumes cause the primary danger from burning materials. Smoke production is often measured in an enclosed chamber or in a collection flume. The density of the smoke is obtained by the reduction in light passing through the chamber between a light source and a photometer. The smoke generated is different for smoldering samples, after ignition and extinction of the flame, than if a specimen is burning.

There are certain flammability characteristics that have been tabulated for organic materials for comparative purposes. Certain property variables can be related to the phases of burning: pyrolysis, ignition, flame propagation extinguishing. These are materials properties rather than ratings assessed against the potential use of the material in specific applications.

The pyrolysis rate is measured in mass of material consumed, per unit time, during burning. Pyrolysis is a function of several parameters: the heat of gasification (ΔH_{vap}) for the polymer at ambient temperature and the amount of heat radiated back to the environment (q_{rad}) by a hot polymer, usually a function of temperature. These are properties of a material, which can be evaluated using a calorimeter. The input heat flux (q_{in}) is the measured input. The measured outputs include the yield of products produced during burning, the rate of combustion product formation for non-flaming combustion, and the mass loss rate. Combustion products include CO_2, CO, hydrocarbons, and smoke.

The ignition properties include a critical heat flux and a thermal response parameter. Below the critical heat flux, no ignition can occur. There can be different critical heat fluxes for auto-ignition and pilot flame ignition. The critical heat flux is generally near to the heat of surface radiation loss for the polymer. The thermal response parameter is a cluster of variables:

$$K = \rho C_p t \Delta T_{ignition} \qquad (10.19)$$

ρ is the density, C_p is the heat capacity, t is the thickness, and $\Delta T_{ignition}$ is the temperature elevation for ignition. This relation holds for thin polymers, and requires modification for thick polymers. The criterion for thick is material-dependent. For thick polymers, the thickness is replaced by the thermal conductivity of the material.

Tabulated propagation parameters include the limiting oxygen index and the flame propagation index. The limiting oxygen index is detailed in ASTM D-2863. LOI is measured on a vertical sample, using downward flame propagation, with air flowing upward. The concentration of oxygen in the air stream required to sustain the flame is the LOI. The lateral ignition and flame spread test is chronicled in ASTM E 1321-90.

The vertical flame propagation at an external heat flux 5 kW/m² higher than the critical heat flux is recorded as the rate of advancement in the pyrolysis front. The pyrolysis front can also be categorized as propagating, slowly propagating, decelerating, and non-propagating.

Ultraviolet Radiation

Some polymeric materials will absorb ultraviolet light. Higher energy *uv* light will be absorbed by the primary bonds, potentially resulting in broken bonds. Materials with extended resonance structures can absorb light energy ranging down into the visible portion of the spectrum. If light energy is absorbed, the energy may be transferred into a photochemical reaction. However, many systems can eliminate the energy in other ways, including quenching, fluorescence, or phosphorescence. In quenching, energy is transferred from excitatory vibrations into rotations, translations, and other heat. In fluorescence and phosphorescence, the energy is reemitted as a photon of a characteristic *hv*, and thus does not cause damage. These alternative energy sinks will block chemical reactions. Reinforcing particles can modify the absorption characteristics of *uv* energy.

Some characteristic *uv* properties of organic materials include the cutoff frequency, location of an absorbance maximum, relative stability in solar radiation, and the type of damage caused. The cutoff frequency is the frequency of light that will be absorbed by a given, usually small, thickness of the material. Table 10.5 shows the wavelengths for some common polymers. Note that the resonance structures in materials such as PET and polycarbonate increase the energy of light that may be absorbed by the matrix material. However, materials with resonance structure can dissipate the energy easily, and so tend to be more stable to *uv*

TABLE 10.5 Spectral range, stability and property damage under uv light.

Polymer	Relative sensitivity	Sensitivity range (nm)	Property showing damage
polypropylene	extremely sensitive	315–330	extensibility
polyvinyl chloride	very sensitive	310–325	yellowing, tensile
polyamides	sensitive		chain scission, flexure
polycarbonate	very sensitive	280	yellowing
polymethyl methacrylate	insensitive	300	chain scission

From Polymer Handbook.

exposure. Damage to materials by *uv* light ranges from yellowing to chain scission. Extreme damage may affect the mechanical properties. Composites and nanocomposites affect *uv* degradation in at least two ways. First, if the reinforcement absorbs light, less can penetrate the material to cause damage. Contrariwise, if the particles reflect light, then energy will pass through the same region of material twice, increasing the opportunities for degradation. Second, the properties of the composite depend on load transfer between polymer and reinforcement. Deterioration of the polymer may have a disproportionate effect on the load transfer. As with many properties, the expected behavior for a polymer matrix or ceramic reinforcement alone must be compared with caution when considering a composite material, especially a nanocomposite, where the large surface area of the particles can increase the total number of polymer subunit/reinforcement atomic interactions.

Solvents

The organic phase of a nanocomposite shows sensitivity to a solvent environment in different ways. Some solvent-related properties of the matrix can provide valuable information for identification or characterization purposes.

Solubility

The response to a liquid environment is also highly characteristic of a composite. The organic component can have one of three specific responses to a solvent: dissolution, swelling, or insolubility. The inorganic component can be soluble, slightly soluble or insoluble. Of course, thermoset and network polymers are insoluble, and even good solvents will only swell the network. Many thermoplastics are soluble only in a limited range of solvents. The basic approach is to identify the solvent class that may allow solubility, such as polar solvents, e.g., water, alcohol, ketone, ether, or nonpolar solvents, such as hydrocarbons, aromatic, or chlorinated solvents. Some polymers, such as the nylon family, are insoluble in most common solvents. As an alternative to immersion, the surface of the polymer may be rubbed with the solvent. The result can be dissolution, scratching, swelling, or other noticeable changes characteristic of specific families.

Swelling is a specific, measurable response to solvent exposure. In a good solvent, a polymer will certainly dissolve, unless cross-linking prevents this. Cross-linked polymers may fracture in good solvents if the swelling pressure exceeds the strength of the bonds. Polymers will not dissolve appreciably in poor solvents. In moderate or Θ solvents, the polymer will swell and dissolve, albeit very or immeasurably slowly. Measuring the swelling parameter, α, the increase in dimensions with the addition of a solvent, provides information of value to understanding the thermodynamic properties of the material.

Solubility may result in partial mass loss, even in fully cross-linked resins. Short chains, unreacted monomer, processing aids, or soluble additives trapped by the polymer will experience a large increase in diffusion mobility out of the solid, commensurate with the swelling parameter.

Residual monomer or processing aids are solvents, also. During the service life of a material, these may outgas. This can sometimes be sufficient to create a detectable odor. Odor is therefore usually significant for inadequately polymerized or extremely modified matrixes. Quantification techniques available for this type of condition include mass loss and chromatography. Mass loss is measured before and after gentle heating. The off gasses may be collected and passed through a gas chromatograph, to identify the products. This may provide insight into, for example, material failure.

Solids and Non-Volatiles

In many circumstances, analyzing the liquid resin used to create the nanocomposite may be instructive. A liquid used to prepare, for example a nanocomposite coating, is often a suspension of dissolved or emulsified polymer mixed with the reinforcing particles. The particles should be removed for separate characterization. The remaining solution of polymer and solvent can be characterized for percent solids, the weight of polymer per total weight of the solution, by evaporating off the solvent thoroughly. As the solvent evaporates, the solution will coagulate and drying efficiency will be observed to slow, as previously described in Chapter 6.

The nanoparticles will not tend to settle out of solution. Observing the long-term stability of the emulsion, for example over the space of a day, can provide quick feedback on the particles. Filtering the nanoparticles is difficult, as described earlier. Centrifugation may allow more rapid separation of the particles from the polymer, and the polymer from the solution, than is practical or possible with the drying technique. The mixture may well separate into individual component layers under centrifugation.

Blocking

Blocking is the tendency of an object to transfer somewhat to the surface of the composite. This effect is caused by a slight flow in the object's surface when another object is placed upon it, even at low temperatures and forces. Blocking is common on softer and stickier materials. Therefore, hardness and adhesion properties will govern whether a material tends to block. Composite reinforcements will often limit blocking, and can be added, particularly at the surface of a material as "anti-blocking" agents. Nanocomposites will generally experience less blocking than bulk polymers. The extent to which an object is affected by blocking is generally a subjective assessment.

MECHANICAL

Tensile Testing

Tensile tests are useful for assessing changes in bulk mechanical properties due to added reinforcement. While the bulk properties often are unchanged by microscopic reinforcements, nanocomposites often show increased modulus, particularly above the glass transition temperature. Most part designs require that performance remain within the proportional limit, where the elastic response is linear and governed by the modulus. Tensile testing provides the necessary information about the modulus and both intrinsically and extrinsically defined yield stress and strain for designing within these limits. A standard test method for tensile measurement in general materials is detailed in ASTM D 638. There are also special test methods for specific materials or configurations, such as D882 for thin plastic sheets, D 1708 for microtensile samples, D 2290 for tubular plastics, D 2343 for glass fibers, and D1623 for cellular plastics, as a limited sample. In addition, a tensile test can provide information on plastically deforming materials above the yield conditions, including the ultimate stress and strain, the toughness, and the characteristic deformation profile, such as brittle, ductile, or work hardening.

Designing and Mounting a Specimen

The standard shape for tensile testing is the dumbbell or dogbone specimen (Figure 10.18). The part may be machined, cast or cut into a cylin-

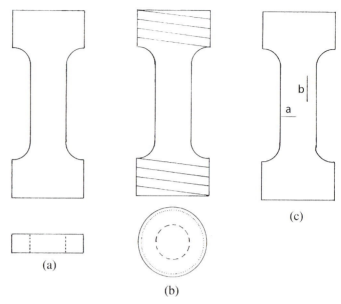

FIGURE 10.18 Examples of dogbone samples for mechanical tests: (a) a flat plate, and (b) a threaded cylinder. (c) Surface cracks may have different effects on testing. a. Cracks perpendicular to the load can increase the local stress. b. Cracks aligned parallel to the load have less impact on the test.

der or into a flat specimen. While in principle a simple block of material can be tested directly, the distribution of force within the sample becomes complicated at the grip points. Saint Venant's Principle (Figure 10.19) states that sufficiently far from the grip, the force distribution field is uniform. In order to insure that the force field is not only uniform, but also that any material response occurs only within a focused region of interest, a narrow gauge length is formed. The force concentration in the gauge length allows precise calculations of dimension changes and property calculations. Table 10.6 shows the stress concentration effect at the narrowing point for various fillets. Some important features of the gauge length are that the width in the gauge must be uniform and the radius of curvature during the transition from the mounting area to the gauge must be precisely cut.

Mounting the specimen can also be a challenge. In order to record accurate data, the grips must be in perfect alignment. Even a slight misorientation can create a normal force, resulting in unexpected, and usually unmeasured, flexure, shearing or bending of the specimen. Modern grip systems are self-aligning. The jaws are mounted on bearings that

TABLE 10.6 The stress concentration in fillets and around holes.

$$C = \frac{\sigma_{max}}{\sigma_{ave}}$$

r/w	C	d/D	r/d	C
0.1	2.7	1.1	0.05	1.85
0.2	2.5		0.1	1.6
0.3	2.35		0.15	1.5
0.4	2.25		0.2	1.4
0.5	2.15	1.3	0.05	2.5
0.6	2.1		0.1	2.05
0.7	2.05		0.15	1.85
			0.2	1.7
		1.5	0.05	2.7
			0.1	2.25
			0.15	1.95
			0.2	1.85

r = radius of hole,	r = radius of fillet,
w = width of sample minus hole	d = characteristic width of gauge
	D = characteristic width of grip region

From Mechanics of Materials.

will align under low load, allowing the grips to move in order to maximize the force. As the load increases, friction increases and locks the grips in place.

If the sample slips, all of the test data for that specimen becomes suspect. The grips are usually serrated in order to prevent the sample from slipping under load. Even during loading, however, an incompressible material may experience dimension changes under the grips. The grips, therefore, must be periodically tightened. A hydraulic grip system, where the grip force is kept constant, can automatically adjust for dimension changes. The specimen may provide separate mounting challenges. A weak material may crush under the normal load applied during grip tightening. Many investigators find that gluing aluminum foil on the gripping ends of the specimen provides sufficient surface mechanical stability to prevent crushing in the grips. Unusual sample geometry, such as that of films or fibers, may require special grip designs (Figure 10.21).

Measuring Load and Displacement

A tensile test, and most mechanical testing, is usually conducted on an

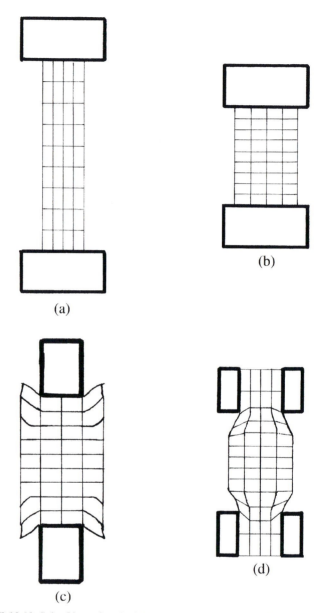

(a)

(b)

(c)

(d)

FIGURE 10.19 Saint Venant's principle expresses the effect of the grips on the stress and strain distribution in the specimen. Illustrated is compression testing: (a) The initial piece can be divided into blocks. (b) When force is applied along the full face of the specimen the blocks are evenly deformed. (c and d) If the force is applied by, for example, anvils or clamps, the regions directly in contact with the grips are compressed more than neighboring zones. Thus, there is uneven deformation near the grips. Far from the grips, the deformation evens out.

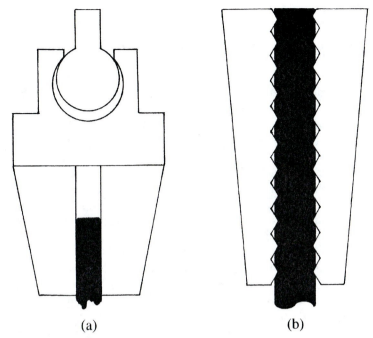

(a)   (b)

FIGURE 10.20 Tensile grips: (a) A self-aligning grip. Under applied load, the grip will adjust to align the force along the axis of the specimen. (b) Flat grips are often cut with ridges to insure that the specimen does not slip under load.

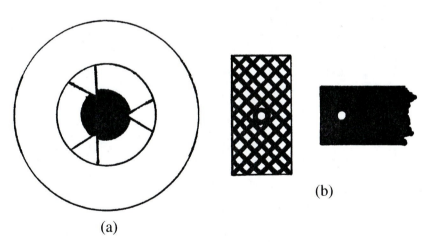

(a)   (b)

FIGURE 10.21 Specialized mounts: (a) Fibers or small rods can be gripped using a three-point, pin vise grip. (b) Thin films are particularly susceptible to slipping in the grips. A hole in the end of the sample, mounted over a corresponding pin in the grip, can keep the specimen in place.

automated load frame. The basic load frame consists of a strong, rigid frame. The frame must be of sufficient size to allow loads that will deform a specimen of interest without deforming in response. Therefore, load frames are specified by the maximum load capacity. One of the load frame bars, top or bottom, is usually adjustable but immobile, and is used to mount a grip connected to some device for measuring load. The other bar is adjustable, with a drive to control the location. This bar is moved to apply a deformation to the specimen while monitoring the load. Combined with a recording device, today almost ubiquitously a computer, this simple device can be used in many different configurations to test tension, compression, shear, flexure, bending, fracture, Poisson's ratio, creep response, stress relaxation and cyclic loading given appropriate specimen design and appropriate instrumentation.

The load cell is possibly the heart of mechanical testing. Three design configurations are typical. The most common is the shear load cell. In this load cell, the displacement of a shearing beam inside the load cell is measured. These cells are usually good for high load capacity. The cell

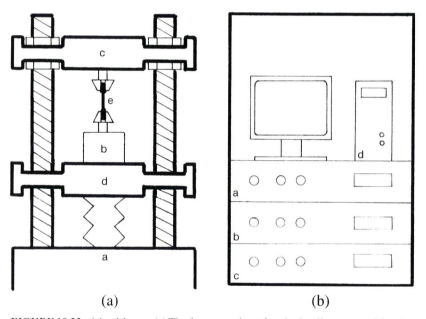

(a)   (b)

FIGURE 10.22 A load frame: (a) The frame consists of a. a hydraulic or screw drive, b. a load cell, c and d. torsion plates, e. the grips. (b) The detector system includes three signal receivers for, a. load, from the load cell, b. displacement, from the rotation or displacement of the lower grip, and c. strain, from an extensometer. These signals are output to d. a recording system, in this case a computer.

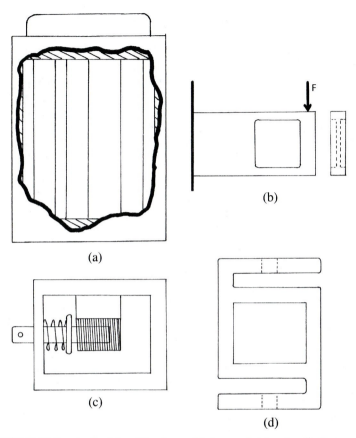

FIGURE 10.23 Load cells: (a) multi-column. (b) shear web. (c) spring displacement. (d) piezoelectric.

has low sensitivity to location of the loading force, and is stable to side loads. Designs using shear load for small loads are less common because the shear web required is small and easily damaged. The load cell capacity reports the measuring range, the safe load limit, the ultimate overload limit, and the maximum safe side load limit. The measuring range is the linear response range, the safe load limit is the maximum force allowed without causing permanent damage to the load cell, thus requiring recalibration, the ultimate overload is the force that will destroy the load cell, and the safe side load is the load that can be applied at 90° to the axis of displacement.

Load cells for low-stress applications sometimes use displacement against the resistance of a spring. Displacement moves a conductive element into an inductive coil, changing the impedance of the coil. These load cells are linear in the elastic range of the spring. Through design, including hard stops in the load cell, these can be made very stable to overloading.

The newest load cells use piezoelectric crystals mounted in the center portion of an S-shaped bending load cell. These can be designed so that small loads can cause large deformations, thereby providing a wider load range for devices of this sort. The current generated by deforming the piezoelectric crystal provides a linear-measured response over a large range.

The load cell is usually calibrated using one or more static calibration weights, applied at the grip. The cell output should match the applied load.

Displacement may be provided by a screw drive. The grips are pulled apart or forced together at a rate determined by the rotational velocity and pitch of the screw. Displacement may also be driven pneumatically or hydraulically. Load control is easier using a hydraulic displacement system rather than a screw-driven system. Creep and cyclic load testing can only be conducted using load-controlled displacement. Screw systems normally cannot respond quickly enough to force measurements to give reproducible results. The displacement is checked using a calibrated measuring device.

An accurate measure of the specimen's deformation is provided by an extensometer, some examples of which are illustrated in Figure 10.24. The knife blade extensometer measures the separation of the blades using an electronic circuit, providing an expensive but reusable system for very accurate measurement of displacement. The maximum measurable displacement is typically small, and therefore not generally used to test ductile composites. The three-lobed circuit extensometer uses a printed circuit that deforms with the part to change the conductivity of the circuit. These circuits are cheaper, but can be less accurate. The multiple orientations of the three circuits provide two-dimensional data on extension, allowing simultaneous measurement of length and width deformation. Laser displacement measurements also are becoming more accessible with the dropping prices of lasers. For materials that plastically deform to a large degree, such as composite elastomers and composites with ductile matrixes, the separation rate set by the screw, the crosshead speed, may be the only measure of the applied displacement. By conservation of mass, the displacement of the crosshead will determine the elongation of the sample. Assuming an affine deformation in the gauge length allows direct correlation between crosshead separation and change in gauge length.

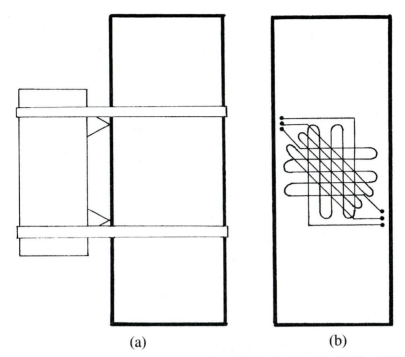

(a)   (b)

FIGURE 10.24 Extensometers: (a) knife blade. (b) 3 circuits oriented at 0°, 45°, and 90°.

The Poisson effect is often ignored in standard tensile tests. However, the Poisson ratio can be measured during standard tensile testing. Measurement requires simultaneously recording deformation in at least two directions. The three-lobed circuit is particularly effective for providing information on compressibility. Measuring the deformation simultaneously along three axes can contribute to a more accurate determination, and is critical in composites that have different compressibilities in the x, y, and z dimensions. Rubber and many polymers react as incompressible liquids, showing a ν of 1/2. Metals, ceramics and semicrystalline polymers often show a different ratio, less than 1/2. Composites may exhibit a complex behavior if some recoverable displacement in the reinforcing particles allows partial compliance in the y-z deformation planes.

Even in nanocomposites, polymeric matrixes are likely to exhibit time-dependent, viscoelastic mechanical responses. Creep deformation occurs under a constant load, such as the response of pipes containing flowing water, the wheel of a parked car, or a beam under constant load.

In order to conduct a creep test, a constant load is required. If there is little change in the cross-sectional area, then the load can be applied using a fixed weight. If sufficient control systems are available, the crossheads can be controlled to provide an accurate load. Displacement as a function of time at the constant load provides a picture of the creep response, including the relaxation time and the response character.

Another viscoelastic response of some importance is stress relaxation, the decay of a stress induced by a deformation that is not removed. Stress relaxation is seen after a forming process, for example when a pipe is bent. Unlike a creep test, a stress relaxation test can be applied using a standard test frame and tensile configuration. A fixed displacement is applied as quickly as reasonable given the test specimen and the load frame. The time response of the load is then monitored. From this the relaxation time and the overall response character can be determined.

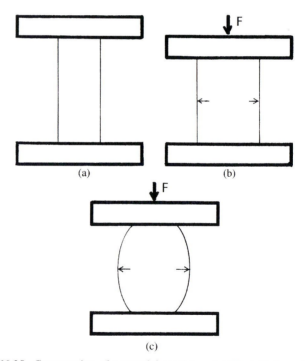

FIGURE 10.25 Compression of a material specimen: (a) The initial specimen under compressive force F, (b) will expand in the x and z dimensions in response to the y compression. If the ends of the specimen are constrained, preventing free expansion, barreling will occur.

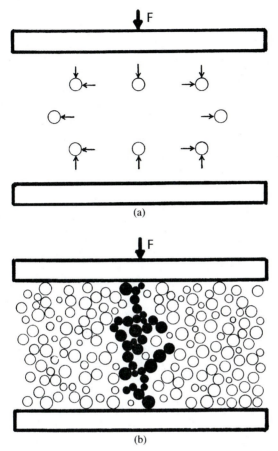

FIGURE 10.26 Compression behavior of a composite material. (a) Under applied load, composites with dilute loading follow the local distortion of the test piece. Along the vertical centerline, the particles move closer together. Along the horizontal centerline, the particles move further apart. In the four quadrants, the particles experience combined displacement. (b) In a densely loaded composite follow a similar distortion pattern. However, direct interactions between the particles form a direct line of communication for the force across the test piece.

Compression

The compressive response of a material is typically measured as the mirror image of a tensile test. Instead of two grips, however, the force may be applied between two anvils, two flat, parallel surfaces. As in tensile testing, the surfaces must be perfectly aligned to avoid creating a normal component that can add shear or bending deformation components. Compression measurements are further complicated by the corresponding expansion implied by Poisson's ratio. The faces of the material must be able to freely expand. If a face sticks, the measured properties will be erroneous, because isostatic compressibility will contribute to the measured load.

The importance of the compressive performance varies with material type and application. Fibrous composites may exhibit less compression resistance, while concrete and cement composites exhibit little tensile strength. For foam cushions, the compressive properties are dominant, while the compressive behavior of foam used as the internal layer of a composite laminate may be less so.

Composite behavior is often very different in compression compared to tension. In tension, the load is carried by the continuous phase, moderated by load transfer to the reinforcement phase. The capability of the interface to carry the load transfer across the interface becomes a dominant property. In compression, the initial load may be completely supported by the matrix, if the volume fraction of the reinforcement is below that required for critical path length formation. As compression loads the continuous phase, the load transfer to the reinforcement can become more effective due to plastic deformation. Once a critical path forms, the ceramic can carry part of the force directly. Final failure begins to occur when the continuous phase cannot restrict the normal forces that are generated by resolution of the force vector from the top surface to the bottom surface.

Nanocomposite behavior can be even more complex. The load transfer across the boundary is governed by curvature. At low loading typical of flake reinforcements, the particles may be widely separated, or perhaps aggregated into small clumps. While common wisdom would assert that aggregated reinforcements will not perform as well as if they were well distributed, the net effect may be stronger than expected.

Shear, Equibiaxial Shear and Pure Shear

The shear modulus is related to the elastic modulus through Poisson's ratio. The shear behavior of a nanocomposite may again show a very different character from that of a bulk atomic solid. The challenge in evaluating the shear properties of materials is applying a pure shear force. Combined modes of load will create a system that is challenging to interpret. Figure 10.27 shows three sample designs for accomplishing shear tests. The most common test for bulk materials is the Iopescu-notched shear sample. The force is applied over the small area defined by the 90°

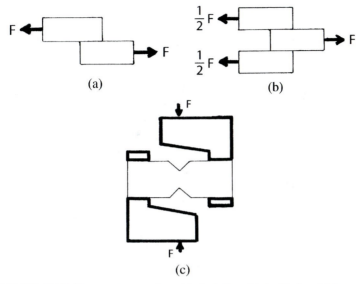

FIGURE 10.27 Shear test approaches: (a) single lap. (b) double lap. (c) Iopescu.

notches. Lap shear, especially double lap shear, of pins can be another reliable way to test. With appropriate specimen design, shear stress and strain data can be collected and subject to the typical theoretical analysis for comparison with expected behavior.

For thin sheets of extensible materials, such as reinforced rubber, an alternative to bulk shear testing is available. By clamping a test specimen using an oval fixture, as shown in Figure 10.28, force can be applied using pneumatic or hydraulic pressure, like blowing up a balloon. Analysis shows that in the center part of the specimen, far from the curved edges, the material will be under pure shear deformation. The displacement can be measured as a function of inflation pressure using glue-on extensometer circuits or a laser extensometer. In the earliest experiments, the separation of fiducial lines was measured. Physical measurement of surface deformation still has value, particularly in specimens expected to deform extensively or non-affinely. Tests in equibiaxial extension, with simultaneous equal displacement in a film or fabric in the x and y directions, may be applied in the same way, using a circular clamping fixture instead of an oval fixture.

Flexure and Torsion

The performance of structures is often defined by their flexure and torsion. Flexure is the result of applying a bending moment along a structural member. For a material under flexure, the inner, concave, surface is under compression while the outer, convex, surface is in tension. Somewhere within the specimen lies a neutral axis, experiencing no deformation. How to obtain the flexural modulus, yield stress, and yield strength is described in ASTM D 790. The test can be conducted in 3-point or 4-point mode. The 3-point bending test is usually sufficient for brittle, rigid materials, but the 4-point test is often considered to provide pure bending data, even for materials with significant compliance during the test. The relationship between the tensile modulus, shear modulus, and the flexural modulus is a function of specimen geometry.

Torsion is the result of counter-twisting the ends of a test specimen. The material experiences no torsion in the center of the part, but the torsion increases as a function of the distance from the center. Thus, surface points at the ends of thicker and longer cylinders will experience more torsional displacement than those at the ends of thinner and shorter cylinders. Again, a "torsional modulus" will be a function of the tensile modulus, shear modulus, and specimen geometry.

Flexure and torsion put a material in a complex distribution of the load,

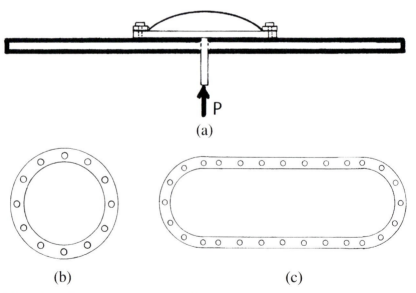

FIGURE 10.28   (a) Inflation can be used to apply shear to thin films. Pressure applies the force to a thin film clamped around the edges. (b) A circular clamp creates an equibiaxial stress state. (c) A cylindrical clamp with rounded ends results in a pure shear state in the center of the part. The stress state closer to the ends is not pure shear.

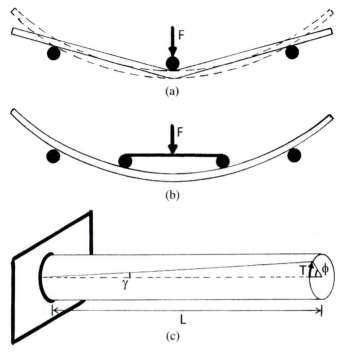

FIGURE 10.29 Mechanical member tests: Bending, using (a) 3-point, and (b) 4-point fixtures. The three-point bend fixture requires that the specimen be absolutely rigid. The piece may buckle at the center point instead of bending smoothly. This is less likely with the 4-point fixture. (c) In torsion testing, a rotational moment is applied to the ends of a test member. The specimen will distort along the length, resulting in a position-dependent stress and strain field.

so the measured properties reflect the test piece geometry. Since nanocomposites are by nature composed of distinct phases distributed along the part, it is likely that differences between the properties of the nanocomposite and the bulk matrix properties will be reflected in the performance at even lower loads than observed for parts in pure shear or tension.

Fracture

Reinforcement particles can interfere with crack growth, thereby modifying fracture behavior. Fracture testing has been conducted in various ways, both scientific and semi-quantitative. The fracture mechanics described in Chapter 3 typically use a compact tensile specimen on a load frame set up for tensile testing. Edge tear and other test methods are possible, giving different interpretations of values for K_{IC} and G_{IC}. A fracture test must break in plane strain. That is, as shown in Figure 10.30, if the specimen is not thick enough to prevent lateral distortions, the fracture test results cannot be interpreted easily.

A particularly interesting facet of the compact tensile sample test may come out of analysis of the fracture surface created. Many ductile polymers display a brittle fracture zone, succeeded by ductile fracture along the remainder of the fracture surface. The length of the shiny, brittle fracture zone correlates with the inherent flaw size observed for polymers and composites.

Of similar interest are impact tests, which use a semi-quantitative approach. The two most common impact tests are described in ASTM D256. Method A describes the Izod test, Method B describes the Charpy test, illustrated in Figure 10.31. In traditional applications, the force is applied using a swinging pendulum. The energy absorbed by the impact is recorded as the difference between the energy stored in the pendulum, and the energy remaining in the pendulum after impact. Electronic sensors on the pendulum to measure the force of impact can provide an impact curve, with more information than just the amount of energy

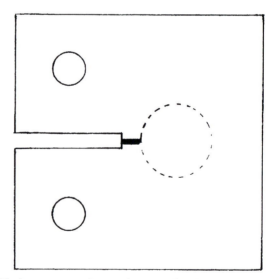

FIGURE 10.30 A compact tension specimen provides fracture data, including K_{IC} and G_{IC}. However, the test piece must be thick enough to insure plane strain conditions. The required thickness depends on modulus. If the specimen dimples, manifesting as a circular distortion field at the tip of the crack, the conditions have not been met.

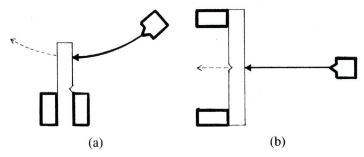

FIGURE 10.31 Impact test configurations: (a) Izod. (b) Charpy.

absorbed by fracture. Both tests provide qualitative information about the fracture energy. These test methods have both flexure and shear contributions to the fracture mode, yielding a difficult analysis if results beyond the Izod or Charpy impact energy are required.

Falling ball fracture provides an even more qualitative result. Balls of increasing mass are dropped from a controlled elevation, and the mass resulting in fracture is recorded.

Drape and Flexibility

Flexible and semi-flexible sheets of material will exhibit flexural behavior that can be described by their drape. The character of drape is particularly important in film, coating, and textile applications. Qualitative tests have been developed in these industries to provide some objective assessment. For example, a sheet of the film or fabric will be slid across a 90° edge. The length of the material that can be extended over the edge before folding is reported as the drape. In another method, the film may be forced onto a die surface, prepared with a marking material such as paint. The area of the film making contact with the underlying dye can assess the flexibility. A slight modification to this approach is recording the force required to bend a film across a die with a known angle.

Composite materials are often less flexible than pure resins. Therefore, they will usually exhibit less drape and flexibility, especially if the reinforcement has directional character, such as fibers or plates. However, the addition of nanoparticulate pigments, such as carbon black, will drive up bulk stiffness but may have less impact on film flexibility.

Hardness

Hardness is another usual property to test. In hardness testing, an in-

TABLE 10.7 Hardness tests.

Test	Indenter	Load	Formula	C
Brinell	10 mm steel (or tungsten carbide) ball	F kg	$HB = \dfrac{2F}{\pi D(D - \sqrt{D^2 - d^2})}$	D = Ball diameter d = indentation diameter (in mm)
Vickers	Diamond pyramid	F kg	$HV = 1.854 \dfrac{F}{l^2}$	l = length of the side of an indentation (in mm)
Knoop	Diamond pyramid (with aspect ratio)	F kg	$HK = 14.2 \dfrac{F}{a^2}$	a = length of the long axis of an indentation (in mm)
Rockwell A	Diamond cone	60 kg		
Rockwell B	1/16 inch steel ball	100 kg		
Rockwell C	Diamond cone	150 kg		
Rockwell E	1/8 inch steel ball	100 kg		

denting tip is applied with a controlled load or displacement. The shape of the tip determines the nature of the hardness test. Table 10.7 shows the most common tip shapes. The round ball hardness tests, since they cover a large area of the specimen, are the most likely to give consistent results for nanocomposites. The area selected inherently provides an average measurement.

Hardness is often highly correlated with wear and scratch resistance, but rubber can be more wear resistant than many harder polymers and ceramics. Hardness is a conflation of the resistance to plastic deformation and the elastic modulus. The elastic modulus can be calculated from hardness data for perfectly elastic, brittle ceramics. Ductile or compliant materials cannot be interpreted this way. Interpreting the results for composite materials is particularly complex. The mechanical properties used for continuum mechanical calculations break down at scales on the order of features in the microstructure, including grains and reinforcements. At any of these features, the force field is distorted, and Saint-Venant's principle will not hold. On a macroscopic scale, an average of properties is observed. The nature of the average is complex, as described earlier.

SURFACE MECHANICAL

The surface properties of materials are often different from their bulk behavior. Boundary conditions at the surface of a material are different. The surface interacts with the environment in ways that the bulk of the material does not. For example, where a bulk material may experience compression, governed by the elastic modulus, the surface will undergo an indentation, governed by the hardness. Where a bulk specimen is under shear load, governed by shear modulus, the surface will be experiencing load transfer by adhesion, as well as surface damage caused by wear or scratching. Surface effects and surface properties govern coatings, an important field of application for nanocomposites.

Adhesion

Nanomaterials have an impact on adhesives and adhesion. The decrease in permeability of the base polymer may make an otherwise water-sensitive adhesive useful for marine or groundwater contact bonding problems. Increased modulus may provide better structural adhesion at the expense of tackiness. Adhesion to a nanocomposite is influenced by the surface energy of interaction. If the nanoreinforcement is well distributed, the surface properties are likely to be considerably different than if the reinforcement is larger, creating surface zones with material properties very different from those of the matrix.

Adhesion, in the sense of a thin layer of material bonding two layers together, is a task common to macromolecules. Adhesion can be classified by use, chemistry, end use or even application technique. Use may be as a structural adhesive or as a pressure-sensitive material. A structural adhesive will generally form a hard coating when cured or dried, while a pressure-sensitive material will remain "tacky" for extended periods. However, some pressure-sensitive adhesives are meant to form structurally strong bonds after contact. A pressure-sensitive adhesive will always exhibit, at some point during the lifecycle, a tacky surface, which is a uniquely polymeric property. Common adhesive chemistries include epoxy, rubber, cyanoacrylate, polyvinyl alcohol, and silicones, to name a very few. The chemistry may provide more information about how bonds form or how long the material may be worked before application, and so forth. Application techniques include waterborne, solvent-based, or hot melt. Low-viscosity adhesive solutions may be sprayed on or painted. The polymer may be cured in place, heat set, or have other applicable descriptive processing steps. The application technique provides information about the liquid state of the polymer: dissolved or molten, low viscosity or high. End uses include laminating, marine, construction, repair, or wood adhesives. Even the application provides information, particularly about the design requirements for the material.

Properties of adhesives include tack, peel, and shear strength. Peel strength describes the force required to remove an adhered coating from a bonded surface. Shear strength describes the force required to drive the adhered coating to shear relative to the bonded surface. Tack, or stickiness, is a permanent property of pressure-sensitive adhesives and a transient property of structural adhesives as they cure.

Peel and shear strength can both be measured using either dynamic or static tests. Dynamic tests are conducted using a standard load frame, set up for the appropriate test. Static tests use a fixed load and record the time required to reach a specific failure criterion. Particularly critical in measuring adhesion is to block the experiment for confounding variables. Variables with a significant effect on measurements, in addition to the standard rate dependence for viscoelastic materials testing, are: the stiffness of the coating substrate, the thickness of the coating, the force applied to create the bond, and the time that the force remains in place during bonding.

The substrate is the material to which the standard test sample is applied. The standard test specimen is usually 1-inch wide, and the value is reported as a force per linear inch, or per the equivalent metric units. If

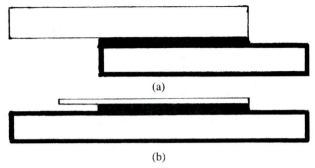

(a)

(b)

FIGURE 10.32 Specimens for testing adhesion: (a) an adhesive material binding two rigid substrates. (b) an adhesive material binding a flexible to a rigid substrate.

the force is too low to measure accurately with available equipment, the width of the specimen is sometimes doubled. If the force is too high, the width may be halved. The area of bonding is also only halved. Lower forces are required to remove waterborne or pressure-sensitive adhesives, while extensive force may be required to remove structural adhesives.

The substrates are typically prepared to a specific quality of surface finish, most frequently electropolished metal. The metal surface is normally degreased with methanol prior to testing. A structural adhesive may be applied and tested between two such substrates. Pressure-sensitive adhesives are usually bonded to a flexible substrate, such as polyester film, and tested against a second, rigid substrate.

Peel

The peel strength of an adhered coating is one valuable characteristic. Peel strength is a combination of tensile and shear response, usually dominated by the surface effects neglected in bulk properties testing. The peel can be measured in several ways, including 90°, 180° or by a tear-off technique.

Many peel tests require that one or both adherent substrates must be flexible. Peel tests use a standard 1-inch test specimen, and the value is reported as a force per linear inch. The 90° peel test is not common. The coating is pulled up so that the force is at 90° to the surface. This requires that the location of the substrate or the mechanical pull point be constantly adjusted as the displacement changes. Test rigs are manufactured to account for the displacement of the debonding zone, and these provide reproducible results. More common is the 180° geometry, where a flexible film is pulled past itself, as shown in Figure 10.34 The T-peel test, described in ASTM D1876, is also a 180° peel geometry.

A particular problem with peel tests is controlling the angle of peel. For the 90° test, the displacement of the specimen will change the angle. For an 180° peel test, the exact angle along which the force is applied is controlled by the stiffness of the material and flexible substrate relative to the peel strength. Using a very thin substrate or a fabric with a high compliance improves the geometry of the test and enhances the accuracy and reproducibility of the test. Controlling the angle of applied force for the T-peel is impossible.

One semi-quantitative peel test is the adherent area. A coating is scored into small squares, cut down to the substrate. A piece of adhesive tape is applied over the scored region, and then removed. The number of squares remaining behind after removing the tape corresponds to the adhered area. This method may be particularly useful in detecting debonded areas of the coating, but is otherwise not as useful as other peel tests, especially for effective, strongly adhered coatings.

Shear

The shear strength of an adhesive governs the resistance to

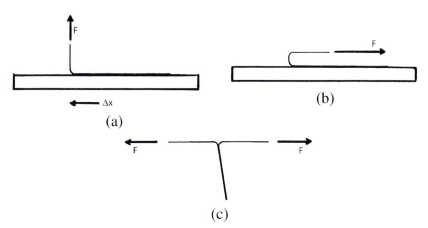

(a)

(b)

(c)

FIGURE 10.33 Peel test geometries: (a) 90° peel – The force is applied to the flexible substrate at a 90° to the rigid substrate. The rigid substrate must be translated to maintain a fixed angle. (b) 180° peel – The force is applied to the flexible substrate at 180°, against the adhesive line. (c) T peel – Force is applied to the ends of two flexible substrates. This results in a 180°, on average, test geometry. A small load may be attached to the bottom end of the test strip to prevent instantaneous, spontaneous movement of the strip.

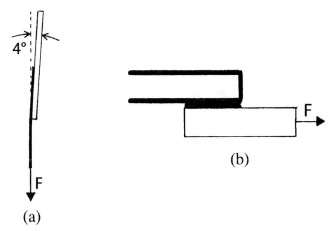

FIGURE 10.34 Shear test geometries: (a) 180° shear. If one substrate is flexible, the rigid substrate panel is mounted at 4° to vertical to eliminate the potential for mixed mode shear. (b) Single lap shear can be applied to test pieces with adhesive bonded to two rigid substrates.

delamination of the coating under a shearing force. Delamination can occur along a boundary created by layers of particulate reinforcement. The coating is in plane stress, rather than the plane strain conditions usually designed for bulk materials. The shear stress techniques used are chosen based on application. The standard test specimen is 1-inch wide and 1-inch long, and the shear is reported in units of pressure. Structural adhesives are usually tested using the lap shear geometry. Pressure-sensitive adhesives are usually tested using the standard shear geometry.

The lap shear test, described in ASTM D1002, uses the sample geometry shown in Figure 10.34. When pulling the lap specimen, the specimen is placed under both tension and shear. The test is usually conducted until failure occurs. The shear modulus and failure stress and strain are reported. Some materials demonstrate yield, even in plane stress.

A 180° shear test is used for pressure-sensitive adhesives in preference to a lap test. The substrate is usually given a 4° back angle. Mounting the substrate at an exact perpendicular to level is functionally impossible. If the angle is slightly forward of perpendicular, then the force will be a combination of shear and tension, causing underestimation of the shear strength. With a back angle, the specimen will be under pure shear. The back angle reduces the effective force applied by a calculable amount. For pressure-sensitive adhesives, the shear strength is usually reported as a shear holding power, which is reported as the length of time the adhesive will hold a standard weight before failing.

Cleavage

Cleavage strength is measured according to ASTM D3433. The standard test uses a 1-inch specimen, mounted between two flat, rigid plates, creating the double-cantilever beam shown in Figure 10.36. The cleavage test is sensitive to small differences in thickness and can be sensitive to the length and width of the cantilever plates. This test produces a value for the fracture modulus.

An alternative configuration is the push test. In this approach, two layers of solid substrate are glued together with the test adhesive. A groove is cut in the top substrate to create a circular block. The groove should penetrate down to, but not into, the bottom substrate, and through the adhesive layer. A second cut is made through the bottom substrate, smaller in diameter to the upper perforation, penetrating down to, but not through, the adhesive layer. The plug created by these concentric perforations is then pushed out. The force to debond per unit area of adhesive bond is reported. One challenge with this test method arises if the substrates are not rigid enough to sustain the applied load for debonding the adhesive without deforming.

Tack

Tack is a permanent feature of a pressure-sensitive coating and can be a temporary and transient property of structural adhesives. For structural adhesives, the interval during which the material is tacky and therefore can be used for bonding, is the open time or the pot life. Thus, some adhe-

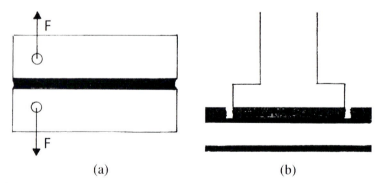

(a) (b)

FIGURE 10.35 Cleavage tests: (a) The cleavage test uses the cantilever beam geometry. (b) A plug shear cleavage test can be applied to a cylindrical plug adhered to a rigid substrate.

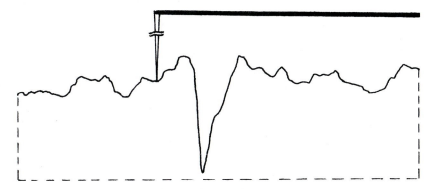

FIGURE 10.36 After a scratch is applied in a specimen, the scratch depth can be measured using profilometry.

sives report a tack time or an open time. The tackiness of the coating can be assessed using qualitative, semi-quantitative, or quantitative tests.

The qualitative or semi-qualitative tests include the finger tack test and rolling ball tack. Finger tack describes whether the material "feels" tacky, and is therefore completely subjective. To measure rolling ball tack, a calibrated ball bearing is rolled down a specified incline from a specified height. The distance the ball rolls is the rolling ball tack.

Applying and removing a probe generates a quantitative value of tack. A probe with a known composition and area is brought into contact with the adhesive using a measured, standard force. The force required to remove the probe is then recorded. The result is reported as the tack.

Scratch Resistance

Scratch tests can be conducted as, for example, documented in ASTM D 5178. A scratch is made in the sample using an applied, fixed load. The test apparatus should be blocked to eliminate important variables such as the angle of the needle, and the rate of scratching. There are specialized scratch and mar test devices with various fixtures and automation to improve the repeatability of the scratch application. A scratch is created by causing localized damage, a sort of plough force applied to the surface of the material. The scratch depth will be affected by the proximity of the testing temperature to the glass transition temperature of the matrix, the amount of crystallinity or lack thereof in the specimen, and the presence, quantity, and effectiveness of any reinforcing particles. The scratch depth is measured using surface profilometry, as described above.

Do It at Home 10.1

Use a mechanical pencil on polymer containers. Try scratching a piece of polystyrene and a piece of PET. Apply the scratch on a hard substrate, such as a piece of metal, and on a soft substrate, such as wood. Use a piece of felt, as well, if one is available. One good source of PS is the clamshell containers used at delicatessens. The top half is clear and the bottom half is black, due to the addition of finely divided carbon black. Compare the scratch resistance of the two parts.

By preference, scratch properties should be determined on bulk materials. The scratch resistance of coatings is often of considerable practical interest, as well. The scratch depth will be influenced by the hardness of the material and proximity to a substrate with very different hardness. If the coating thickness is not repeatable by processing, then the coating must be thick enough to eliminate substrate influences. The required thickness is material dependent, and must be determined using a series of coating thicknesses. The required minimum thickness occurs where the test scratch ceases to change as a function of specimen thickness. A closer match in hardness between coating and substrate may also improve the accuracy of the measurement.

Like hardness, the scratch depth of a composite can become a complicated problem. The scratch is a point application of force, and if the point size drops near to or below the size of the reinforcing particles, the force can interact directly with individual particles and change the load distribution in an unpredictable and difficult to interpret fashion.

There has been some interest in using the area of material removed by the scratch applicator as an estimate of the toughness of the material. Such an approach may produce comparative information. Extrapolating the toughness gathered using scratch testing to that obtained by fracture testing can be challenging.

Coefficient of Friction

The coefficient of friction (COF) is a measure of the relative ease with which two surfaces slide against each other. The COF is the ratio of the force required to slide the surfaces to the force perpendicular to the surfaces, and is therefore dimensionless. The value obtained for the friction coefficient will depend on both the load pressure and the velocity. The friction coefficient is also a function of both surfaces. Polymers are usu-

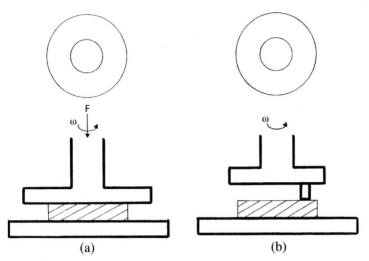

(a)   (b)

FIGURE 10.37 The coefficient of friction can be measured using a thrust washer or pin on disk tester. (a) In the thrust washer, a thin, annular test piece is mounted between two flat plates. The upper plate is rotated, under compressive force. (b) A pin on a disk tester uses a pin of specific composition, typically steel, as a wear point. A disk of the sample material is spun in contact with a steel pin.

ally tested versus polished steel, and this result is reported as an attribute of the polymer. The static COF is recorded at the point of incipient motion. The dynamic COF is measured instead at a constant velocity. The difference between these two is the "slip-stick" condition. Smoother and or slipperier, lubricated, or non-interactive surfaces have less resistance to sliding motion. Polymers with a low slip-stick are well suited for parts that undergo back-and-forth or stop-and-go movements. Materials are usually tested dry, with no additional lubricant, against polished steel. ASTM 3702 uses the thrust washer geometry. Alternative methods are described in ASTM D1894. Figure 10.37 shows some test geometries for measuring COF.

Wear Resistance

Rather than applying a point force to a surface to apply damage, wear testing uses a distributed, continuous force to inflict damage. Nanocomposites have shown particular improvements in wear resistance above the bulk polymer properties. Flowing water, blown particulates, or repeated physical contact may all cause wear. These may also serve as the basis for wear testing.

Pin on Disk

Sliding pin-on-disk wear testing is described in ASTM G 99. The test is composed of a disk spinning with a controlled rotational velocity, in contact with a pin applied with a controlled surface load. A number of cycles of the disk are applied, after which the wear area may be measured using a profilometer. The number of disk revolutions may be defined by

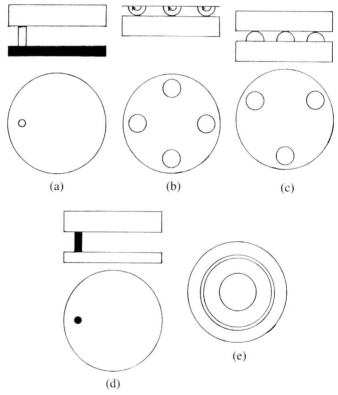

FIGURE 10.38 Examples of disk wear testing configurations: (a) A disk, either coated with or comprised entirely of the test material, is spun in contact with a fixed steel pin. (b) A disk, either coated with or entirely constituted of the test material, is spun in contact with four rolling, steel bearings. (c) A disk, either coated with or entirely constituted of the test material, is spun in contact with three fixed, steel bearings. (d) A steel disk is spun in contact with a pin, either coated with or constituted entirely of the test material. (e) After a predetermined number of revolutions, the result of test configurations (a)-(d) is a disk with a worn-in groove. The cross sectional area of the wear groove is characteristic of the wear resistance of the material.

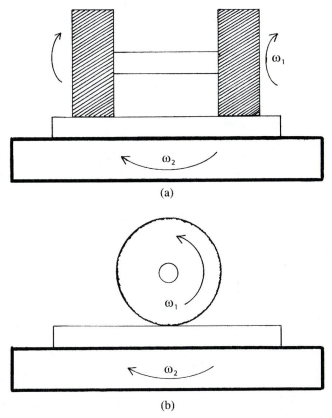

FIGURE 10.39 (a) In a Taber tester, a disk, either coated with or constituted entirely of the test material, is mounted on a turntable. The turntable spins the specimen in contact with two spinning grinding wheels. Thus the specimen experiences alternating co-rotating and counter-rotating wear events. The area removed after a predetermined number of revolutions of the specimen turntable is characteristic of the wear resistance of the material. (b) End on view of the Taber tester.

standard requirements. The number of revolutions may also be chosen for convenience, based on the amount of damage done. The test configuration depends on the test situation. If only a small amount of test material is available, the stationary pin may be made from the composite, and the rotating disk may have an abrasive surface. The usual form of the test is a rotating disk composed of or coated with the composite specimen. The abrading counter body is usually a stainless steel ball of high hardness. As with scratch testing, the temperature and the film thickness can affect the recorded result. When recording the results of pin on disk testing, the normal force, surface velocity, and number of rotations should be reported.

Taber

An alternative wear machine is the taber tester. A disk of the composite coating is placed on the device, again a rotating turntable. The rotational velocity, number of revolutions, and the applied normal force are controlled. The abrasion is applied using two wheels, counter rotating to the specimen, as shown in Figure 10.39. The wheels have a standard roughness. A particular difficulty with Taber testing polymeric or polymer-containing materials is accumulating material on the abrasion surface. If the abrasive becomes clogged, the test values will show variability.

Abrasive Grit

A semi-qualitative test is possible using grit blasting. A standard grit powder can be used in a grit blasting cabinet. The specimen is prepared by masking part of the surface. The specimen is then subjected to the abrasive air stream for a finite, controlled time. The amount of material eroded, measured as a step change relative to the masked portion, provides comparative data for the wear resistance of a material. This test may be useful in aircraft or architectural design, where resistance to airborne particles is critical.

Surface Tension and Surface Free Energy

The surface tension and surface energy required to create a free surface, are related. Thus, measures of the surface energy are useful for mea-

TABLE 10.8 Surface energies of materials.

Liquid	γ (mN/m)	γ_p (mN/m)	γ_d (mN/m)
Water	72.8	51.0	21.8
Formamide	58.2	18.7	39.5
Methylene Iodide	50.8	2.3	48.5
Glycerol	63.4	26.4	37.0
α-Bromonaphthalene	44.6	0	44.6
n-Hexadecane	27.6	27.6	0

From Fundamentals of Adhesion.

suring chemical changes to surfaces and for predicting property changes from crystallinity, or the quantity of reinforcing particles at the surface, and so forth. Table 10.8 shows the liquid surface tension for several polymers. Notice that the surface tension becomes higher as the polarity of the polymer increases.

Surface tension is often calculated through contact-angle measurements. The basic relation describing the free energy is based on the density:

$$\gamma = \gamma_\beta \rho^\beta \qquad (10.20)$$

where γ_β and β are materials constants. For many polymers, β is $3.0 - 4.5$. This equation makes some sense in that the number of functional groups at the surface of the material should be proportional to the density, and the energy effect of these functional groups can have an exponential effect. The surface energy is temperature dependent:

$$\gamma = \gamma_0 \left(1 - \frac{T}{T_c}\right)^{11/9} \qquad (10.21)$$

where γ_0 is the surface tension at 0 K and T_c is the critical temperature, which is a little difficult to measure for polymers. T_c is approximately 1000 K for most polymers, and the surface tension is a weak function of temperature. Some polymer surface tensions have stronger temperature dependence.

The surface tension also is dependent on molecular weight:

$$\gamma^{1/4} = \gamma_\infty^{1/4} - \frac{k}{M_n} \qquad (10.22)$$

where γ_∞ is the surface tension at infinite molecular weight and k is a constant. At a molecular weight of between about $2000 - 3000$, the surface tension is very near the maximum attainable. There is also a relationship between the surface tension in different material states, exemplified by the change from glass to crystalline structure, as shown below:

$$\gamma_c = \left(\frac{\rho_c}{\rho_a}\right)^\beta \gamma_a \qquad (10.23)$$

The c subscript refers to the crystalline state, and a to the amorphous. This is clearly related to the density change due to the change in phase.

The surface tension is composed of two contributions, due to the dispersive force interactions, d, and the polar interactions, p:

$$\gamma = \gamma_d + \gamma_p \qquad (10.24)$$

These components determine the interfacial tension between two polymers:

$$\gamma_{12} = \gamma_1 + \gamma_2 - \frac{4\gamma_{1,d}\gamma_{2,d}}{\gamma_{1,d} + \gamma_{2,d}} - \frac{4\gamma_{1,p}\gamma_{2,p}}{\gamma_{1,p} + \gamma_{2,p}} \qquad (10.25)$$

The subscripts 1 and 2 refer to the two materials in contact.

The Young equation relates the contact angle of a drop of liquid to the surface energies of the liquid and substrate. A drop of liquid is applied to the surface of a flat, polished specimen. An equilibrium drop is one that forms a natural shape. Alternative techniques use the contact angles of advancing and receding drops. The contact angle is measured at the intersection of the drop with the surface. The liquid used should not spread over the surface, and the angle must be above 10° in order to take measurements.

EXPOSURE

Environmental factors play a role in the lifecycle of many products. The composite should be durable in the context of use, whether that is outdoors, in a chemical plant, or for use as a biomaterial to be implanted in a living creature.

Environmental Resistance

Materials that will be exposed to sunlight during service can be tested using a high-energy *uv* light box. Salt fog testing exposes the sample to high humidity and a saltwater environment, and provides information about corrosion stability. Weather chambers can subject a material to cycles of light and dark, precipitation and aridity. Conditioning ovens can cycle a material through hot-cold cycles, under any level of humidity. After exposure to the elements of interest, often under intense and accelerated cycles, materials properties are evaluated as the experimental variables, with cycles or exposure time as the control variable. If weathering has only a small measurable effect on the mechanical properties, for example, then the material's environmental stability is high. Environmental resistance tests are often important for architectural, clothing, infrastructure and transportation applications.

Most plastics and composites are very resistant to moisture. However, one of the additional values of composites for exterior applications is chemical resistance. Resistance to chemical attack is a function of structure. Polyesters and nylons show good resistance to organic liquids, polyolefins show good resistance to polar solvents. Polymers tend to be resistant to bases, even strong caustics. They are preferable for storing bases above metals, which corrode, or glasses, which are etched by caustic liquids. Polymers are moderately resistant to acids, though they will oxidize and degrade at varying rates, depending on permeability to the liquid and the susceptibility of the backbone. Polyolefins with no polar groups are the most stable to acidic attack.

Polymers can be susceptible to the phenomenon of environmental stress cracking. Under the combined forces of temperature cycling, exposure to solvents, and cyclical pressures, a crack can initiate and propagate, causing failure. Some stress cracking can be accelerated by surfactants. Reinforcement particles may hinder ESC by restricting diffusion and improving surface properties.

Biodegradability and Biocompatibility

Creating biodegradable nanocomposites requires generating sufficiently small reinforcing particles that are biodegradable over a reasonable time scale, but also are stable enough to process into useful parts. Biodegradability is not a single property. Environmental biodegradability implies that the material is able to degrade under reasonable environmental conditions. In a physiological system, on the other hand, biodegradability refers to the ability of the material to be absorbed by the body without an adverse physiological response. Biocompatibility is a related issue referring to the suitability of a material for use in a physiological system.

Environmental biodegradability is reported by dissolution time in a given environment. Typically, the material will be immersed in hot water, and the mass or strength as a function of time will be recorded. Nanoreinforcements may interfere with the degradation process by blocking the path for moisture infiltration, or they may increase the open, internal surface area for moisture transport if they are poorly adherent with the matrix.

Biodegradability is measured in vitro using water or simulated body fluid (SBF), an aqueous, ionic solution. Again, the mass loss or deterioration of properties may be the evaluation variable. *In vivo* measurements in animal models, for example rats, guinea pigs or dogs, requires surgery to implant a specimen, and a second surgery on the animal, often a sacrifice of the animal, to recover the material for analysis. Animal studies require appropriate animal care facilities, and can be quite expensive. Tests on animals must pass an institutional review process. In an institutional review, the experiment protocol, the expected outcomes and the animal care protocols are peer reviewed to insure the humane treatment of the test subjects and the value of the experiments. *In vivo* testing is expensive and takes considerable time to accomplish.

Biocompatibility describes the ability of the material to work in a physiological system. There are several classes of compatibility. Incompatible materials are inimical to cells. PVC is the gold standard of this behavior, proving completely cytotoxic (lethal to cells). "Stealth" materials stimulate no discernible biological reaction from the host organism. This is a very rare trait, exhibited by polyethylene oxide polymers. More typically, materials stimulate varying degrees of a wound healing cascade process. The first step of this process is attracting an acellular, fibrous capsule to wall the material away from the surrounding, native tissue. A typical biocompatible material will continue without further damage to the surrounding tissue beyond the formation of the fibrous capsule. A typical incompatible material will generate further steps in the process, with inflammation, infiltration by giant cells and macrophages as the body attempts to remove the offending material. A truly biocompatible material will be integrated by the body and then remodeled into a natural, repaired tissue, called granulation tissue. Collagen is an example of a compatible material. The body consumes collagen, leaving little damage in the process. Compatibility seems to be a combination of chemical, structural, and mechanical compatibility with the surrounding tissue.

Techniques are used to evaluate the compatibility *in vitro*, and to improve the predictability of success with expensive *in vivo* testing. Cell cytotoxicity, attachment, and spreading are three such tests. Tissue-culture-treated polystyrene is the gold standard for cell attachment, binding strength, and spreading. Cell cytotoxicity is measured by inoculating a well containing the material with a certain number of cells of a specific type of interest. The number of surviving cells is counted after specific intervals. Materials that kill an unexpected number of the cells are cytotoxic. Cell attachment follows a similar process, but in this case, the number of cells attaching to the surface of the material is counted. Cell spreading can also be evaluated from the same test. Cell spreading is regarded as a sign of cell health and viability. There are additional *in vitro* methods for evaluating cell interactions with materials.

Unfortunately, the ability of *in vitro* testing to predict *in vivo* performance may not generalize across materials. Cytotoxic materials such as PVC usually do show catastrophic performance in living systems. However, many materials show excellent cell adhesion properties *in vitro*, but do not succeed as well in living systems.

Liquid Stability

Many *in situ* polymerization suspensions containing nanocomposites have a limited stability. Liquid resins, whether unreacted monomers, oligomers or polymers in solution, are environmentally unstable. The stability is reported as the length of time that the material can be stored until it becomes unusable. A safety factor is usually included if the material is being sold to a demanding market. The stability testing can be accelerated using thermal cycling and elevated temperatures. Accelerated testing seldom gives an exact prediction of the eventual shelf life. If a sample

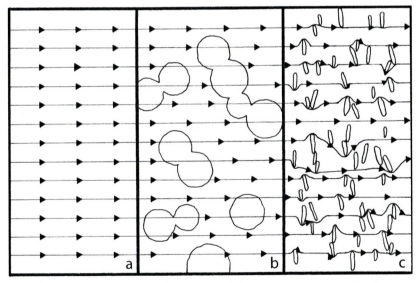

FIGURE 10.40 The presence of pores or particles affects the average diffusion path through the specimen. (a) The virgin material has a fixed diffusion rate, dependent on the thickness. (b) In a porous specimen, diffusion through the pores is more rapid. The average net diffusion rate is greater. (c) In a specimen containing particles, here, for example, oriented plates, part of the specimen is blocked to diffusion. As a result, the average net diffusion rate decreases.

passes accelerated testing, the exact shelf life will not be determined. However, a minimum stable time may be determined.

Instability manifests itself in various forms. Waterborne materials are unstable to freezing, as the water crystallization process will exclude the resin from the solution. Liquid monomers and oligomer solutions will polymerize slowly, becoming too viscous for use over time. Adding an inhibitor will slow this premature polymerization, but will not prevent it. Solvent borne polymers may lose solvent over time. If the solvent loss process is caught early enough, additional solvent may redissolve the polymer, but if the polymer loses enough solvent, the solid may not dissolve in a reasonable amount of time.

BARRIER PROPERTIES

Polymer coatings, particularly nanocomposite coatings, provide enhanced barrier properties in many important applications. Coatings are applied to prevent liquids and vapors from penetrating to the protected surface. Therefore, the permeability of the coating is one of the key performance properties.

Permeability

A decrease in cross-sectional area caused by reinforcing particles contributes to the dramatic improvements in permeability for composites. The permeability is increased by tortuosity, and decreased by increased porosity in the coating. Permeability describes the ability of a membrane to pass vapors and gasses through the thickness. Usually of more interest is the resistance to gas and vapor permeation. The permeability, P, is proportional to the diffusion rate, D, and the solubility, S:

$$P = D \cdot S \qquad (10.26)$$

The diffusion rate and solubility at a single temperature are constant for low concentrations of gas, or if the gas or vapor is insoluble in the polymer phase. At higher concentrations or for solvents soluble in the polymer, the membrane will be plasticized by the vapor, and the permeability will increase in a nonlinear, even exponential, manner. Solubility and diffusion are each dependent on temperature in the standard Boltzmann's activation energy form, and therefore so is permeability.

Polymers exhibit a different susceptibility to permeation by polar and non-polar organic liquids.

In addition, water vapor permeability is a key factor in protective coating performance. The permeability of a coating can be determined using

ASTM D 1653. In this method, a shallow metal Payne cup is covered with a film of the composite film. Payne cups are available with different fixtures to seal the film across the opening of the cup. The method describes two protocols for conducting the test, wet and dry cups. In the wet cup method, a precisely measured mass of water is placed inside the cup. The cup is stored in a desiccator. The mass loss of the cup is recorded as a function of time. In the dry cup method, desiccant is placed in the cup, which is then kept in a controlled humidity environment. The weight gain again is measured as a function of time. The film must be of well-characterized and of uniform thickness. The porosity must also be characteristic of the test material. If a stand-alone film cannot be obtained, films can be tested on a porous paper backing. This is not the preferred approach, however.

The mass transport is a function of time, film thickness, and area of exposure. The mass flux will go through an initial transient before establishing steady state. The known film thickness and area can be used to convert the water mass flux to the vapor transmission rate. The wet cup procedure is preferred for evaluating materials for use in high-humidity applications. The wet cup permeability test also can be applied for sol-

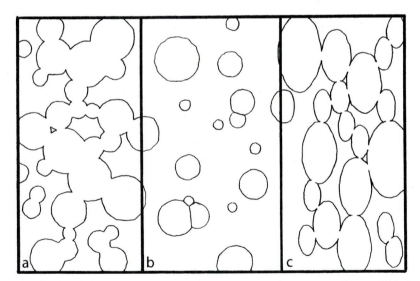

FIGURE 10.41   Various pore geometries: (a) Spherical, interconnected pores. This is the typical case, described well by typical theories. (b) Spherical, closed pores. Closed porosity cannot be measured by vapor or liquid penetration. (c) Compressed, oval, open porosity. The narrow passages between pores can restrict liquid infiltration, resulting in challenges for accurate porosimetry.

vents, by placing a quantity of solvent inside the cup, sealing the coating across the lid, and measuring mass loss with time in the same way previously described.

Porosity

The permeability through a thin, contiguous film is diffusion controlled. Many routes to films produce coatings with some amount of porosity. Porosity can allow rapid convection of vapors and gasses through the protective layer, with a corresponding increase in permeability. Closed porosity has the least impact on liquid transport. Small closed pores do not permit convection cells, but do have a faster diffusion constant than a solid. Large pores may allow convection cells to form, further enhancing transport beyond just the improvements due to the low-density gas. Open porosity allows direct penetration of liquids and vapors into the specimen through the surface of the coating. If the porosity is above a critical concentration, the pores may penetrate the entire thickness of the specimen. The effects of open porosity may be modified by the size of interconnections between the pores. Narrow openings are not much barrier to small, rapidly moving gas molecules but exclude liquids with an incompatible surface energy. Porosity can be measured in many ways, and depends somewhat on the nature of the pores.

Two example techniques for measuring closed porosity are density and image analysis. Density measurements using the Archimedes principle, as described earlier, will measure the actual mass and displacement of the material. A porous material will be less dense than the theoretical solid. The difference in volume between the measured and predicted values provides an estimate of the pore volume. Analysis of an image of a cross-section of a coating can provide information about the porosity as well. Figure 10.42 shows three stages of a typical image-analysis protocol for extracting porosity data. A filter removes background noise. Assuming that the distribution of gray levels will be normal allows one to set a threshold for the black level so that it represents porosity. The porosity is then the ratio of the thresholded pixel area to the total image area.

Open porosity can also be measured by image analysis of an image cross-section. Open porosity is often measured by liquid or gas porosimetry, as well. Mercury is a non-wetting liquid on the majority of polymers, and will not penetrate a pore until sufficient external pressure is applied to drive the liquid in against the surface energy. The pressure is governed by the Washburn equation:

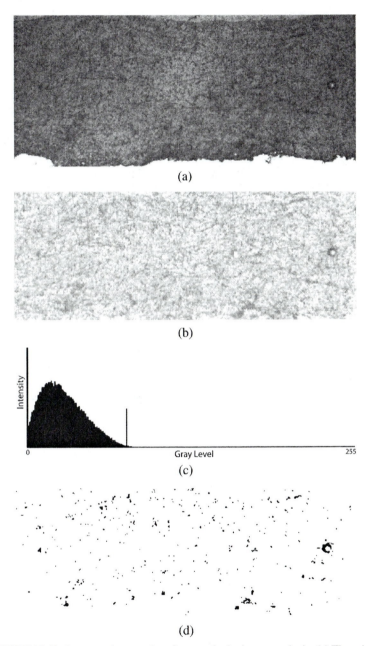

(a)

(b)

(c)

(d)

FIGURE 10.42 An example procedure for porosity by image analysis: (a) The original image. (b) The image after applying a rolling ball filter to remove background illumination irregularities. (c) The histogram after filtering. Image filtering reduces the available gray levels. (d) Threshold set at the point where the gray level distribution deviated from a Gaussian distribution. The features identified were classified as porosity.

523

$$P \cdot d = -4\gamma \cos\theta \qquad (10.27)$$

P is the applied pressure, d is the minimum diameter that may be penetrated at that pressure, γ is 480 dyne/cm, the surface tension of mercury, and θ is the contact angle of the mercury with the pore wall, usually reported as 140°. The sample is immersed in mercury in the porosimeter. As the pressure is increased, the mercury level will drop as an increased number of pores are penetrated. When further increases in pressure do not drive further penetration, the size of the smallest pores have been identified. Mercury porosimetry cannot penetrate closed porosity. Further, the porosity measured is based on the diameter of the pore opening, not necessarily the diameter of the pore.

A measure of the total open porosity content can be made by gas sorption. When gas encounters a solid, there will be interactions. The weak interactions characteristic of dispersive forces will cause weak adsorption, while chemical interactions will allow stronger, permanent interactions. The interaction between gas and surface creates a reversible but stable monolayer, and will thus lower the ambient pressure of gas in the surroundings. This is used for the BET (Brunauer, Emmett and Teller) technique. A quantity of gas is introduced to a calibrated chamber containing a solid specimen. The pressure in the chamber is measured. The reduction in pressure caused by gas adsorption can be checked against the ideal gas law:

$$V \approx \frac{V_m \dfrac{P}{P_0} e^{\left(\frac{E_1 - E_L}{RT}\right)}}{1 - \dfrac{P}{P_0}} \cdot \frac{1}{1 + \dfrac{P}{P_0} e^{\left(\frac{E_1 - E_L}{RT}\right)}} \qquad (10.28)$$

where V is the volume of gas adsorbed, V_m is the volume of the monolayer, is the fugacity or partial pressure divided by the partial pressure above a saturated solution, E_1 is the heat of adsorption of the gas in the monolayer, and E_L is the heat of condensation for the gas. This allows measurement of the quantity of gas adsorbed onto a monolayer on the surface of the specimen. From the cross-sectional area of the gas molecule, the total surface area can be calculated. As the gas pressure increases, the gas will fill the available pores. The equilibrium gas volume compared to the relative saturation pressure can be converted to a pore size distribution. As the gas is removed, the gas will leave the pores. The gas adsorption and desorption phase will usually follow different pres-

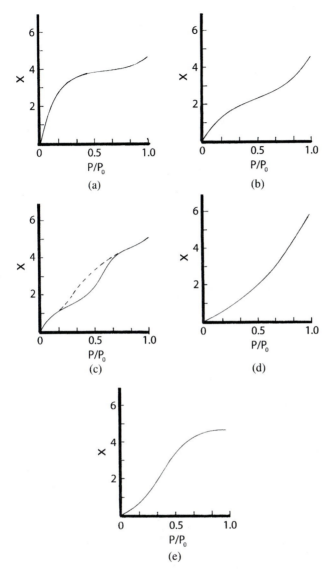

FIGURE 10.43 Physical adsorption isotherms for porous material beds plot the amount of vapor absorbed vs. the partial pressure of the vapor relative to the saturated partial pressure at a constant temperature. (a) An adsorption typical for materials such as carbon, where the external surface of the particle is small relative to the surface area of the internal pores. (b) An inverted S typical of large, non-porous particle beds. (c) An adsorption curve with hysteresis between adsorption, solid line, and desorption, dashed line. Particles with very small pores resist penetration, and resist desorption as well. (d) An adsorption typical for liquids in contact with chromatography substrates, such as silica gel. (e) An S shape, sometimes observed for polar liquids interacting with non-polar substrates, or vice-versa. Figure after Cadle.

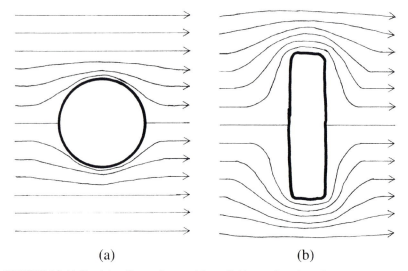

(a)   (b)

FIGURE 10.44 Particles disrupt flow and force fields as a function of particle size, concentration and shape. (a) Round particles exert an influence modeled by moderate diversion of flow lines, and concentration of flow lines around the particle. (b) At an equivalent volume, a plate-like particle can exert an influence modeled by severe diversion of flow lines, with concentration of flow lines particularly at the edges of the plates. This can result in an anomalous decrease in diffusion rate and permeability in the direction perpendicular to the plate, but a much smaller decrease parallel to the plate.

sure histograms. This allows some evaluation of the pore shape, as well as the energy of sorption. The amount of gas adsorbed per unit weight of the powder follows several characteristic behaviors, as seen in Figure 10.43.

Tortuosity

The permeability of a composite film is influenced by tortuosity, which describes the shortest average distance a particle must travel to traverse the film. Tortuosity can be measured as the average distance a molecule must travel to get through a film, divided by the thickness of the film. This quantity is also called the mean free path. In a pure, flawless, amorphous, polymer film the shortest distance traveled is the thickness of the film. Within a semicrystalline matrix, the crystallites block direct travel, and the diffusion path must travel farther to pass through the film. The reinforcement causes a similar increase in tortuosity. However, nanoparticulates will have a lower impact on tortuosity than will larger

This is a controlled document. Only the latest revision, which appears in the company database, may be used as a working document. Any printed copy is uncontrolled, unless otherwise identified. The user is responsible to verify the validity before use.

Scope and Purpose

Nanocomposite powders are a valuable precursor for many products. This procedure is used to prepare 200 g batches of a nanocomposite Nylon 11/Silica feedstock powder with 10% by volume fumed silica content.

Responsible Authority

Manufacturing is responsible for all processes. Personnel are responsible for following documented procedures.

Approvals:

Quality Manager: _____

Manufacturing: _____

Related and Support Documents

C-0257 Balance procedure

Materials

Nylon 11, Rilsan D-60 or equivalent   Batch #: _____

Silica, Degussa Aerosil A380 or equivalent   Batch #: _____

Top Loading Balance, Ohaus Adventurer 3100 or equivalent

500 g calibration weight

Disposable Aluminum Weighing Pan

Spatula

Ball Mill, Amoco Ball Mill or equivalent

Ball Jar, 4 liter capacity, Amoco or equivalent

500g +/- 10 g Alumina Milling Media or equivalent

#10 Mesh Screen Sieve

Round Wide Mouth Bottle, HDPE

Procedure	Initial
Calibrate the Ohaus balance.	_____
Load the Ball Jar with alumina milling media.	_____
Weigh 40 g +/- 0.1 g of silica on the top loading balance.	_____
Amount: _____	_____
Load silica into ball jar. Dispose of weighing pan.	_____
Weigh 110 g +/- 0.1 g of nylon 11 on the top loading balance	_____
Amount: _____	_____
Load nylon 11 into ball jar. Dispose of weighing pan.	_____
Close the ball jar, carefully closing the screw clamp lid.	_____
Place the ball jar on the ball mill.	_____

Initials: _____

FIGURE 10.45 Written test methods are prepared in various formats. (a) In an SOP, very specific instructions are given to maximize repeatability and reproducibility. (b) A recipe provides ingredients and steps, in order to lead to a useful result. There is often significant freedom in applying the recipe. (c) In an experimental procedure description, all the information necessary to repeat the experiment to achieve a similar result is provided. However, many of the specifics are left unstated.

527

Materials
Nylon 11
7 nm hydrophobic silica
Top loading balance
Ball mill
4 liter ball jar
500 g Alumina milling media
Metal particle sieves
Storage jar

Procedure
Load the ball jar with alumina milling media.
Accurately weigh 40 g of silica. Add to the ball jar.
Accurately weigh 110 g of nylon 11. Add to the ball jar.
Place the ball jar on the mill. Run the mill for 48 hours. Record the milling time to the nearest 20 minutes.
Separate the milling media from the composite powder using a sieve.
Store the composite powder in a plastic or glass storage jar for later use. Be certain to label the jar with the contents, including the true volume fraction of silica.

(b)

Samples of nylon 11 and various sizes of nanoscopic silica were mixed for ball milling. The mixtures ranged from 10 - 50 %$_{vol}$ silica with particle sizes ranging from 7 nm to 10 μm. Mixtures were milled using an Amoco 115 Volt ball mill for up to 48 hours. The materials were milled with 500 ± 10 g of Al$_2$O$_3$ balls at a speed of 60 rpm. Samples were withdrawn from each mixture at geometric time intervals, beginning after thirty minutes. Samples were characterized to determine the amount of silica present by ashing and scanning electron microscopy.

(c)

FIGURE 10.45 **(continued)** Written test methods are prepared in various formats. (a) In an SOP, very specific instructions are given to maximize repeatability and reproducibility. (b) A recipe provides ingredients and steps, in order to lead to a useful result. There is often significant freedom in applying the recipe. (c) In an experimental procedure description, all the information necessary to repeat the experiment to achieve a similar result is provided. However, many of the specifics are left unstated.

particles. Open porosity can exhibit tortuosity, as well. Some of the impact on tortuosity of reinforcement is shown in Figure 10.44.

RECIPES AND STANDARDS

Physical properties testing is a wide ranging activity, with deep, historic roots. As discussed in Chapter 2, the periodic table was derived originally using materials properties in part. The current goal of quality engineering, as applied to manufacturing, is to assure that the properties and performance meet or exceed the requirements of the application. Property testing is rooted, in origin and modern application, in ascertaining or assuring that a material will meet a stipulated design goal. The performance requirements depend on the application. Will the material be used in bulk or as a coating? Are the properties improved, degraded, or unchanged by the addition of reinforcement? How does one determine if a coating will survive 20 years of weathering, without waiting 20 years? What are the minimum requirements for the material chosen as the frame of a pair of glasses?

A casual example may help to illustrate the point. A civil engineer, tasked with picking a coating for an expensive bridge, has a list of performance criteria to meet based on the structure of the bridge. The building costs are all significant: the cost of making and applying the coating, the cost of making repairs to an unsatisfactory application, the costs associated with legal responsibility should the coating fail. The coating should be the best for the application, but must at least be adequate. Based on years of experience with coatings, manuals for bridge coatings, and other sources, a set of performance characteristics known to have a high probability of success in the application is set. The coating will be designed, specified and tested to these characteristics. The tests are almost exclusively standardized. ASTM, a professional organization for testing materials, prepares standards, e.g., for coating adhesion, wear resistance, solvent resistance, water uptake, weatherability, gloss, haze and myriad other performance metrics. The results of standardized tests provide a systematic set of data for discriminating among material choices.

On the other hand, an engineer first characterizing the same nanocomposite coating may have only limited ideas of what properties to expect. The coating might provide protection from the elements, mechanical strength, or improvements in appearance. Each of these features could play a role in the eventual utility of the composite coating. Since the goal at the invention and development level is to create a product that consumers want to buy, a variety of characterization techniques may be applied. Often, the techniques available in standard methods will not apply directly to a new material, in a new configuration, with a different chemistry. For example, one standard may call for a flexibility measure, assuming that most coatings are flexible. A nanocomposite coating may be too stiff to test in the standard way, which will entail test modifications. Further, the engineer serving in such a creative role may hope that the material will fill functions yet undeveloped, and may have to design tests that illustrate the positive features for this application.

FIGURE 10.46 A representation of testing as a continuum, with maximum repeatability at the left end and maximum experimental freedom at the right end. The SOP is meant to provide exacting specifications, down to the type of glassware to be used. The experiment has no specific rules, requiring only attention in assuring the accuracy and precision of measurements, and careful reporting.

Still differently, a coating manufacturer must comply with industry standards for characterization and performance. An engineer working for such a manufacturer may be collecting reference test results in order to establish whether the new material meets current performance expectations. Further, though, the engineer will often have to uncover performance advantages providing a market advantage.

Properties arise from the underlying structure of the composite. Structural characterization was considered thoroughly in the last chapter. The physical properties, however, can be treated as arising from a black box. Commercial consumers of materials are more interested in the properties a product has than the properties it might have. Tables of materials properties and standard methods for obtaining them help narrow the field of choices. From there, testing to ascertain true performance according to a specific test method can provide a narrow but clear picture of the expected behavior in service of a material. This focused response to an applied question and answer is often quite satisfactory. While a lack of structural information may make the steps for property modification less obvious, modifying the properties is often irrelevant for many applications.

The standards organizations mentioned previously provide a large contingent of standard test methods for measuring physical properties. ASTM International in particular is dedicated to standards for materials properties testing. The tests are designed to assess performance as it pertains to function in a particular application. A test may be extremely specific to the application, moderately specific, or non-specific. The standard might provide a process for collecting results, or describe an approach to making accurate measurements. Professional standard methods provide results that, in principle, can be compared with all other test results from that method.

Corporations making a product according to a set process usually test the product during processing and at completion according to methods defined in standard operating procedures, or SOPs. The SOP provides systematic instructions for conducting tests that lead to reproducible and comparable results. The SOP for a test method should include all the details necessary to faithfully test the material and provide comparable test results: details of equipment and supplies needed, along with suitable substitutions when appropriate; a detailed sequence of steps that can be followed by anyone with appropriate training; and defined reporting and decision requirements. An SOP is designed to remove decisions about material performance from the experiment, where temptation may lead an investigator to bias a report toward some desired outcome. Decisions about performance requirements reside with the individual designing the SOP, and with the managers making decisions based on the reports.

For educational settings, experimental work, creating large batches of test solutions, or for guiding performance short of setting dogmatic procedures, a recipe may be a more appropriate format for describing a test method. A recipe provides a list of supplies, some general guidance on equipment, and a series of steps that usually leads to an interesting or useful result. There is room for variation within a recipe, because the goal of the process is not necessarily a specific result. The output should be reproducible if proper note taking is accomplished.

Standard methods are usually targeted at a specific property, for a specific bulk material, under specific conditions. However, professional standards sometimes must be modified in order to provide specific results that are more appropriate for a material that falls slightly outside of the test parameters. Composites often do not provide consistent results from standard test methods, because of the combination of materials or the structural features they exhibit. Nanocomposites, in particular, can show behavior not anticipated in setting a standard method. Making slight modifications to the standard method procedure may sometimes accommodate this. At other times, an entirely new method may be required. Any new method developed for testing a nanocomposite should be based on sound judgment and scientific principles, careful consideration of positive and negative controls, assessment of accuracy and precision, and with attention to the testing goals, just as is characteristic of methods developed by standards-producing organizations such as ASTM.

Testing is often done without a formal procedure provided from an external resource. Informal test procedures are not carelessly designed. They must be as free as possible from sources of error, factors that might introduce an artifact to the results. A test must also, formal or informal, be repeatable. They are not as formal, however, because the approach is defined exclusively by the needs of the investigator. Standards and blanks are included for comparison, to insure that the test is accurate and properly applied. Certain features specific to composites testing may be evaluated that might not be in a generic test design. In order to make the test more formal, a round-robin method evaluation and other formal procedures may be implemented.

The flexibility provided by test types ranges from none, in an SOP, to complete for informal test design. Comparison between results of tests on different materials in different laboratories is more likely to be acceptable using an SOP or standard test.

REFERENCES

Mark, J., ed., Physical Properties of Polymers Handbook, Springer-Verlag, New York, 1996.

Scheirs, J., Compositional and Failure Analysis of Polymers, John Wiley & Sons, Chichester, 2000.

Hunter, R. and R. Harold, The Measurement of Appearance, 2nd ed., John Wiley & Sons, Chichester, 1987.

Callister, W. Materials Science and Engineering, 6th ed., John Wiley & Sons, Hoboken, 2003.

Mathot, V., ed., Calorimetry and Thermal Analysis of Polymers, Hanser, Munich, 1994.

Young, R. and P. Lovell, Introduction to Polymers, Chapman Hall, Cambridge, 1991.

Kocman, V. and P. Bruno, Thermal stability of paper fillers: a strong case for a low temperature paper ashing, TAPPI Journal, March 1996, p. 303.

Fortner, B and T. Meyer, Number by Colors, Springer-Verlag, New York, 1997.

Index

Biography

Dr. Twardowski has been working in nanocomposites since 1997 and in composites since 1987. He has diverse research experience in composites, adhesives, biomaterials and science education. Dr. Twardowski has taught courses in materials, textiles, chemical, and mechanical engineering. He currently provides teaching and research support to various institutions in the Delaware Valley region, including Widener and Villanova Universities. Dr. Twardowski is developing the Materials Workshop as a resource for materials education and research.